Lecture Notes in Computer Science 15903

Founding Editors

Gerhard Goos
Juris Hartmanis

AF172554

The series Lecture Notes in Computer Science (LNCS), including its subseries Lecture Notes in Artificial Intelligence (LNAI) and Lecture Notes in Bioinformatics (LNBI), has established itself as a medium for the publication of new developments in computer science and information technology research, teaching, and education.

LNCS enjoys close cooperation with the computer science R & D community, the series counts many renowned academics among its volume editors and paper authors, and collaborates with prestigious societies. Its mission is to serve this international community by providing an invaluable service, mainly focused on the publication of conference and workshop proceedings and postproceedings. LNCS commenced publication in 1973.

Michael H. Lees · Wentong Cai ·
Siew Ann Cheong · Yi Su · David Abramson ·
Jack J. Dongarra · Peter M. A. Sloot
Editors

Computational Science – ICCS 2025

25th International Conference
Singapore, Singapore, July 7–9, 2025
Proceedings, Part I

 Springer

Editors
Michael H. Lees ⓘ
University of Amsterdam
Amsterdam, The Netherlands

Siew Ann Cheong ⓘ
Nanyang Technological University
Singapore, Singapore

David Abramson ⓘ
The University of Queensland
Brisbane, QLD, Australia

Peter M. A. Sloot ⓘ
University of Amsterdam
Amsterdam, The Netherlands

Wentong Cai ⓘ
Nanyang Technological University
Singapore, Singapore

Yi Su
Institute for High Performance Computing
A*STAR
Singapore, Singapore

Jack J. Dongarra ⓘ
The University of Tennessee
Knoxville, TN, USA

ISSN 0302-9743 ISSN 1611-3349 (electronic)
Lecture Notes in Computer Science
ISBN 978-3-031-97625-4 ISBN 978-3-031-97626-1 (eBook)
https://doi.org/10.1007/978-3-031-97626-1

This Springer imprint is published by the registered company Springer Nature Switzerland AG
The registered company address is: Gewerbestrasse 11, 6330 Cham, Switzerland

If disposing of this product, please recycle the paper.

Preface

Welcome to the 25th International Conference on Computational Science (ICCS - https://www.iccs-meeting.org/iccs2025/), held on July 7–9, 2025 at Nanyang Technological University (NTU), Singapore.

This 25th edition in Singapore marked our return to a fully in-person event. Although the challenges of our present times are manifold, we have always tried our best to keep the ICCS community as dynamic, creative, and productive as possible. We are proud to present the proceedings you are reading as a result.

ICCS 2025 was jointly organized by Nanyang Technological University, the A*STAR Institute of High Performance Computing, the University of Amsterdam, and the University of Tennessee.

Considered one of the most developed countries in the world, the island country of Singapore is a major aviation, financial, and maritime shipping hub in Asia. Singapore is multilingual, multiethnic, and multicultural, and as such a very popular, safe tourism destination.

NTU Singapore is a public university ranked among the world's best, with 35,000 students, and home to the world-renowned autonomous National Institute of Education and S. Rajaratnam School of International Studies. In addition to many research institutes and centers at the university, college, and school levels, NTU also hosts two National Research Foundation (NRF) and Ministry of Education (MOE) Research Centers of Excellence, namely the Singapore Center for Environmental Life Sciences Engineering (SCELSE) and the Institute for Digital Molecular Analytics & Science (IDMxS), and 11 Corporate Labs in partnership with various industries. ICCS 2025 took place on the One-north campus.

The Institute of High Performance Computing (IHPC) is a national research institute under the Agency for Science, Technology and Research (A*STAR), dedicated to advancing science and technology through computational modeling, simulation, AI, and high-performance computing. With a multidisciplinary team of scientists and engineers, IHPC drives innovation across sectors such as advanced manufacturing, microelectronics, sustainability, maritime, and biomedical sciences. It leads Singapore's national efforts in hybrid quantum-classical computing and digital twin platforms, and partners extensively with industry and government agencies to translate deep tech into real-world impact.

The International Conference on Computational Science is an annual conference that brings together researchers and scientists from mathematics and computer science as basic computing disciplines, as well as researchers from various application areas who are pioneering computational methods in sciences such as physics, chemistry, life sciences, engineering, arts, and humanitarian fields, to discuss problems and solutions in the area, identify new issues, and shape future directions for research.

The ICCS proceedings series has become a primary intellectual resource for computational science researchers, defining and advancing the state of the art in this field.

We are proud to note that this 25th edition, with 23 workshops (the Workshops on Computational Science), one co-located event (the Asian Network of Complexity Scientists Workshop), and over 300 participants, kept to the tradition and high standards of previous editions.

The theme for 2025, "**Making Complex Systems tractable through Computational Science**", highlighted the role of Computational Science in tackling the complex problems of today and tomorrow. This conference was a unique event, focusing on recent developments in scalable scientific algorithms; advanced software tools; computational grids; advanced numerical methods; and novel application areas. These innovative novel models, algorithms, and tools drive new science through efficient application in physical systems, computational and systems biology, environmental systems, finance, and others.

ICCS is well known for its lineup of keynote speakers. The keynotes for 2025 were:

- **Johan Bollen**, Indiana University Bloomington, USA
- **Jack Dongarra**, University of Tennessee, USA
- **Mile Gu**, Nanyang Technological University, Singapore
- **Erika Fille Legara**, Center for AI Research|Asian Institute of Management, Philippines
- **Yong-Wei Zhang**, Institute of High Performance Computing, A*STAR, Singapore

This year, the main track of ICCS registered 162 submissions, of which 64 were accepted as full papers, and 52 as short papers. There were on average 2.4 single-blind reviews per submission.

We would like to thank all committee members from the main track and workshops for their contribution to ensuring a high standard for the accepted papers. We would also like to thank *Springer, Elsevier,* and *Intellegibilis* for their support. Finally, we appreciate all the local organizing committee members for their hard work in preparing this conference.

We hope you enjoyed the conference and the beautiful country of Singapore.

July 2025

Michael H. Lees
Wentong Cai
Siew Ann Cheong
Yi Su
David Abramson
Jack J. Dongarra
Peter M. A. Sloot

Organization

Program Committee Chairs

Peter M. A. Sloot	University of Amsterdam, The Netherlands
Jack J. Dongarra	University of Tennessee, USA
Michael H. Lees	University of Amsterdam, The Netherlands
David Abramson	University of Queensland, Australia
Wentong Cai	Nanyang Technological University, Singapore
Cheong Siew Ann	Nanyang Technological University, Singapore
Su Yi	Institute for High Performance Computing, A*Star, Singapore

Local Program Committee at NTU Singapore

Ee Hou Yong	Nanyang Technological University, Singapore
Kang Hao	Nanyang Technological University, Singapore

Publicity Chairs

Leonardo Franco	University of Málaga, Spain
Muhamad Azfar Ramli	Institute for High Performance Computing, A*Star, Singapore

Impact Chair

Valeria Krzhizhanovskaya	University of Amsterdam, The Netherlands

Outreach Chair

Alfons Hoekstra	University of Amsterdam, The Netherlands

Program Committee Chair – Workshops on Computational Science

Maciej Paszynski   AGH University of Krakow, Poland

Program Committee – Workshops on Computational Science

Amanda S. Barnard   Australian National University, Australia
Yongjie Jessica Zhang   Carnegie Mellon University, USA

Reviewers

Julen Alvarez-Aramberri   University of the Basque Country, Spain
Philipp Andelfinger   Nanyang Technological University, Singapore
Adrian Bekasiewicz   Gdańsk University of Technology, Poland
Nik Brouw   University of Amsterdam, Netherlands
Roland V. Bumbuc   University of Amsterdam, Netherlands
Wentong Cai   Nanyang Technological University, Singapore
Pedro J. S. Cardoso   Universidade do Algarve, Portugal
Eddy Caron   ENS-Lyon/Inria/LIP, France
Lock-Yue Chew   Nanyang Technological University, Singapore
Ana Cortes   Universitat Autònoma de Barcelona, Spain
Daan Crommelin   CWI Amsterdam, Netherlands
Carlo Cunha   Northern Arizona University, USA
Bartosz Czaplewski   Gdańsk University of Technology, Poland
Venkata Rupesh Kumar Dabbir   Google LLC, USA
Eric Dignum   University of Amsterdam, Netherlands
Vitor Duarte   Universidade NOVA de Lisboa, Portugal
Mariusz Dzwonkowski   Medical University of Gdańsk, Poland
Nahid Emad   Paris-Saclay University, France
Roberto R. Expósito   Universidade da Coruña, CITIC, Spain
Ruy Freitas Reis   Universidade Federal de Juiz de Fora, Brazil
Wlodzimierz Funika   AGH University of Krakow, Poland
Victoria Garibay   University of Amsterdam, Netherlands
Paweł Gepner   Warsaw University of Technology, Poland
Alex Gerbessiotis   New Jersey Institute of Technology, USA
Maziar Ghorbani   Brunel University London, UK
Konstantinos Giannoutakis   University of Macedonia, Greece
Jorge González-Domínguez   Universidade da Coruña, Spain
Yuriy Gorbachev   Soft-Impact LLC, Russia
Michael Gowanlock   Northern Arizona University, USA

George Gravvanis   Democritus University of Thrace, Greece
Derek Groen   Brunel University London, UK
Loïc Guégan   UiT the Arctic University of Norway, France
Rafiazka Hilman   University of Amsterdam, Netherlands
Cillian Hourican   University of Amsterdam, Netherlands
Neil Huynh   Institute of High Performance Computing,
                                     A*STAR, Singapore
Alireza Jahani   Brunel University London, UK
Song Jie   Institute of High Performance Computing,
                                     A*STAR, Singapore
Zhong Jin   Computer Network Information Center, Chinese
                                     Academy of Sciences, China
David Johnson   Uppsala University, Sweden
Takahiro Katagiri   Nagoya University, Japan
Sotiris Kotsiantis   University of Patras, Greece
Sergey Kovalchuk   Huawei, Russia
Valeria Krzhizhanovskaya   University of Amsterdam, Netherlands
Michael Kuhn   Otto von Guericke University Magdeburg,
                                     Germany
Jaeyoung Kwak   Nanyang Technological University, Singapore
Michael Lees   University of Amsterdam, Netherlands
Malcolm Low   Singapore Institute of Technology, Singapore
Lukasz Madej   AGH University of Science and Technology,
                                     Poland
Tomas Margalef   Universitat Autònoma de Barcelona, Spain
Paula Martins   University of Algarve, Portugal
Pedro Medeiros   Universidade Nova de Lisboa, Portugal
Isaak Mengesha   University of Amsterdam, Netherlands
Marianna Milano   Università Magna Græcia di Catanzaro, Italy
Dhruv Mittal   University of Amsterdam, Netherlands
Francisco J. Moreno-Barea   Universidad de Málaga, Spain
Marcin Paprzycki   IBS PAN and WSM, Poland
Giulia Pederzani   Universiteit van Amsterdam, Netherlands
Alberto Perez de Alba Ortiz   University of Amsterdam, Netherlands
Dana Petcu   West University of Timisoara, Romania
Jolan Philippe   IMT Atlantique, France
Dirk Pleiter   University of Groningen, Netherlands
Alexander Pyayt   EPAM Systems, Russia
Rick Quax   University of Amsterdam, Netherlands
Muhamad Azfar Ramli   Institute of High Performance Computing,
                                     A*STAR, Singapore
Amir Raoofy   Technical University of Munich, Germany

Sophie Robert   University of Orléans, France
Daniel Rodriguez   University of Alcalá, Spain
Bertil Schmidt   University of Mainz, Germany
Martin Schreiber   Université Grenoble Alpes/Inria/Laboratoire Jean
                            Kuntzmann, France
Md. Shalihin Othman   D-SIMLAB Technologies Pte. Ltd., Singapore
Joaquim Silva   Nova School of Science and Technology - NOVA
                            LINCS, Portugal
Mateusz Sitko   AGH University of Science and Technology,
                            Poland
Sucha Smanchat   King Mongkut's University of Technology North
                            Bangkok, Thailand
Alexander Smirnovsky   SPbPU, Russia
Yong Sheng Soh   National University of Singapore, Singapore
Ryszard Tadeusiewicz   AGH University of Krakow, Poland
Daisuke Takahashi   University of Tsukuba, Japan
Gary Tan   National University of Singapore, Singapore
Wen Jun Tan   Nanyang Technological University, Singapore
Vítor V. Vasconcelos   University of Amsterdam, Netherlands
Lars Wienbrandt   Kiel University, Germany
Yani Xue   Brunel University London, UK
Xin-She Yang   Middlesex University London, UK
Felix Zhu   IHPC, Singapore

Contents – Part I

ICCS 2025 Main Track Full Papers

Backtranslation and Paraphrasing in the LLM Era? Comparing Data Augmentation Methods for Emotion Classification

Łukasz Radliński[✉][iD], Mateusz Guściora, and Jan Kocoń[iD]

Department of Artificial Intelligence, Wrocaw University of Science and Technology, Wyb. Wyspiaskiego 27, 50-370 Wroclaw, Poland
lukasz.radlinski@pwr.edu.pl

Abstract. Numerous domain-specific machine learning tasks struggle with data scarcity and class imbalance. This paper systematically explores data augmentation methods for NLP, particularly through large language models like GPT. The purpose of this paper is to examine and evaluate whether traditional methods such as paraphrasing and backtranslation can leverage a new generation of models to achieve comparable performance to purely generative methods. Methods aimed at solving the problem of data scarcity and utilizing ChatGPT were chosen, as well as an exemplary dataset. We conducted a series of experiments comparing four different approaches to data augmentation in multiple experimental setups. We then evaluated the results both in terms of the quality of generated data and its impact on classification performance. The key findings indicate that backtranslation and paraphrasing can yield comparable or even better results than zero and a few-shot generation of examples.

Keywords: Data Augmentation · Large Language Models · GPT · data scarcity · class imbalance

1 Introduction

Improvements in the quality of the results of artificial intelligence (AI) systems in recent years have led to their more widespread application and increasing interest. Deep learning (DL), as the main engine of these changes, requires vast amounts of input data to produce good output. Thus, data collection and processing have become crucial. One branch of AI is Natural Language Processing (NLP), which deals with textual data, among others. DL models that handle such data require a large amount and quality of them. However, many domains still have a shortage of such data.

Data augmentation (DA) is one technique for tackling this challenge. It has had a positive impact on computer vision and audio processing, and similar attempts are being made to augment text data. The manipulation of such data

© The Author(s), under exclusive license to Springer Nature Switzerland AG 2025
M. H. Lees et al. (Eds.): ICCS 2025, LNCS 15903, pp. 3–17, 2025.
https://doi.org/10.1007/978-3-031-97626-1_1

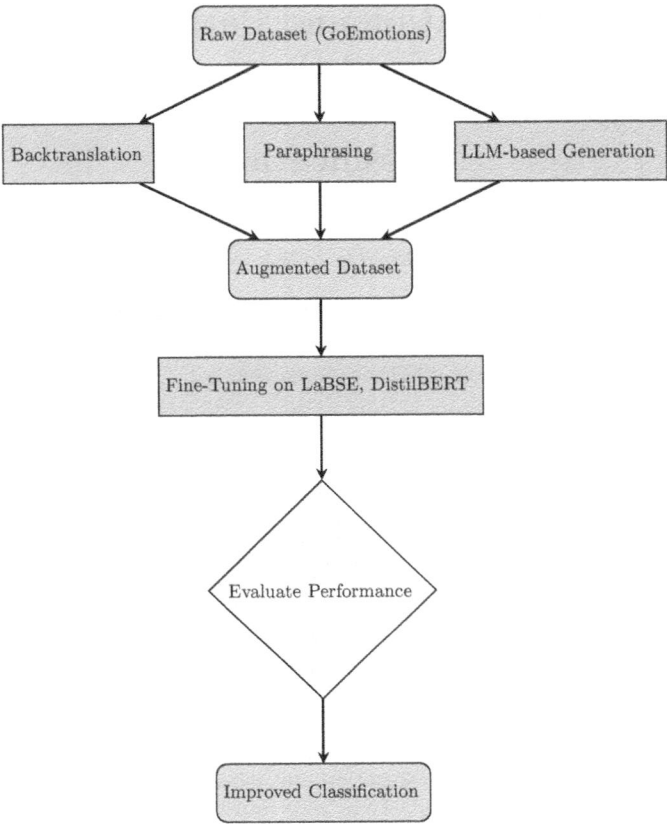

Fig. 1. Data augmentation for emotion classification.

can vary in complexity and sophistication. Methods range from simple techniques like random insertion of characters to more advanced approaches, such as employing the generative power of large language models (LLMs) for paraphrasing entire data samples. LLMs, such as ChatGPT, are becoming more popular and accessible, making them valuable tools. Their generative capabilities enable the creation of extensive and coherent text samples. Although improvements in DL and NLP have been substantial, domain-specific tasks continue to face challenges due to data scarcity and the resulting class imbalance. The use of LLMs in NLP presents a promising new approach to DA. In this paper, we conducted a comprehensive comparison of multiple DA techniques on the GoEmotions classification dataset [8]. We chose the most suitable categories to augment the dataset. Then we compared the conventional LLM generative approach to more traditional methods, i.e. backtranslation and paraphrasing, performed using language models. We then conducted a detailed analysis of the lexical diversity and semantic fidelity of the generated data. In order to verify the impact of the augmentation, we then used the augmented dataset to fine-tune two popular transformer models and compared the increase in performance across experimental setups, see Fig. 1. The analysis of results gives key insights into how the new generation of language models can effectively leverage traditional approaches to data augmentation. All the code and prompts used to carry out the experiments are publicly available on GitHub[1].

2   Related Work

Looking at a survey published last year on data augmentation [37], two main groups of data augmentation techniques are commonly used. The first group that saw a significant increase in popularity over the past years is the generated content-based approaches. Generative models have been utilized for data augmentation ever since they appeared. Works such as [11] and [19] showed that transformer models could effectively increase data diversity while preserving semantic fidelity. The works of [22] have shown that such augmentation can directly increase classification performance. The release of GPT-3 cemented the role of generative models in data augmentation, as multiple studies have shown that they can be utilized to increase performance in classification tasks [2]. Various techniques were used to increase the effectiveness of these models in data augmentation, such as fine-tuning the model [36] or using reinforcement learning [22]. However, one of the most successfully employed techniques has been, without a doubt, Few Shot Learning (FSL). Numerous studies showed its effectiveness for models such as GPT-3 [35] and ChatGPT [7, 12, 14–18, 23, 24, 30, 31]. However, looking at the survey dedicated to text data augmentation with Large Language Models [5], one of the crucial challenges in using LLM is to ensure the diversity and quality of generated data.

The second, less popular, but still common group of methods listed in [37] are label-based methods. These largely focus on using dataset labels or their embeddings to improve sampling quality or text generation. Although often slightly more complex than generation-based approaches, They have been proven to substantially improve the quality of generated data [6, 29, 32] and even allow for explainability in data augmentation [20].

Another commonly employed technique for data augmentation that has not been listed in [37] since it predates Large Language Models is backtranslation. The core idea to augment data by translating text into a foreign language and then back to the original has been used effectively long before generative devices such as BART or GPT were released [9]. It was limited by the quality of the translation tools. However, only a few years later, models from the Workshop on Machine Translation (WMT models) [33] and then the seq2seq transformers [3] would show greater and greater improvements in the achievable results. Although the backtranslation technique predates the Large Language Models, it is still utilized [34] and was even effectively used to increase the quality of instructions for Large Language Models [21, 25, 26].

---

[1] https://github.com/marentoo/data_augmentation_text_classification_task.git.

3   Dataset – GoEmotions

We decided to conduct our experiments on the GoEmotions dataset. This extensive dataset comprises textual comments labeled with specific emotions. Collab-
oratively developed by researchers from Google and Amazon, the dataset focuses
on emotion-related tasks. It uses data from Reddit from 2005 to 2019, meticu-
lously curated from subreddits with at least 10,000 comments, and exclusively
in English [8]. GoEmotions is a notably large dataset that benefits from human
annotation, resulting in high ground truth accuracy. These attributes are particularly valuable given the scarcity of similar public datasets and the relative
expense of human annotation. However, the data set has some obvious drawbacks that researchers point out as well. The dataset contains biases and is not
representative of global diversity due to taking data specifically from Reddit.
Notably, it is a multi-label dataset with 27 emotional labels. Therefore, in our
research, we decided to focus on five labels that are least represented in the
dataset. Looking at the class distribution of the labels, Fig. 2, there is a visible
class imbalance in the dataset. It exhibits very limited representation in certain
emotional categories. The five emotions with the least representation are:

- embarrassment (291 samples),
- nervousness (156 samples),
- relief (145 samples),
- pride (105 samples),
- grief (75 samples),

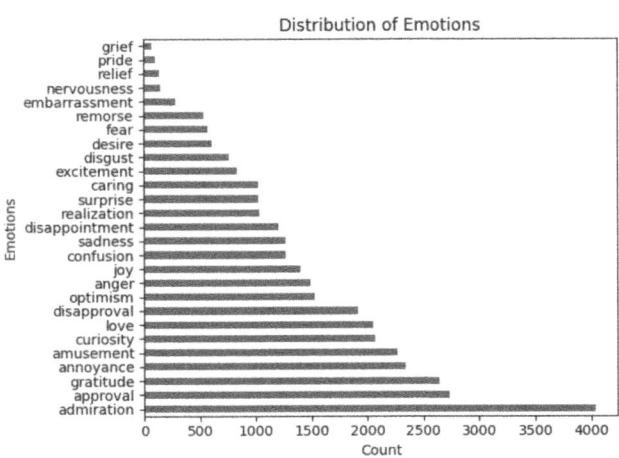

Fig. 2. Emotion Labels Distribution.

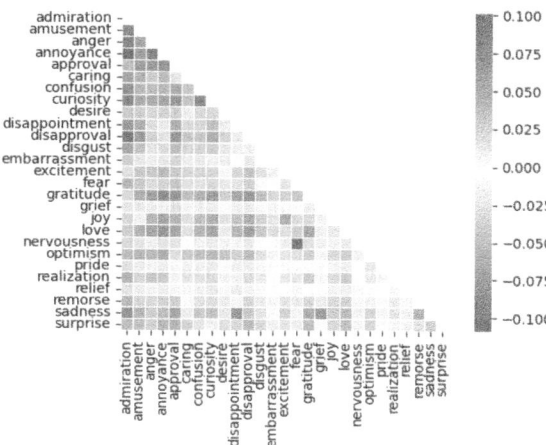

Fig. 3. Correlation Heatmap of labels.

Furthermore, looking at the correlations between individual emotions, Fig. 3, there are visible dependencies between some of the labels. These include a high correlation between annoyance and anger, sadness and grief, remorse and sadness, disappointment and sadness, fear and nervousness, desire and optimism, excitement and joy, as well as confusion and curiosity. This means that many models may struggle to distinguish between these labels, even more so given the significant imbalance of the dataset. These qualities make the GoEmotions dataset very suitable for the data augmentation task. In our experimental setup, we generated additional samples for each of the five least represented classes and then used the modified data set to fine-tune the models used for evaluation.

4 Data Augmentation Methods

In our study, we wanted to focus on novel approaches to data augmentation using LLM. We decided to conduct our research using four data augmentation techniques:

– Oversampling with dummy copies - baseline method
– Paraphrasing via prompting on a sentence level
– Generating via prompting with zero-shot learning and few-shot learning
– Backtranslation

Our goal is to verify both the quality of the generated data and its impact on the utility of the dataset.

4.1 Baseline - Oversampling

Oversampling is a method used to address the problem of class imbalance in machine learning. It involves artificially manipulating the dataset to mitigate this issue. One of the most well-known techniques in this context is the Synthetic Minority Oversampling Technique (SMOTE). Oversampling is used to mitigate class imbalance by increasing the instances of the minority class. In its fundamental form, it involves randomly duplicating instances of this class or classes. We use this approach as a baseline for the research as it provides no linguistic diversity while maintaining maximum semantic similarity to the original dataset. These two criteria are the main aspects of the quality of the generated data that we wanted to evaluate. Therefore, oversampling becomes the obvious choice as a baseline method. We compare two datasets augmented through oversampling. One is modified with three samples of each example in augmented classes, while the second is modified with five.

4.2 Paraphrasing

Paraphrases are texts that have been rewritten or rephrased using different words, structures, or forms. While the semantic meaning of a paraphrased sentence remains the same, its lexical and syntactic structures may differ. Paraphrasing is a highly effective and valuable method for data augmentation in NLP problems because it maintains semantic fidelity while introducing lexical diversity. Through such an enhancement, the model can learn new relationships and expand its vector space. By enhancing the scope of wording, the model can also prevent overfitting to specific linguistic patterns. This may have a wide range of applications, such as language understanding, summarization, generation, and translation [3]. In our experiments, we utilize GPT-3.5 and GPT-4.0, which have already shown great potential for data augmentation [7,23]. Both models are used in three prompt configurations:

- Prompt 1 - iteratively asking the model for a single paraphrase of each sample multiple times
- Prompt 2 - Nmax - asking models for generating multiple paraphrases at once for each sample and introducing as many as possible to the dataset
- Prompt 2 - Nbal - asking models for generating multiple paraphrases at once for each sample and introducing an equal number of samples from each class to the dataset

4.3 Generating via Zero-Shot and Few-Shot Generation

Zero-shot learning (ZSL) and few-shot learning (FSL) are classes of methods used in machine learning to increase the generalizability of a model, Fig. 4. These techniques have been widely adopted in classification problems where data is limited. The idea is to use either a few examples, in the case of FSL, or no examples at all, in the case of ZSL, to employ a trained deep learning model for a chosen new task. The task requires a model to infer classes that it has never seen during training (zero-shot learning) or has only seen a small number of times (few-shot learning). Methods designed to address data scarcity and class imbalance typically prioritize low-resource approaches. Zero-shot and few-shot learning are very resource effective. When applied using language models, these methods can be particularly effective as they also benefit from in-context learning. In terms of performance and results, FSL often yields better results than zero-shot, but it requires samples and more resources. FSL can achieve results comparable to those obtained through fine-tuning, and in certain tasks, these methods are even capable of outperforming human-level performance [4, 13,27]. In our experiments, we used both GPT-3.5 and GPT-4 in zero-shot and 5-shot scenarios. This approach cannot be easily compared to backtranslation and paraphrasing in terms of data diversity and semantic fidelity. However, it is important to include it since it is arguably the simplest and most resource-effective LLM-based augmentation method.

Example - Zero Shot Learning

Prompt: "Generate N different sentences in various forms that express a strong emotional sentiment for the following emotion:"

System message: "You are a helpful assistant. Output sentences separated by newline in reply to <prompt>. Sentences should vary in type, slang, length, structure, tone and style, sentences such as comments, responses, opinions, and facts. Its not necessary to often use the emotions name in every sentence. Do not number output or use bullet point for the output."

Example output: ["I'm trying to laugh it off, but inside I'm dying of embarrassment.", "I can't believe I just did that!", "I still cringe thinking about that time I tripped and fell in front of everyone.", "Embarrassment can make you wish the ground would open up and swallow you whole.", "You know that feeling when you accidentally send a text meant for your friend to your boss? Yeah, pure embarrassment.", "Embarrassment can turn your face as red as a tomato."]

Fig. 4. Example ZSL, input and output for GPT model.

4.4 Backtranslation

Backtranslation (BT) is a specific type of paraphrasing commonly used to generate semantically similar text with different lexical features. This process requires the source text to be translated into another language and then translated back into the source language. The paraphrases generated in this process are essentially by-products of the translation process by a model. Using various languages to generate new text content results in diversity, potentially enhancing model performance and expanding the dataset. The produced text exhibits diversity because different languages have unique structures, syntax, and linguistic features. Furthermore, imperfect translations can lead to numerous variations in the translated text [9]. Researchers have tested different variations of BT. These variations include employing a single pair of languages, using multiple languages at different levels, and making architectural modifications to diversify the output [1,33]. With this method, certain challenges also arise. Often, translation models tend to favor high-frequency words, which can lead to repetition and limit diversity. This can result in overfitting the data and losing subtle details. Furthermore, the quality of the results is highly dependent on the quality of the translation model [20,22]. Translation models tend to perform with various translation qualities depending on the target language. Using only one language for back-translation also vastly limits the number of paraphrases generated. Therefore, to increase the number of models that can be compared to each other and reach a sufficient number of generated examples, we decided to perform backtranslation on multiple languages for each model and then aggregate all the examples into one dataset. In this study, experiments involving backtranslation were conducted using the following models (translating into the following languages):

- DeepL (Russian, Polish, Finnish, Japanese, Chinese, Bulgarian, Spanish, Hungarian, Greek, Turkish)
- GPT-3.5 (Polish, Chinese, Russian, Hindi, Hungarian, Finnish, Spanish, Japanese, Turkish, Arabic)
- GPT-4 (Polish, Chinese, Russian, Hindi, Hungarian, Finnish, Spanish, Japanese, Turkish, Arabic)
- GPT-4-turbo(Polish, Chinese, Russian, Hindi, Hungarian, Finnish, Spanish, Japanese, Turkish, Arabic)
- MarianMT model family (Hindi, Polish, Hungarian, Finnish, Russian, Chinese, Spanish, Japanese, Turkish, Arabic)

Most of the languages used for the backtranslation, but some of them differ due to a limited common group of sufficiently foreign languages to English available for all solutions.

5   Results Evaluation

We evaluate the increase in data in terms of both the quality of the generated data and the increase in classification performance achieved through augmentation. The evaluation of data quality is performed by measuring the linguistic diversity and the semantic fidelity achieved. The evaluation involved analyzing every fourth sentence pair between the reference and the generated text. The results were recorded for each pair, followed by calculating the average for the entire set. Such actions were repeated for every generated set. Assessment of the increase in classification is achieved by fine-tuning two reference models and comparing the results across the experimental setups.

5.1   Linguistic Diversity

There are multiple known metrics that are used to measure lexical diversity. In order to receive more nuanced information, we decided to utilize four metrics for measuring diversity:

- word count and word count ratio
- Jaccard dissimilarity
- information entropy
- ratio of the Type Token Ratios (TTR)

The word count ratio is calculated as the ratio of average word counts between original sentences and paraphrases. Jaccard dissimilarity (1) is calculated as follows:

$$J_{\text{dissimilarity}}(A, B) = 1 - \frac{|A \cap B|}{|A \cup B|} \tag{1}$$

where A and B are two sentences to compare. These are reference sentences and generated sentences. Information entropy is calculated as the ratio between the information entropy (Fig. 2) of the generated and reference sentences. The calculation of the information entropy for one of these sentences is as follows:

$$H(S) = -\sum_{i=1}^{n} P(w_i) \log_2 P(w_i) \tag{2}$$

TTR ratio is calculated as the ratio between TTR (3) of the generated set of sentences and divided by the original. Each TTR is calculated as follows.

$$TTR = \frac{\text{types}}{\text{tokens}} \tag{3}$$

5.2 Semantic Fidelity

Many techniques and studies emphasize measuring semantic fidelity, which is the similarity in meaning between the new text and the original. For data augmentation and, consequently, for subsequent machine learning tasks, maintaining semantic fidelity is crucial. Preserving the meaning and the label is key to determining the effectiveness of a model. There are two most common approaches to measuring semantic fidelity. The first is to use distance measures such as cosine similarity, dot product similarity, or Euclidean distance to calculate the distance between vector representations of the latent space of text. The other is BERTscore, which employs BERT embeddings to grasp semantic relations between texts. In our experiments, we decided to use both types of measures:

– cosine similarity for embeddings with Hugging Face model: 'paraphrase-MiniLM-L6-v2'
– BERTScore employing DistilBERT, which returns an F1 score

Cosine similarity (4) is calculated as follows:

$$\text{Cosine Similarity}(\mathbf{A}, \mathbf{B}) = \frac{\mathbf{A} \cdot \mathbf{B}}{\max(\|\mathbf{A}\|, \varepsilon) \cdot \max(\|\mathbf{B}\|, \varepsilon)} \tag{4}$$

Where A and B are two vectors, specifically the embedding representations of the reference and generated text sentences, the max function and epsilon parameter are used to avoid division by zero or very small numbers.

5.3 Classification Improvement

We measure the improvement in classification by fine-tuning two models, LaBSE [10], and distilBERT [28], both showing impressive performance in multiple classification tasks, including emotion classification. Then, we compare the results with the performance of models before fine-tuning. We measure the change in the f-1 measure for both the entire dataset and the f1-macro measure for the augmented classes.

6   Results

Table 1. Average results of Lexical Diversity for paraphrasing methods

Data aug.	Word Original	Word Generated	Word Ratio	Jaccard Dissimilarity	Entropy	TTR Ratio
Prompt 1 GPT-3.5	12	16	**1.4146**	**0.8190**	**1.1523**	**1.2167**
Prompt 2 GPT-3.5 Nmax	11	13	1.3148	0.8069	1.1250	1.1876
Prompt 2 GPT-3.5 Nbal	12	15	1.3900	0.8117	1.1410	1.2164
Prompt 1 GPT-4	12	14	1.2479	0.8229	1.1010	1.1733
Prompt 2 GPT-4 Nmax	11	12	1.2419	0.8233	1.1055	1.1858
Prompt 2 GPT-4 Nbal	12	14	1.2414	0.8184	1.1027	1.1862
BT with MarianMT	12	14	**1.9729**	**0.7992**	**0.9157**	**1.1748**
BT with DeepL	13	13	1.0093	0.5036	1.0006	1.0016
BT with gpt-3.5	13	14	1.1380	0.6152	1.0482	1.0554
BT with gpt-4	13	14	1.0861	0.5757	1.0303	1.0462
BT with gpt-4-turbo	13	21	**2.3324**	**0.9636**	**1.3250**	**1.4611**

Table 2. Average results of Semantic Fidelity for paraphrasing methods

Data aug.	Cosine Similarity	Bertscore-F1
Prompt 1 GPT-3.5	**0.7031**	**0.8365**
Prompt 2 GPT-3.5 Nmax	0.6903	0.8324
Prompt 2 GPT-3.5 Nbal	0.6774	0.8299
Prompt 1 GPT-4	0.6960	0.8461
Prompt 2 GPT-4 Nmax	0.6774	0.8354
Prompt 2 GPT-4 Nbal	0.6928	0.8400
BT with MarianMT	0.4758	0.7848
BT with DeepL	**0.8733**	**0.9248**
BT with gpt-3.5	0.8092	0.8989
BT with gpt-4	0.8465	0.9127
BT with gpt-4-turbo	0.1516	0.6892

6.1   Lexical Diversity and Semantic Fidelity of Generated Data

Looking at the diversity measures of the paraphrasing methods (Table 1), GPT-3.5 created longer sentences and had a higher TTR ratio than GPT-4. However, the Jaccard dissimilarity indicates that the GPT-4 methods were better at introducing lexical diversity. BT techniques (Table 1) introduced less diversity overall, except when using MarianMT models and the GPT-4-turbo, where the results indicated a high diversity. The best fidelity of meaning for paraphrasing (Table 2) was achieved with prompt 1 for GPT-3.5-turbo and GPT-4, both for cosine similarity and BERTScore. BT techniques (Table 2) received better results of generated text than paraphrasing methods, except for Hugging Face models and GPT-4-turbo. Using DeepL and GPT-4 presented the highest results.

Table 3. Comparison of performance change for experimental setups.

Data aug	FT Model	F1-macro (all Cls)	%Change (all Cls)	F1-macro (aug Cls)	%Change (aug Cls)	F1-macro (othr Cls)	%Change (othr Cls)
Baseline	LaBSE	0.467	0.00	0.211	0.00	0.522	0.00
	distilBERT	0.458	0.00	0.174	0.00	0.520	0.00
Oversampling 3x	LaBSE	0.477	2.19	0.330	56.67	0.509	-2.59
	distilBERT	0.474	3.43	0.229	31.34	0.527	1.39
Oversampling 5x	**LaBSE**	**0.484**	**3.62**	**0.327**	**55.39**	**0.517**	**-0.92**
	distilBERT	**0.481**	**5.06**	**0.307**	**76.12**	**0.519**	**-0.11**
Prompt 1 GPT-3.5	LaBSE	0.473	1.35	0.276	30.94	0.516	-1.25
	distilBERT	0.472	3.10	0.286	64.06	0.513	-1.34
Prompt 2 GPT-3.5 Nmax	**LaBSE**	**0.489**	**4.74**	**0.388**	**84.15**	**0.511**	**-2.23**
	distilBERT	0.474	3.56	0.266	52.81	0.520	-0.03
Prompt 2 GPT-3.5 Nbal	LaBSE	0.479	2.66	0.285	35.41	0.521	-0.21
	distilBERT	0.476	3.86	0.262	50.23	0.522	0.49
Prompt 1 GPT-4	LaBSE	0.486	4.22	0.313	48.32	0.524	0.35
	distilBERT	**0.484**	**5.70**	**0.307**	**76.23**	**0.523**	**0.56**
Prompt 2 GPT-4 Nmax	LaBSE	0.478	2.36	0.276	30.80	0.522	-0.14
	distilBERT	0.466	1.77	0.221	26.69	0.520	-0.05
Prompt 2 GPT-4 Nbal	LaBSE	0.481	3.11	0.256	21.45	0.530	1.50
	distilBERT	0.466	1.72	0.213	22.22	0.521	0.23
0-shot gpt-3.5	**LaBSE**	**0.492**	**5.38**	**0.324**	**53.68**	**0.528**	**1.14**
	distilBERT	0.472	3.10	0.255	46.56	0.520	-0.07
0-shot gpt-4	LaBSE	0.474	1.61	0.243	15.09	0.524	0.42
	distilBERT	0.482	5.17	0.280	60.79	0.526	1.12
5-shot gpt-3.5	LaBSE	0.485	4.03	0.326	54.53	0.520	-0.40
	distilBERT	**0.488**	**6.55**	**0.309**	**77.55**	**0.527**	**1.38**
5-shot gpt-4	LaBSE	0.489	4.72	0.310	47.08	0.527	1.00
	distilBERT	0.479	4.50	0.289	66.07	0.520	0.01
BT with DeepL	**LaBSE**	**0.497**	**6.47**	**0.377**	**78.74**	**0.523**	**0.13**
	distilBERT	**0.494**	**7.77**	**0.387**	**121.99**	**0.517**	**-0.55**
HuggingFace models	LaBSE	0.490	5.10	0.340	61.22	0.523	0.18
	distilBERT	0.473	3.27	0.298	71.30	0.511	-1.68
BT with gpt-3.5	LaBSE	0.487	4.41	0.293	38.82	0.530	1.40
	distilBERT	0.486	5.66	0.348	99.83	0.516	-0.79
BT with gpt-4	LaBSE	0.493	4.39	0.341	62.03	0.526	0.71
	distilBERT	0.483	5.46	0.313	79.39	52.02	0.07
BT with gpt-4turbo	LaBSE	0.482	3.40	0.309	46.99	0.520	-0.41
	distilBERT	0.479	4.75	0.313	79.97	0.516	-0.72

6.2 Impact on Classification

Looking at the results of fine-tuning LaBSE and distilBERT models on augmented datasets (Table 3), there is a visible increase in performance with regard to augmented classes, as well as the entire dataset across all experimental setups. There is also no significant observable decrease in the performance of non-augmented classes, which means that improvement in the augmented classes does not happen at the cost of the rest of the dataset. Even baseline oversampling improved the model's ability to generalize. The mere addition of copies of the examples to the training set to increase results in a gain of more than 76% with distilBERT in the augmented classes and 5% on the entire dataset with negligible impact on the classification of other classes. Oversampling with more samples predictably yielded better results.

When it comes to paraphrasing, the smallest increase was approximately 1.3% and 30% increase on the dataset and augmented classes with prompt 1 using the LaBSE model. However, the highest results were achieved with Prompt 1 GPT-4 with DistilBERT and Prompt 2 GPT-3.5 nmax with the LaBSE model, resulting in increases of more than 76% and 84%, respectively. In particular, Prompt 2 GPT-3.5 Nmax setup achieved the largest decrease in performance in other classes simultaneously, although it was only 2.23%. The results with LaBSE were slightly better in all methods. Looking at the purely generative approach, the highest gain was reported for 5-shot GPT-3.5, showing a 6.5% and 77.5% increase with DistilBERT, and for 0-shot GPT-3.5 with LaBSE. These two experiments also both resulted in a slight increase in the performance of the non-augmented classes. No significant difference was between the results achieved by 0-shot and 5-shot learning. BT achieved better results than the previous methods (Table 3). The best gain was reported with backtranslation using DeepL with disilBERT, showing a 7. 7% increase in the F1-macro in the entire data set and a greater 121% increase in the augmented classes that decreased only by 0. 55% in the other classes.

It is worth noticing that there is significant variance in results achieved by particular experimental setups for each method. All the methods compared to the oversampling yielded the results underperforming 3x oversampling on both models and the results beating 5x oversampling. Another interesting observation is that GPT-4 achieved the best results for a particular augmentation method in only one instance and only on the distilBERT model. This leads to the conclusion that while modern Language models can effectively leverage the traditional approaches to data augmentation, their overall performance largely depends on the models used for augmentation and specific configuration elements such as prompt and example selection.

7 Limitations and Future Research

Data augmentation using large language models seems effective in improving performance and mitigating class imbalance, as shown with the GoEmotions dataset for multi-label classification. The techniques presented in this research have the potential for a broader application across various datasets and problems, but further experiments need to be performed to confirm these findings. Future research and experiments should focus on:

- methods using different models and datasets,
- experimenting with different parameters of the model,
- evaluation of generated text with LLM-as-a-judge,
- specific setups for backtranslation and multiple prompts for both paraphrasing and generation,
- evaluating the impact of LLM and in-context learning biases on the data augmentation,
- addressing cultural and global biases resulting from sourcing data from Reddit, e.g. by enriching data with other datasets,

8 Conclusions

We conducted experiments on multiple different approaches for data augmentation utilizing LLM. All methods produced semantically similar but distinct samples. Paraphrasing showed greater lexical diversity, while backtranslation maintained better semantic similarity, except for HuggingFace models and GPT-4-turbo. All methods improved the classification results. Although the improvements were modest, they are significant given the augmented classes' underrepresentation. The best F1-macro scores were achieved using the GPT-4 model for paraphrasing. Zero-shot and few-shot learning generally outperformed paraphrasing, with the best results from GPT-3.5. Backtranslation produced the best overall results, with the highest classification outcome using the DeepL model.

The best F1 scores across all classes were achieved using backtranslation with LaBSE and DistilBERT via DeepL, GPT-4 on LaBSE, and zero-shot learning with GPT-3.5 turbo. These methods and setups achieved an F1-macro score above 0.49. Analyzing the results of classifying only augmented classes, the greatest gains were observed with backtranslation. The F1 score increased from 0.17 up to 0.38. Assessing different methods and setups should include the costs of data augmentation. The main costs include the complexity of creating augmented data, API usage fees, the time required for fine-tuning the model on the augmented training set, and the computational resources needed. While DistilBERT produced slightly worse results than LaBSE, it required significantly less training time and computational resources, making it a more suitable model when resources are limited. Zero-shot learning and few-shot learning require fewer input tokens compared to paraphrasing and backtranslation. Oversampling slightly mitigated the class imbalance problem, but data scarcity remains. It is the least demanding method for acquiring new data in terms of complexity and time.

Acknowledgements. Financed by: (1) the National Science Centre, Poland (2021/41/B/ST6/04471); (2) CLARIN ERIC (2024âĂŞ2026), funded by the Polish Minister of Science (agreement no. 2024/WK/01); (3) CLARIN-PL, the European Regional Development Fund, FENG programme (FENG.02.04-IP.040004/24); (4) statutory funds of the Department of Artificial Intelligence, Wroclaw Tech; (5) the Polish Ministry of Education and Science ("International Projects Co-Funded" programme); (6) the European Union, Horizon Europe (grant no. 101086321, OMINO); (7) the EU project "DARIAH-PL'n', under investment A2.4.1 of the National Recovery and Resilience Plan. The views expressed are those of the authors and do not necessarily reflect those of the EU or the European Research Executive Agency.

References

1. Abonizio, H.Q., Paraiso, E.C., Barbon, S.: Toward text data augmentation for sentiment analysis. IEEE Trans. Artif. Intell. (2021)
2. Balkus, S.V., Yan, D.: Improving short text classification with augmented data using gpt-3. In: Natural Language Engineering, pp. 1–30 (2022)
3. Beddiar, D.R., Jahan, M.S., Oussalah, M.: Data expansion using back translation and paraphrasing for hate speech detection. Online Social Networks and Media (2021)
4. Brown, T., et al.: Language models are few-shot learners. Adv. Neural. Inf. Process. Syst. **33**, 1877–1901 (2020)
5. Chai, Y., Xie, H., Qin, J.S.: Text data augmentation for large language models: a comprehensive survey of methods, challenges, and opportunities (2025)
6. Chen, J., Yang, Z., Yang, D.: Mixtext: linguistically-informed interpolation of hidden space for semi-supervised text classification. arXiv preprint arXiv:2004.12239 (2020)
7. Dai, H., et al.: Auggpt: leveraging chatgpt for text data augmentation. arXiv preprint arXiv:2302.13007 (2023)
8. Demszky, D., Movshovitz-Attias, D., Ko, J., Cowen, A., Nemade, G., Ravi, S.: Goemotions: a dataset of fine-grained emotions. arXiv preprint arXiv:2005.00547 (2020)
9. Fadaee, M., Bisazza, A., Monz, C.: Data augmentation for low-resource neural machine translation. arXiv preprint arXiv:1705.00440 (2017)
10. Feng, F., Yang, Y., Cer, D., Arivazhagan, N., Wang, W.: Language-agnostic bert sentence embedding (2022)
11. Feng, S.Y., Gangal, V., Kang, D., Mitamura, T., Hovy, E.: Genaug: data augmentation for finetuning text generators. arXiv preprint arXiv:2010.01794 (2020)
12. Ferdinan, T., Kocoń, J.: Fortifying nlp models against poisoning attacks: the power of personalized prediction architectures. Inf. Fusion **114**, 102692 (2025)
13. Gilardi, F., Alizadeh, M., Kubli, M.: Chatgpt outperforms crowd workers for text-annotation tasks. Proc. Natl. Acad. Sci. **120**(30) (2023)
14. Kazienko, P., et al.: Human-centered neural reasoning for subjective content processing: hate speech, emotions, and humor. Inf. Fusion **94**, 43–65 (2023)
15. Kochanek, M., et al.: Improving training dataset balance with chatgpt prompt engineering. Electronics **13**(12), 2255 (2024)
16. Kocoń, J.: Deep emotions across languages: a novel approach for sentiment propagation in multilingual wordnets. In: 2023 IEEE International Conference on Data Mining Workshops (ICDMW), pp. 744–749. IEEE (2023)

17. Kocoń, J., et al.: Chatgpt: jack of all trades, master of none. Inf. Fusion (2023)
18. Koptyra, B., Ngo, A., Radliński, Ł., Kocoń, J.: Clarin-emo: Training emotion recognition models using human annotation and chatgpt. In: International Conference on Computational Science, pp. 365–379. Springer, Heidelberg (2023). https://doi.org/10.1007/978-3-031-35995-8_26
19. Kumar, V., Choudhary, A., Cho, E.: Data augmentation using pre-trained transformer models. arXiv preprint arXiv:2003.02245 (2020)
20. Kwon, S., Lee, Y.: Explainability-based mix-up approach for text data augmentation. ACM Trans. Knowl. Disc. Data (2023)
21. Li, X., et al.: Self-alignment with instruction backtranslation (2024)
22. Liu, R., et al.: Data boost: text data augmentation through reinforcement learning guided conditional generation. arXiv preprint arXiv:2012.02952 (2020)
23. Møller, A.G., Dalsgaard, J.A., Pera, A., Aiello, L.M.: Is a prompt and a few samples all you need? Using gpt-4 for data augmentation in low-resource classification tasks. arXiv preprint arXiv:2304.13861 (2023)
24. Ngo, A., Kocoń, J.: Integrating personalized and contextual information in fine-grained emotion recognition in text: a multi-source fusion approach with explainability. Inf. Fusion, 102966 (2025)
25. Nguyen, T., et al.: Better alignment with instruction back-and-forth translation (2024)
26. Qi, Y., Peng, H., Wang, X., Xu, B., Hou, L., Li, J.: Constraint back-translation improves complex instruction following of large language models (2024)
27. Radford, A., Wu, J., Child, R., Luan, D., Amodei, D., Sutskever, I.: Language models are unsupervised multitask learners (2019)
28. Sanh, V., Debut, L., Chaumond, J., Wolf, T.: Distilbert, a distilled version of bert: smaller, faster, cheaper and lighter (2020)
29. Sun, L., Xia, C., Yin, W., Liang, T., Yu, P.S., He, L.: Mixup-transformer: dynamic data augmentation for nlp tasks. arXiv preprint arXiv:2010.02394 (2020)
30. Woźniak, S., Kocoń, J.: From big to small without losing it all: text augmentation with chatgpt for efficient sentiment analysis. In: 2023 IEEE International Conference on Data Mining Workshops (ICDMW), pp. 799–808. IEEE (2023)
31. Woźniak, S., Koptyra, B., Janz, A., Kazienko, P., Kocoń, J.: Personalized large language models. arXiv preprint arXiv:2402.09269 (2024)
32. Wu, X., Lv, S., Zang, L., Han, J., Hu, S.: Conditional BERT contextual augmentation. In: Rodrigues, J., et al. (eds.) ICCS 2019. LNCS, vol. 11539, pp. 84–95. Springer, Cham (2019). https://doi.org/10.1007/978-3-030-22747-0_7
33. Xie, Q., Dai, Z., Hovy, E., Luong, T., Le, Q.: Unsupervised data augmentation for consistency training. Adv. Neural Inf. Process. Syst. (2020)
34. Xu, Q., Hong, Y., Chen, J., Yao, J., Zhou, G.: Data augmentation via back-translation for aspect term extraction. In: 2023 International Joint Conference on Neural Networks (IJCNN), pp. 1–8 (2023)
35. Yoo, K.M., Park, D., Kang, J., Lee, S.W., Park, W.: Gpt3mix: leveraging large-scale language models for text augmentation. arXiv preprint arXiv:2104.08826 (2021)
36. Zheng, C., et al.: Augesc: dialogue augmentation with large language models for emotional support conversation. arXiv preprint arXiv:2202.13047 (2022)
37. Zhou, Y., Guo, C., Wang, X., Chang, Y., Wu, Y.: A survey on data augmentation in large model era (2024)

Explainable Artificial Intelligence for Bioactivity Prediction: Unveiling the Challenges with Curated CDK2/4/6 Breast Cancer Dataset

Adam Sułek[1], Jakub Klimczak[1], Jakub Jończyk[1,2], Tomasz Kosciolek[1], Tomasz Danel[1,3(✉)], and Barbara Pucelik[4(✉)]

[1] Sano Centre for Computational Medicine, 30-054 Kraków, Poland
[2] Department of Medicinal Chemistry, Faculty of Pharmacy, Jagiellonian University Medical College, 30-688 Kraków, Poland
[3] Faculty of Chemistry, Jagiellonian University, 30-387 Kraków, Poland
tomasz.danel@uj.edu.pl
[4] Łukasiewicz Research Network, Kraków Institute of Technology, 30-418 Kraków, Poland
barbara.pucelik@kit.lukasiewicz.gov.pl

Abstract. In recent years, the interplay between machine learning (ML) and cheminformatics has driven advancements in bioactivity prediction. However, the challenge of model explainability remains a significant barrier to adopting these approaches in drug discovery. This study addresses critical shortcomings in existing modeling techniques by examining the assumptions of feature independence and contribution additivity that are the foundation of traditional explainability methods. We investigate fingerprint-based and molecular graph models within quantitative structure-activity relationship modeling. While these models demonstrate impressive predictive performance, they offer limited actionable insights for medicinal chemists. To assist researchers in developing useful and interpretable activity prediction models, we propose a new benchmark based on the pharmacophore concept, commonly used in preliminary compound filtering. Furthermore, we introduce PharmacoScore, a novel evaluation metric designed to assess whether ML-based explanations prioritize essential pharmacophore components over non-critical features. Our findings highlight a crucial misalignment between ML model explanations and established pharmacophore principles, revealing a pressing need for innovative interpretability strategies in cheminformatics. This work not only offers a valuable resource but also sets the stage for future research, enhancing the transparency of ML in drug discovery.

Keywords: explainable artificial intelligence · machine learning · cheminformatics · bioactivity prediction · pharmacophores

M. H. Lees et al. (Eds.): ICCS 2025, LNCS 15903, pp. 18–32, 2025.
https://doi.org/10.1007/978-3-031-97626-1_2

1   Introduction

Explainable artificial intelligence (XAI) is a rapidly expanding field, mostly due to the need for understanding predictions of machine learning (ML) models in high-risk applications such as medical image analysis, fraud detection, and autonomous vehicle decision-making [1]. In drug discovery, XAI methods help medicinal chemists identify novel therapeutic molecules with improved bioactivity, which is the most crucial property in the early stages of drug discovery. Bioactivity refers to the effect that a compound has on the organism, usually caused by the activation or suppression of the selected protein targets. Nowadays, ML models are employed to predict bioactivity from the chemical structure to expedite the search for new potent molecules and reduce the cost related to unsuccessful experimental trials. Explainability techniques provide additional insights about model predictions that can be leveraged by medicinal chemists to propose more effective molecular designs [4].

In bioactivity prediction, simple ML models were classically used to predict the activity of small molecules within a series of compounds sharing the same core, e.g. by fitting a multiple linear regression to a few molecular descriptors. This process is called quantitative structure-activity relationship (QSAR) modeling [4]. Currently, more advanced ML models are utilized for predicting activity across more diverse compound libraries obtained from large databases. One of the predominant approaches is applying models like random forest (RF) or support vector machines (SVM) to predict activity from molecular descriptors or fingerprints, which are feature vector representations. The other approach uses graph neural networks (GNN) that work directly on chemical structures represented as molecular graphs [13]. Inspired by current advancements in natural language processing (NLP), text representations like SMILES are sometimes used [9]. All of these techniques are purely ligand-based and provide an alternative to more computationally demanding structure-based methods like molecular docking. However, the quality of predictions of such methods depends on the quality of the training data, and most of these models fail to generalize to novel chemical spaces, providing accurate predictions primarily for the compounds close to the training dataset [7]. That is also why XAI methods should be employed to better understand the knowledge gained by these models and avoid overfitting to certain chemical structures.

Many XAI methods for molecules have been adopted from other domains, such as computer vision and NLP, where they proved to produce explanations that are effective and comprehensible for the users. However, the adaptation of these methods to the molecular domain is not always straightforward and can provide misleading insights if applied or interpreted incorrectly. For example, Local Interpretable Model-agnostic Explanations (LIME) [10] and SHapley Additive exPlanations (SHAP) [6] are explainability methods used for feature vector representations, but these techniques assume the features are independent, which is rarely held for typical molecular representations like calculated descriptors or fingerprints [2,5]. Additionally, LIME assumes that model behavior is locally linear, which may not be true for bioactivity prediction models

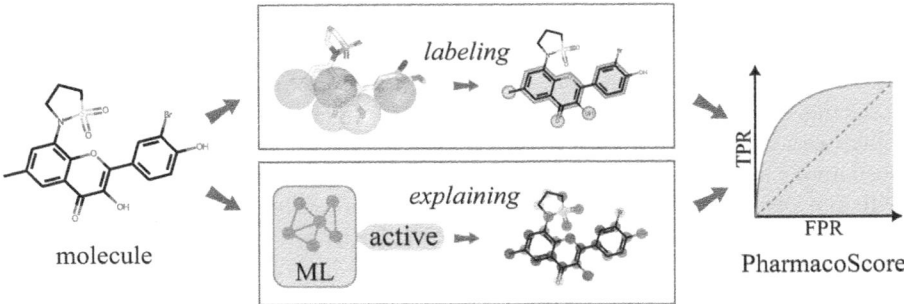

Fig. 1. Overview of our benchmark. ML methods predict compound activity, but only inherently interpretable models or post-hoc XAI techniques can identify the atoms that are significant for predictions. With our new PharmacoScore metric, we compare atom-based explanations against fragments that match a pharmacophore model used for data labeling.

due to so-called activity cliffs, which are pairs of similar molecules with huge differences in activity. SHAP assumes that predictions can be decomposed into a sum of effects attributable to each feature, but this may also be inaccurate for highly non-linear relations between compound structure and its activity. In the context of molecular graphs, saliency map techniques like Class Activation Mapping (CAM) [18] and Vanilla Gradients (VG) have been adopted from computer vision due to similarities in image and graph convolutions [1]. However, these methods are not particularly effective for elucidating how individual atoms contribute to the final prediction. For example, CAM explains predictions by multiplying atom activations after the last graph layer, which may not provide any meaningful signal due to the graph oversmoothing problem of GNNs.

We hypothesize that effective QSAR models, to accurately predict activity across diverse datasets of compounds, must be capable of identifying more general features that ensure binding to their macromolecular targets. One such category of features is a pharmacophore, which describes key points in three-dimensional space that will improve binding when a molecule matches that description [15]. Cyclin-dependent kinases (CDKs), a well-known class of anti-cancer drug targets, exemplify this by possessing a distinctive pharmacophore. For instance, CDK4/6 inhibitors, when combined with endocrine therapy, have become a milestone in managing hormone receptor-positive breast cancer (HR+BC) [3]. However, the latest data revealed that the aberrant CDK2 plays a key role in driving resistance to CDK4/6 in HR+BC. Their ATP-competitive ligands typically contain a heterocyclic core with a hinge-binding motif and a hydrophobic component in the gatekeeper sub-pocket [9].

To confirm that models can learn these high-level features, XAI methods can be applied since the models should prioritize these key features over more fine-grained structural modifications. We introduce a new dataset and an evaluation metric based on pharmacophores to facilitate the testing of bioactivity prediction model generalization and the robustness of XAI methods (Fig. 1). The dataset includes a primary collection of 2,131 molecules with experimentally measured CDK2 activity, along with 2,252 for CDK4 and 679 for CDK6, extended by an additional test set of generated decoy compounds. To simplify the complex problem of predicting activity, we propose a proxy task in which molecules are labeled by matching their structure to a pharmacophore hypothesis constructed from our activity dataset. Through this new benchmark, we make the following observations.

1. The evaluated XAI methods produce inconsistent explanations on our dataset that fail to highlight only the structures that should directly affect the predicted label.
2. Many commonly used ML models, such as RF, XGB, MLP, and GNN, face challenges in effectively learning the concept of a pharmacophore.
3. Activity-trained models share some common XAI patterns with pharmacophore-trained models, but they also reveal that activity data does not always align with the pharmacophore hypothesis.

2   Related Work

The emergence of GNNs has prompted the creation of explainability methods for graph data. **GNN-Explainer** [16] is a model-agnostic approach designed to provide explanations for predictions made by GNNs. It does this by identifying a subgraph and node features that are most important for a given prediction. The approach formulates explanation generation as an optimization problem, where the goal is to find a compact subgraph that maximizes mutual information with the original model's prediction. **SubgraphX** [17] is an explainability method designed for GNNs that generates subgraph-based explanations in a self-interpretable manner. Instead of focusing on individual nodes or edges like traditional attribution-based methods, SubgraphX identifies the most influential subgraphs that contribute to the model's predictions. It employs a Monte Carlo Tree Search strategy to efficiently explore possible subgraphs and rank them based on their contribution to the final prediction. Unlike post-hoc methods that approximate model behavior, SubgraphX provides more stable and meaningful explanations by maintaining structural integrity within graphs. This makes it particularly effective for tasks in chemistry, biology, and social network analysis, where understanding group-wise interactions is critical.

In parallel, intrinsically interpretable graph-based models are being proposed. For example, **ProGReST** [11] achieves interpretability by integrating prototype learning, soft decision trees, and GNNs. This architecture enables predictions to be explained through learned prototypical parts, which serve as reference points for molecular structures. Unlike post-hoc explainability methods, ProGReST is inherently interpretable, as it directly links its decision-making process to identifiable molecular substructures. Additionally, the model ensures interpretability through tree-based reasoning, where each decision node is associated with meaningful prototypes that can be analyzed and validated by experts.

Various reports highlight the use of XAI in drug discovery projects. Wong et al. [14] present a novel model that combines GNNs with explainable graph algorithms to uncover chemical substructures linked to antibiotic activity. By integrating explainability, the model allows researchers to interpret the key molecular features driving antibiotic properties, paving the way for a more rational and efficient approach to antibiotic discovery. The **EvoGradient** [12] model combines explainable deep learning with virtual evolution to predict and optimize antimicrobial peptides (AMPs) in drug discovery. Utilizing LSTM, Transformer, and gradient-based analysis, the model achieves high predictive performance in AMP classification and potency prediction. EvoGradient outperforms traditional models by identifying key amino acids linked to bioactivity. However, like attention-based models, its explanations are sparse, highlighting broad regions rather than specific antimicrobial features. The **BiLAT** model [9], leveraging BiLSTM and Transformer-based encoding layers, achieves high predictive performance for CDK activity, surpassing traditional ML methods in identifying the CDK2 hinge motif, though its explanations remain sparse, highlighting large molecular regions.

3 Methods

This section summarizes the ML modeling techniques and explainability methods employed in this study. Following this, we address the dataset construction, detailing how the pharmacophore hypothesis was created for labeling our dataset. Lastly, we outline the evaluation metrics established in our benchmark. The code and data are available at https://github.com/AdamSulek/pharmacoscore-benchmark.

3.1 Activity Prediction Models

We trained XGBoost (XGB), Random Forest (RF), and a multilayer perceptron (MLP) on the RDKit-generated ECFP fingerprint (radius 2, 2048-bit), and a graph neural network (GNN) with one-hot encoded atomic descriptors using PyTorch Geometric. Models were tuned using grid search and selected based on validation ROC AUC. The final evaluation was done on a separate test set.

3.2 Explainability Methods

For model explainability, we employed SHAP explanations using the SHAP package for the RF, XGB, and MLP models. Additionally, we used vanilla gradient explanations for MLP and GNN models and Grad-CAM for GNN [8]. From each explanation for fingerprint-based models, we identified the top 5 most important features and then determined which atoms contributed to these bits per molecule using atom fragment mappings. As a result, the top 5 bits often corresponded to more than 5 unique atoms. In contrast, for GNN models, we directly selected the top 5 most important nodes, which always corresponded to exactly 5 atoms. These identified atoms were then used for visualization, sparsity score calculations, pharmacophore-type matching, and PharmacoScore computation.

3.3 Dataset Construction

This study employed three datasets profiling compound activity (IC_{50} and/or Ki) against CDK2 (1432 active, 699 inactive, $1\,\mu M$ cutoff), CDK4 (1102 active, 1150 inactive, $1\,\mu M$ cutoff), and CDK6 (307 active, 372 inactive, $100\,nM$ cutoff). All compound data originated from the ChEMBL database. Data was split into training, validation, and test sets using Bemis-Murcko scaffold-based splitting to ensure structural dissimilarity between sets. This work introduces a new CDK pharmacophore-based benchmark dataset. To vary the dataset's difficulty, we used a five-element pharmacophore for CDK2 and three-element pharmacophores for CDK4 and CDK6. Molecules matching the pharmacophore were labeled '1', others '0', yielding 1057/1074 (pos/neg) for CDK2, 2071/181 for CDK4, and 638/41 for CDK6. We added a second test set to better assess machine learning models' understanding of pharmacophores. The updated dataset includes positive examples reused from the prior test set and newly generated negative examples, where we slightly altered the molecular structures while preserving their overall shapes. Our modifications involved random substitutions of hydrogen donors and acceptors with carbon atoms, removal of aromaticity in the rings, adjustments to the length of the linkers connecting the rings, and replacement of hydrophobic groups with polar ones.

3.4 Pharmacophore Model Preparation

The pharmacophore models were created with Maestro Schrödinger Release 2021-2, based on the structures of human protein complexes CDK2 (PDB ID: 2A4L), CDK4 (PDB ID: 9CSK) and CDK6 (PDB ID: 5L2I). Atom types, hydrogens, and charges were assigned using the Protein Preparation Wizard. OPLS4 and PROPKA optimized the complex's structure and hydrogen bond network. A starting set of pharmacophores was produced with the Phase module, using the ligand-receptor complex and the CDKs ATP-binding site cavity as a basis. For each instance, pharmacophores were generated using the automated E-Pharmacophore method, with a feature limit of ten. A subset of pharmacophores was created using the top features from the initial set of automatically generated pharmacophores. These features focused on ATP-binding pocket interactions, particularly Hinge Region contacts. The optimal pharmacophore model for discriminating activity from inactive compounds was determined through application of the Hypothesis Validation tool to a curated dataset. Phase considers a molecule aligned only if all pharmacophore points match corresponding features within a 2 Å range and generates a single best-fit conformation for each ligand.

3.5 Evaluation

The performance of bioactivity prediction models is conducted using standard classification metrics: ROC AUC, accuracy, and F1-score. The explanations are evaluated using the sparsity and fidelity metrics, defined as follows:

$$\text{Sparsity} = \frac{1}{N}\sum_{i=1}^{N}\frac{|m_i|}{n_i}, \quad \text{Fidelity} = \frac{1}{N}\sum_{i=1}^{N}\left(\mathbb{1}(y_i = f(x_i)) - \mathbb{1}(y_i = f(\bar{x}_i))\right),$$

where N is the number of testing examples, n_i is the number of atoms in the i-th example, $m_i \in \{0,1\}^{n_i}$ is an explanation for the i-th example, $|m_i|$ denotes the number of non-zero elements of m_i, x_i is the input representation of i-th molecule, y_i is its label, $\mathbb{1}$ is the indicator function, and $\bar{x}_i$ is the masked input where important atoms are removed.

We implement two additional metrics to measure the agreement between the ground-truth explanations and model explanations. Feature detection accuracy is defined as the percentage of testing examples in which a particular pharmacophoric feature is detected as important, i.e., at least one atom of this feature overlaps with the model explanation. We also introduce **PharmacoScore** (Ph-Score), an evaluation metric designed to test whether the whole pharmacophore is prioritized over other atoms in a molecule. It is defined as the ROC AUC between the explained atom importances and binary ground-truth atom labels.

4   Results

We trained ML models on our activity dataset. In the following sections, we first explore the quality of predictions. Second, we check if these models can distinguish decoys, which are molecules with small structural changes that break the given pharmacophoric structure. Next, we introduce a new evaluation metric based on pharmacophore alignment to test XAI methods on our dataset. Finally, we compare predictions of models trained on our proxy pharmacophore data with those trained on the original activity data.

4.1   Prediction of Compound Activity

Four ML models were trained on our dataset that was labeled by finding molecules that match the predefined pharmacophore. RF, XGB, and MLP are three models trained on ECFP fingerprints. GNN is a model trained on the molecular graph representation. All neural networks were trained for 20 epochs using the Adam optimizer. The best set of hyperparameters for each model was found using a random search with 32 randomly sampled hyperparameter sets.

Table 1 shows the performance of ML models in predicting both experimental activity and pharmacophore matching. The results indicate that models trained on the pharmacophore matching task achieve higher ROC AUC scores compared to those trained on the activity prediction task. This suggests that the pharmacophore task may be more straightforward or better defined, potentially due to the more homogeneous structural relationships inherent in pharmacophore data. In contrast, activity prediction may involve more complex factors, such as mixed or non-competitive enzyme inhibition, which could introduce greater variability and challenge model performance. Furthermore, methods utilizing fingerprint-based representations consistently outperform GNNs that rely on simple one-hot encoding of atom features. The models demonstrate strong generalizability across all metrics, ROC AUC, accuracy, and F1-score.

Table 1. Model performance measured on the testing set.

Target	Model	experimental activity			pharmacophoric labels		
		AUC	Accuracy	F1-score	AUC	Accuracy	F1-score
CDK2	RF	0.824	0.625	0.641	0.888	0.778	0.778
	XGB	0.813	0.611	0.648	0.900	0.775	0.783
	MLP	0.800	0.602	0.630	0.865	0.789	0.810
	GNN	0.713	0.637	0.628	0.826	0.724	0.734
CDK4	RF	0.918	0.858	0.886	0.939	0.933	0.961
	XGB	0.912	0.860	0.889	0.944	0.931	0.959
	MLP	0.892	0.824	0.856	0.930	0.931	0.961
	GNN	0.916	0.862	0.895	0.904	0.929	0.965
CDK6	RF	0.807	0.800	0.830	0.964	0.911	0.952
	XGB	0.803	0.785	0.820	0.878	0.933	0.963
	MLP	0.804	0.748	0.785	0.927	0.911	0.952
	GNN	0.769	0.704	0.733	0.919	0.975	0.983

4.2 Model Performance in Decoy Detection

Overly optimistic results from evaluating molecular property prediction models on random test sets may arise from the similarity of compounds within a given chemical series. To assess the generalizability of these models, a stratified sampling technique is sometimes employed, grouping compounds with identical chemical scaffolds within the same subset. This approach might also fail if the chemical series includes scaffold modifications.

We propose a more challenging test to learn if the evaluated models discover our pharmacophore hypothesis instead of memorizing molecular fragments in active compounds. Therefore, we create decoy molecules for each test molecule that fit our pharmacophore model. Decoys are used to introduce small structural modifications that should hinder binding to the biological target, removing the critical pharmacophoric features. This new testing set with generated close negatives is used to evaluate our models, and their results are shown in Table 2.

For most types of modifications generated, the accuracy of predictions is generally similar. However, in about 50% of CDK2 cases, models make errors when the distance between aromatic rings is altered. Although all models highly recognized both aromatic rings (Table 3), changing the length of the linker between the rings did not significantly affect the model predictions. As a result, the predictions remained positive despite the absence of pharmacophore matching due to the incorrect distance in three-dimensional space.

Table 2. Model performance tested on the decoy dataset.

Model	CDK2			CDK4			CDK6		
	AUC	Accuracy	F1-score	AUC	Accuracy	F1-score	AUC	Accuracy	F1-score
RF	0.715	0.621	0.482	0.661	0.434	0.586	0.826	0.448	0.618
XGB	0.732	0.631	0.504	0.687	0.484	0.602	0.852	0.630	0.702
MLP	0.635	0.505	0.436	0.568	0.418	0.583	0.580	0.459	0.623
GNN	0.585	0.514	0.416	0.532	0.533	0.612	0.530	0.525	0.650

The degraded performance on the testing set can often be attributed to the models making predictions for the molecules coming from outside the training distribution. This is not the case for our decoy dataset because all molecules are close analogs of positive examples in the original dataset, which is depicted in Fig. 2. The mean Tanimoto distance between decoys and unmodified compounds is less than 0.33. The differences are subtle, often replacing only one atom or changing aromaticity of a ring, which is also depicted in the figure.

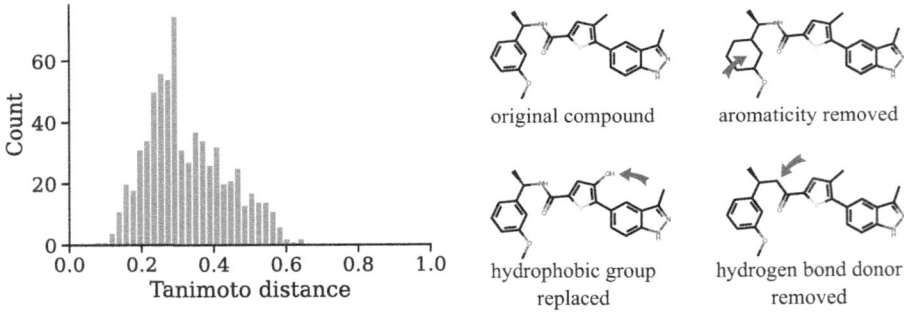

Fig. 2. Tanimoto distance between CDK2 pharmacophore-matching compounds and their decoys, with example decoys shown on the right.

4.3 Assessment of Model Explanations

By matching and aligning molecules with our pharmacophore hypothesis, we obtain ground-truth labels to evaluate XAI methods. In this experiment, we first test whether our models recover parts of the pharmacophore hypothesis and then employ our PharmacoScore metric to quantify the amount of the pharmacophore that is learned and correctly attributed by the XAI methods.

Table 3 shows the percentages of recovered pharmacophoric features by each explainability method. Additionally, sparsity is reported to account for different numbers of atoms highlighted by each method. A higher sparsity value—meaning more atoms are considered important in the explanation—correlates with a higher percentage of correctly covering all important atoms, while not penalizing models for incorrectly marking non-important atoms as important. The central aromatic group (aromatic 1), located near the HBA and HBD groups, was generally easier to identify across all models. Additionally, a high sparsity value is often linked to the presence of five or six atoms within a pharmacophore point, making it more likely that the model will highlight at least one of them.

Table 3. Ability of the models to detect specific CDK2 pharmacophoric features.

Method	Sparsity	HBA	HBD	Aromatic 1	Aromatic 2	Hydrophobic
RF+SHAP	22%	32%	36%	79%	41%	32%
XGB+SHAP	21%	22%	40%	78%	57%	24%
MLP+SHAP	23%	25%	41%	88%	47%	38%
MLP+VG	23%	25%	41%	88%	47%	38%
GNN+Grad-CAM	16%	18%	10%	52%	42%	47%
GNN+VG	16%	8%	6%	47%	18%	42%

Most XAI methods do not provide guidance on how to select important atoms. Instead, they assign each atom an importance weight. This allows us to rank all atoms and measure the model's ability to prioritize the pharmacophore correctly using our PharmacoScore. Table 4 presents the agreement between predicted atom attributions and the ground-truth pharmacophore label. Moreover, we report the fidelity of the XAI technique, which measures how crucial the highlighted structure is to the model prediction. Low fidelity scores in MLP and GNN may suggest that poor alignment with the true pharmacophore may result from errors in the explanation method.

PharmacoScore is a challenging metric, with standard models often struggling to accurately mark the pharmacophore atoms, despite being able to classify the entire molecule label. While the top 5 important fragments frequently highlight some of the pharmacophore atoms, standard models tend to mark atoms randomly, assigning high importance to non-relevant side fragments in the global explanations. As a simple baseline, we labeled all aromatic atoms as 1 and others as 0, achieving a score of 0.75 for a 5-point pharmacophore containing two aromatic groups. This high score reflects the prevalence of aromatic fragments, some of which include HBA or HBD atoms that contribute to the CDK2 pharmacophore. The 'all aromatic' baseline can serve as a reference point for future PharmacoScore evaluations. Notably, PharmacoScore performs better for 3-point pharmacophores, where approximately 50% of atoms are labeled as 1.

We visualize some of the explanations in Fig. 3. The GNN models, which exhibited the lowest sparsity due to the selection of the top 5 nodes that accounted for an average of 16% of the molecule, were able to recognize hydrophobic fragments at a level close to 40%, demonstrating their effectiveness in identifying these key components. In models trained on fingerprints, the

Table 4. Ability of the models to prioritize pharmacophore over less important molecular substructures. The model named "all aromatic" is a simple baseline where all aromatic atoms are marked as important.

Method	CDK2		CDK4		CDK6	
	Fidelity	Ph-Score	Fidelity	Ph-Score	Fidelity	Ph-Score
all aromatic	–	0.75	–	0.67	–	0.60
RF+SHAP	0.20	0.54	0.18	0.53	0.17	0.69
XGB+SHAP	0.30	0.53	0.53	0.51	0.82	0.62
MLP+SHAP	0.09	0.49	0.00	0.51	0.01	0.56
MLP+VG	0.05	0.51	0.01	0.53	0.01	0.64
GNN+Grad-CAM	0.14	0.41	0.35	0.49	0.27	0.54
GNN+VG	0.06	0.48	0.33	0.49	0.23	0.51

sparsity was higher, caused by the presence of several atoms within a single fingerprint feature. The models frequently predicted the NH group between aromatic rings as interacting with the HBD, which often resulted in correct predictions (e.g., CHEMBL482211). However, in some molecules (CHEMBL115220), this atom was not the actual interacting group. The models also struggled with molecules containing three aromatic rings (CHEMBL232735), failing to identify the key rings. The RF and XGB models highlighted all three aromatic rings, while the MLP model marked only two, ignoring one key aromatic ring. However, in many cases, fingerprint and SHAP produce similar XAI results.

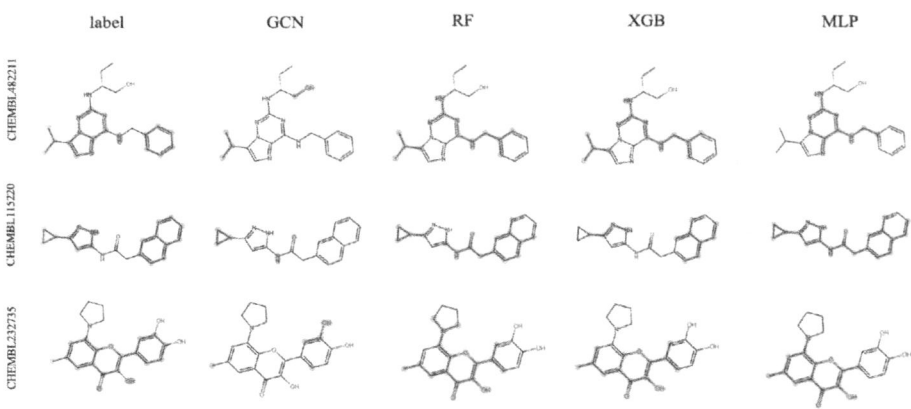

Fig. 3. Example explanations of XAI techniques applied in our proxy pharmacophore prediction problem. Pharmacophore atoms are highlighted: HBA in orange, HBD in green, hydrophobic in yellow, and aromatic in pink. Important atoms identified by Grad-CAM (GCN) and SHAP (RF, XGB, MLP) are shown in pink, highlighting key regions influencing model predictions. (Color figure online)

4.4   CDK2 Inhibitors Case Study

To validate the usefulness of our benchmark in real-world scenarios, we compare the insights derived from the model trained on the experimental data with those from our proxy pharmacophore-based labels. Figure 4 illustrates the differences between explanations of XGB models trained on experimental and pharmacophoric labels. The model trained on experimental labels fails to identify HBA atoms in CHEMBL361833 and the aromatic ring in CHEMBL4297488. However, in some cases, predictions fully align, as observed for CHEMBL497854 and CHEMBL3655766. The overall consistency in highlighting similar structural features can be partly explained by the high precision of the pharmacophore hypothesis (73%) and the notable correlation between the label sets (accuracy 56% and precision 54% for compounds with matching labels).

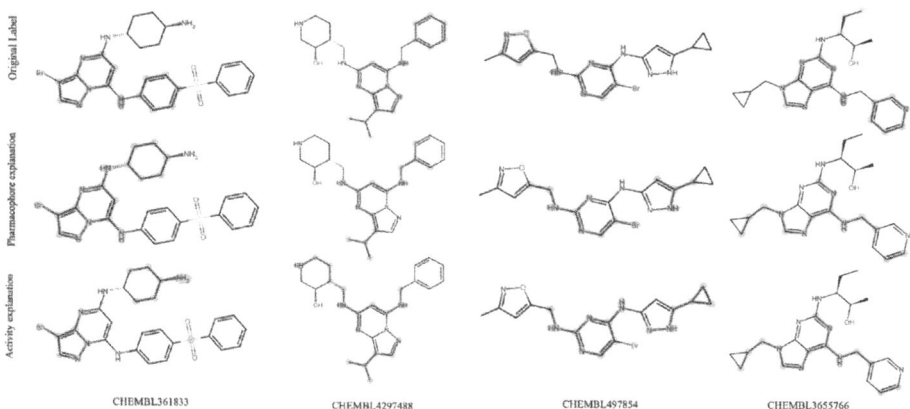

Fig. 4. Comparison of explainability-based pharmacophore classification model and activity classification model against the reference label. The plot illustrates how both models align with the ground truth, highlighting differences in their ability to capture key pharmacophore features.

Interestingly, some misclassification of the activity model can be attributed to finding pharmacophoric features, which is shown in Fig. 5. In the HBD modification, new fragments are highlighted as important, but fragments common with the original molecule prediction are still visible, hence the false positive (FP) prediction. In the HBA modification, the key pharmacophore fragment was removed, so the model does not highlight this fragment as important. However, the presence of remaining important fragments still results in a positive prediction despite the lack of pharmacophore matching. In the example of hydrophobic modification, the original prediction was positive, and the hydrophobic point was not crucial for the positive classification. Therefore, after modifying this site, the prediction remained positive. However, correct prediction of the distance between aromatic rings is crucial to solving the pharmacophore matching issue.

The modification of ring distance shows that the model does not recognize the correct distance, still highlighting both aromatic rings as important despite the reduction of the distance and the lack of pharmacophore matching.

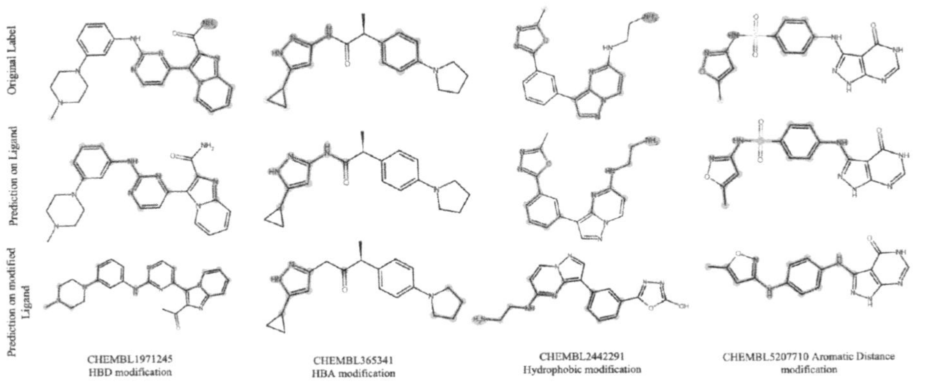

Fig. 5. Visualization of FP cases by the XGB model, showing positive true labels, positive predictions, and modification-based predictions. SHAP explanations highlighting key atoms in the model's predictions.

4.5 Limitations

Although the proposed benchmark spotlights crucial problems with the commonly used QSAR models and XAI techniques, it has limitations that should be considered when drawing conclusions in more general drug discovery setups.

Multiple Pharmacophores. The proposed proxy activity labeling considers only one pharmacophore per target. In reality, molecules can have different binding modes that affect their target. Moreover, some molecules can bind allosterically to a different part of the target protein, making the pharmacophore constructed on a set of typical orthosteric ligands uninformative. Nevertheless, the labels produced using only one pharmacophore prove to be challenging for some models, and more complex models should be used only after this benchmark is solved.

3D Prediction Models. This benchmark does not consider 3D models because they require one particular molecular conformation as input. However, the conformation that matches our pharmacophore might differ from the lowest-energy conformation that can be computed using force-field methods. It is possible that models that use 3D descriptors might be better at understanding high-level pharmacophoric features, which we leave as future work.

5 Conclusions

In this study, we introduced a novel benchmark dataset alongside a new metric, PharmacoScore, to evaluate the explainability of ML models in cheminformatics. Our findings reveal that commonly used classification and regression models, despite their strong performance in traditional tasks, face significant challenges in aligning with the interpretability requirements of pharmacophore-based modeling. Pharmacophores present a notable difficulty for ML models, as even minor structural modifications—such as those introduced by decoys—lead to a significant drop in performance, with ROC AUC scores decreasing about 0.2. This decline highlights the critical need for models that can reliably predict bioactivity for compounds with small modifications, a capability essential for hit-to-lead optimization in drug discovery. Moreover, our results demonstrate that typical explainability methods are not well-suited for bioactivity prediction tasks, as they often fail to accurately identify pharmacophoric interaction sites. This underscores a fundamental gap in the applicability of current ML approaches to cheminformatics, where interpretability is as crucial as predictive accuracy. By providing a benchmark and a robust evaluation framework, we aim to inspire the development of more interpretable and chemically meaningful models, ultimately advancing the field of computational drug discovery.

Acknowledgements. This research has been conducted as part of the FIRST TEAM FENG project No FENG.02.02-IP.05-0029/23, funded by the Foundation for Polish Science (FNP) under the European Funds for a Modern Economy (FENG) program, co-financed by the European Union. The publication was created within the project of the Minister of Science and Higher Education "Support for the activity of Centers of Excellence established in Poland under Horizon 2020" on the basis of the contract number MEiN/2023/DIR/3796. This project has received funding from the European Union's Horizon 2020 research and innovation programme under grant agreement No 857533. This publication is supported by Sano project carried out within the International Research Agendas programme of the Foundation for Polish Science, co-financed by the European Union under the European Regional Development Fund. We gratefully acknowledge Polish high-performance computing infrastructure PLGrid (HPC Center: ACK Cyfronet AGH) for providing computer facilities and support within computational grant no. PLG/2024/017108.

References

1. Ali, S., et al.: Explainable artificial intelligence (xai): what we know and what is left to attain trustworthy artificial intelligence. Inf. Fusion **99**, 101805 (2023)
2. Garreau, D., Luxburg, U.: Explaining the explainer: a first theoretical analysis of lime. In: International Conference on Artificial Intelligence and Statistics, pp. 1287–1296. PMLR (2020)
3. Guven, D.C., Sahin, T.K.: The association between her2-low status and survival in patients with metastatic breast cancer treated with cyclin-dependent kinases 4 and 6 inhibitors: a systematic review and meta-analysis. Breast Cancer Res. Treat. **204**(3), 443–452 (2024)

4. Jiménez-Luna, J., Grisoni, F., Schneider, G.: Drug discovery with explainable artificial intelligence. Nat. Mach. Intell. **2**(10), 573–584 (2020)
5. Li, Z.: Extracting spatial effects from machine learning model using local interpretation method: an example of shap and xgboost. Comput. Environ. Urban Syst. **96**, 101845 (2022)
6. Lundberg, S.M., Lee, S.I.: A unified approach to interpreting model predictions. Adv. Neural Inf. Process. Syst. **30** (2017)
7. Muratov, E.N., et al.: Qsar without borders. Chem. Soc. Rev. **49**(11), 3525–3564 (2020)
8. Pope, P.E., Kolouri, S., Rostami, M., Martin, C.E., Hoffmann, H.: Explainability methods for graph convolutional neural networks. In: Proceedings of the IEEE/CVF Conference on Computer Vision and Pattern Recognition, pp. 10772–10781 (2019)
9. Qian, X., et al.: An interpretable multitask framework bilat enables accurate prediction of cyclin-dependent protein kinase inhibitors. J. Chem. Inf. Model. **63**(11), 3350–3368 (2023)
10. Ribeiro, M.T., Singh, S., Guestrin, C.: "why should i trust you?" explaining the predictions of any classifier. In: Proceedings of the 22nd ACM SIGKDD International Conference on Knowledge Discovery and Data Mining, pp. 1135–1144 (2016)
11. Rymarczyk, D., Dobrowolski, D., Danel, T.: Progrest: prototypical graph regression soft trees for molecular property prediction. In: Proceedings of the 2023 SIAM International Conference on Data Mining (SDM), pp. 379–387. SIAM (2023)
12. Wang, B., et al.: Explainable deep learning and virtual evolution identifies antimicrobial peptides with activity against multidrug-resistant human pathogens. Nat. Microbiol., 1–16 (2025)
13. Wojtuch, A., Danel, T., Podlewska, S., Maziarka, Ł: Extended study on atomic featurization in graph neural networks for molecular property prediction. J. Cheminf. **15**(1), 81 (2023)
14. Wong, F., et al.: Discovery of a structural class of antibiotics with explainable deep learning. Nature **626**(7997), 177–185 (2024)
15. Yang, S.Y.: Pharmacophore modeling and applications in drug discovery: challenges and recent advances. Drug Disc. Today **15**(11–12), 444–450 (2010)
16. Ying, Z., Bourgeois, D., You, J., Zitnik, M., Leskovec, J.: Gnnexplainer: generating explanations for graph neural networks. Adv. Neural Inf. Process. Syst. **32** (2019)
17. Yuan, H., Yu, H., Wang, J., Li, K., Ji, S.: On explainability of graph neural networks via subgraph explorations. In: International Conference on Machine Learning, pp. 12241–12252. PMLR (2021)
18. Zhou, B., Khosla, A., Lapedriza, A., Oliva, A., Torralba, A.: Learning deep features for discriminative localization. In: Proceedings of the IEEE Conference on Computer Vision and Pattern Recognition, pp. 2921–2929 (2016)

Bus Loop Scheduling with Dueling Double Deep Q Network

Andri Pradana$^{(\boxtimes)}$ⓘ and Lock Yue Chewⓘ

Nanyang Technological University, 21 Nanyang Link, Singapore 637371, Singapore
{andri.pradana,lockyue}@ntu.edu.sg

Abstract. In this paper, we investigate the application of a reinforcement learning algorithm known as the Dueling Double Deep Q-Network to discover bus scheduling strategies and compare them against conventional approaches. In particular, we look into real-time control strategies where buses may choose to stay or leave at bus stops. We explore both waiting time and travel time as the optimization objectives. The results for uniform bus frequency show that average waiting time can be reduced by allowing buses to stay longer at stops with higher passengers' arrival rate but at the cost of increased average travel time. This is also supported by our analytical calculation on a theoretical bus loop model. We then apply our method to a model based on a real world bus loop in Nanyang Technological University. The results highlight the potential benefit of reinforcement learning methods to find novel strategies that can be better than conventional approaches.

Keywords: Bus scheduling · Reinforcement Learning · Complex Systems

1   Introduction

Recently, the machine learning technique of reinforcement learning (RL) has been successful in tackling computational problems that are NP-hard. For example, self-RL had become a demonstrably state-of-the-art approach in discovering novel solutions in board games like Chess and Go [1,2], which are NP-hard. In fact, DeepMind's self-RL algorithm known as AlphaZero [3] contributed new theoretical insights into chess playing after only four hours of self-play, even though humans developed extensive theories and principles in chess over centuries. Another class of NP-hard problem where RL finds success is the protein folding problem. Here, AlphaFold uncovered millions of intricate 3D protein structures which closely match laboratory determined experimental structures, leading to the team developing AlphaGO to win the Nobel Prize in Chemistry in 2024.

Bus scheduling is also a NP-hard problem. A network of buses picking-up and delivering commuters at bus-stops is in fact a complex system whose dynamics

M. H. Lees et al. (Eds.): ICCS 2025, LNCS 15903, pp. 33–48, 2025.
https://doi.org/10.1007/978-3-031-97626-1_3

have been analyzed with its bunching behavior being recognized as a synchronization phenomenon [4]. Bus bunching is a form of operational inefficiency as it increases the waiting and travel time of the commuters, leading to a drop in the quality of service of the buses. Bus operators had tried to address the bunching problem with different strategies, such as holding [5], stop-skipping [6], deadheading [7], limiting boarding [8], and dispatching buses with wide doors [9]. In addition, models are also created as a test bed to simulate intervention strategies to overcome the inefficiencies of bus bunching [10]. Recently, research exploring the use of self-RL to overcome the bunching problem interestingly uncovered two new strategies: the no-boarding strategy [11–13], and the semi-express bus configuration [14,15]. For the no-boarding strategy, the bus may leave the bus stop even though there is somebody wishing to board. RL found that a combination of no-boarding and holding strategies creates a staggered bus configuration that avoids bus bunching and minimizes the average waiting time of commuters. The semi-express bus configuration, on the other hand, consists of a combination of normal buses (which pick up and deliver commuters at all bus stops) and express buses (which pick up and deliver commuters at selected bus stops). Such semi-express bus configuration has already been observed in bus networks and it is found to lower the average waiting time of commuters relative to the operation of purely normal buses. It is surmised that the efficiency of the semi-express bus configuration results from its intrinsic chaotic behavior.

In this paper, we investigate into an advanced RL algorithm known as the Dueling Double Deep Q Network (D3QN) to yield bus scheduling strategies that go beyond our previous approaches [11–13]. In particular, we base our evaluation by minimizing the average travel time of the commuters. This differs from our earlier approach and other works where optimization is performed according to bus headway. We compare our results with the case where the buses adopt the holding strategy. The reason for using the holding strategy as a basis of comparison results from its robustness and consistency in granting bus schedules that nearly achieve the optimal efficiency in bus operations.

2   Literature Review

To schedule buses, bus system operators perform planning, control, and operation on the buses after processing diverse sources of historical and real-time information. Their aim is to determine an optimal schedule for buses, where travel time are minimized, fleet utilization is maximized, and commuters face minimal waiting time. To achieve this goal, there is a need for accurate demand and traffic prediction, optimization of route and timetable, and a fleet management system that is efficient.

Our research focuses on the active real-time decision aspect of this optimization process, instead of the passive decisions which are only relevant over a longer time horizon, such as frequency of bus operations, timetabling of buses, and drivers' scheduling and rostering. Specifically, we look into real-time control strategies where buses may choose to stay or leave at bus stops. This is made possible recently with real-time data from Automated Passenger Counter System (APC), Automated Vehicle Location System (AVL), Geographical Positioning System (GPS), and Automated Fare Collection System (AFC) [16]. Although techniques such as Integer Programming, Mixed-Integer Linear Programming, Genetic Algorithms, Simulated Annealing, Particle Swarm Optimization, Ant Colony Optimization, etc. have been used to yield the optimal schedules, the advent of novel machine learning techniques have given rise to a more flexible approach to derive these optimal schedule in real-time [17]. In particular, we desire to determine the best bus scheduling strategies that would minimize either the waiting time or the travel time of the commuters.

In the literature, the machine learning techniques that relate to bus scheduling in real-time performs travel time prediction of the buses. Understanding the travel time of the buses would allow the inference of the travel time of the commuters and their waiting time. For instance, travel time predictions were performed using the k-nearest neighbor and random forest methods with GPS data when the bus travels between consecutive bus stops with traffic intersections in between [18]. When the road traffic exhibits high variability conditions, such as adverse weather, traffic junctions, diverse vehicular types, a Kalman filter cum support vector regression approach becomes applicable [19]. In another approach that predicts bus travel time, projection pursuit regression, support vector regression, random forest, together with dynamic weighting and dynamic weighting with selection in the integration step, are employed with data from AVL [20]. From another perspective, prediction of bus arrival time is also relevant. In this context, the techniques of artificial neural network and linear regression are utilized with a traffic density matrix to perform the prediction [21]. Alternatively, multilayer perceptron and a deep neural network implemented with PyTorch and TensorFlow could be used to predict the arrival time of buses at bus stops [22]. Finally, models capable of estimating travel and arrival time of buses can be built through linear regression, artificial neural network, and long short term memory network model with the usage of historical data from AVL systems, bus routes, and bus stop information as inputs [23].

One of the most successful machine learning techniques for bus scheduling is reinforcement learning. The decision outcome from it invariably relieves bus bunching and gives a staggered configuration that is optimal [24]. It is also applicable for real-time bus scheduling through the dynamical optimization of online schedule which leads to shorter commuters' waiting time and lower operating cost [25, 26]. RL has also been used to optimize the holding durations of each bus by means of a multi-agent framework when the bus system adopts the holding strategy [27]. A model that incorporates reinforcement learning in this sense was built to carry out dynamic holding control in a noisy environment, where the buses are modeled as agents that minimize headway deviations. The use of multi-agent reinforcement learning improves real-time operations compared with previous works which are focused on centralized control. This model adopts a hierarchical approach such that on top of the bus agents, other agents are needed to coordinate, manage and interface the agents with the environment

[28]. In another piece of work, deep RL was applied to maintain bus headway at a prescribed value through holding the bus. The authors used Double Deep Q-learning to minimize deviations from a target headway, bus travel time, and holding time by combining these quantities in a customized cost function. The action is holding the bus at a bus stop dynamically [29]. A recent improvement in dynamic bus control employs proximal policy optimization (PPO) [30] and a joint action tracker to exploit the multi-agent nature of the problem [31]. The technique was found to compare favorably against simpler methods.

A further development in Deep Q-learning techniques comes in the form of D3QN which integrates the Dueling Network Architecture with Double Deep Q-Learning [32]. The combination reduces overestimation bias and improves learning efficiency. Whilst D3QN has been explored in bus scheduling by deciding on the departure time of buses in the timetable [26], it has yet to be investigated on its efficacy in the determination of optimal dynamic bus control strategies, which is of principal interest in this paper.

3   Approaches

3.1   Dueling Double Deep Q-Network

The Dueling Double Deep Q-Network (D3QN) is an advanced RL architecture designed to improve the performance and stability of the standard Deep Q-Network (DQN) [32]. It combines three key techniques: Dueling Network Architecture, Double Q-Learning, and Deep Q-Learning.

Deep Q-Learning uses the DQN algorithm which is a RL algorithm implemented with neural networks to approximate the Q-value function $Q(s, a)$. $Q(s, a)$ gives the expected return for taking an action a in a state s. The network learns to predict the Q-values by minimizing the error between the predicted and target Q-values.

Because the same Q-network is used to select and evaluate actions, Deep Q-Learning can be plagued by overestimation bias. To address this issue, Double Q-Learning employs two separate networks where one is used for selection and the other evaluation. Specifically, Double Q-Learning uses a *Policy Network* as the primary network to choose the action, while it uses a *Target Network* to evaluate the Q-value of the chosen action as follows:

$$y = r + \gamma Q_{\text{target}} \left(s', \arg \max_a Q_{\text{policy}}(s', a) \right) . \tag{1}$$

Furthermore, the Q-value is computed for every action in the output layer in DQN. This, however, may not lead to good decision making. The Dueling Network Architecture improves this by splitting the evaluation of the Q-value function into two separate streams in the network. One stream estimates the *Value Function* $V(s)$, which denotes how good it is to be in state s, independent of the action. The second stream estimates the *Advantage Function* $A(s, a)$, which denotes the relative benefit of each action a in state s, compared to other actions. The two calculations are combined as follows to give the Q-value function of D3QN:

$$Q(s, a) = V(s) + \left(A(s, a) - \frac{1}{|A|} \sum_{a'} A(s, a') \right). \tag{2}$$

3.2   The Holding Strategy

The holding strategy is a bus scheduling technique employed to improve service reliability and manage the adherence of headway, i.e., the spacing between buses. The idea is to have the buses wait at certain control points (usually the bus stops) so as to prevent them from running too early or too close to the preceding bus. Thus, the strategy minimizes bus bunching and serves to maintain a consistent schedule for the commuters.

To implement the holding strategy, holding points need to be defined along the route where buses are instructed to wait. In addition, bus locations and headway are monitored in real-time using GPS or other tracking systems. Information from these devices allows the bus to know the gap ahead with the previous bus, and the gap behind with the following bus. The bus would then decide how long to wait by avoiding (a) being too close to the previous bus and (b) leaving a large gap with the following bus. This decision depends on a target interval between buses (which could be computed in real-time) the system aims to maintain. The consequence is a minimization of waiting time for the commuters at the bus stops.

The advantage of the holding strategy is a reduction of irregular service by preventing bus bunching. It improves service reliability by distributing the buses evenly along the route. Its key disadvantage is an increase in in-vehicle travel time for commuters who are already onboard. To be effective, it requires accurate real-time data and communication systems. It may become ineffective when subjected to delays due to traffic congestion.

4   Methods

4.1   Modeling and Simulation

We model a bus loop system as an isometric (distance-preserving) map on a unit circle, where the location along the loop can be denoted by a phase angle from $0°$ to $360°$. The arrival of passengers at each bus stop is assumed to follow a Poisson process corresponding to a specified arrival rate. Each bus has a natural frequency (rev/s) with which it travels the loop. After dropping all alighting passengers at a bus stop, each bus can choose an action: either *stay* or *leave*. The alighting and boarding rate is assumed to be 1 passenger per second. The simulation advances in discrete time steps with a 1 s simulation time step.

In a naive approach, after all the passengers alight at the bus stop, the bus picks up all the passengers wanting to board the bus and then simply leaves if there are no more passengers wanting to board. This is the approach in which bus bunching commonly occurs. In the holding strategy, after dropping all alighting passengers and picking up all boarding passengers, the bus has to stay or hold if its phase headway with the bus behind it is greater than the perfectly staggered phase, $360°/N_{bus}$, where N_{bus} is the number of buses in the loop. Otherwise, the bus just leaves. In our simulation, once a bus decides to stay/hold, we implement a 5 s holding duration before the bus checks its headway again. If a passenger arrives within this duration, the passenger can board the bus. In the RL approach, the buses still have to drop all alighting passengers but can choose to stay or leave afterward, even when there are passengers wishing to board. However, once a bus decides to stay, we impose the condition that it must pick up all passengers wishing to board before making the decision to stay or leave again. Here, we also implement a 5 s holding duration.

4.2 Reward Function

Suppose that a decision for an action is required at time t and, after the decision is made, the next time a decision is required again is at time t'. The reward corresponding to an action a_t given the state s_t at time t is calculated according to a continuous-time discounted future reward function [33] given by

$$R(s_t, a_t) = \int_t^{t'} e^{-\beta(\tau - t)} r_\tau \, d\tau, \tag{3}$$

where r_τ is the instantaneous reward at simulation time τ and β is the decay rate of the discount factor $e^{-\beta(\tau - t)}$. The instantaneous reward r_τ is a function of the elapsed time from each passenger's perspective, counted from the time of arrival, which is given by

$$r_\tau = -\sum_p \left(\tau - t_p^{\text{arrival}}\right)^k \tag{4}$$

where the sum is taken over all passengers in the bus loop system. There is a negative sign in the reward because the optimization goal is to minimize, not maximize, either the waiting time or travel time. $k \geq 0$ is an exponent that controls the scaling of the reward with respect to passengers' elapsed time. $k = 0$ gives uniform reward regardless of the length of each passenger's elapsed time, whereas $k > 0$ penalizes longer passenger's elapsed time. We obtain the best results using $k = 0$ for optimization of average waiting time and $k = 1$ for optimization of average travel time. For waiting time optimization, the counting of a passenger's elapsed time ends when the passenger boards, whereas for travel time optimization, the counting ends when the passenger alights at the destination stop.

4.3 State Representation

The state of the system encompasses information about the bus stops and buses. For optimization of average waiting time, bus stop information includes the elapsed time of the earliest passenger arriving and still waiting at each stop. For optimization of average travel time, bus stop information includes the number of passengers waiting at each stop. There are N_{stop} entries for bus stop information, where N_{stop} is the number of bus stops.

For each bus, three types of information are included, and each of these information has N_{stop} entries: whether the bus is going to a bus stop or currently at a bus stop (binary), its phase headway, and the number of passengers in the bus going to each bus stop. Although the headway for each bus is a single value, we find that the RL algorithm learns better when it is paired with the information of whether the bus is going to or currently at a bus stop. So, for the N_{stop} headway entries (each entry representing a bus stop), a headway entry can only be nonzero if the bus is going to or currently at a bus stop corresponding to that entry.

The ordering of the buses in the state description follows certain rules depending on the scenarios discussed in the following. The first scenario is one in which the buses have approximately identical natural frequencies. In this scenario, the ordering of the buses in the state description follows the actual ordering of the buses in the loop. The first bus in the state description is always the bus requiring the decision. The second one is the bus in front of the first bus in the loop, the third one is the bus in front of the second one, and so on. There are only two nodes in the output layer of the neural network, corresponding to the actions that this first bus can take. This implies that these buses follow a collective strategy, i.e., given similar situation or circumstances, the different buses make the same decision.

The second scenario is one in which there is *frequency detuning*, where the buses have different natural frequencies. In this scenario, the ordering of the buses in the state description is fixed, regardless of the actual ordering of the buses in the loop. The buses individually follow their own strategies, instead of a single collective strategy as in the previous scenario. Therefore, in the output layer of the neural network, there should be N_{bus} pairs of nodes, where each pair corresponds to each bus's set of actions. Thus, for the bus requiring the decision, the relevant Q-values are contained in the pair of nodes associated with it.

5   Results and Discussion

5.1   Uniform Bus Frequency

Reinforcement Learning Results. We applied our reinforcement learning methods to a simple case of 4 bus stops and 3 buses with identical frequencies. Two bus stops are located at $0°$ and $30°$, each with passenger arrival rate of $0.04/\text{s}$, and the other two are located at $180°$ and $210°$, each with passenger arrival rate of $0.02/\text{s}$. It is assumed that all passengers are traveling to the stop antipodal to their origin stop. The frequency for all the buses is 1 mHz, so that without stopping, each bus can complete the loop in about 17 min. The results are summarized in Table 1. The results for the naive approach (where bus bunching occurs) and the holding strategy are also shown for comparison.

Table 1. The average waiting time (AWT), average time spent traveling in a bus (ABT), and average travel time (ATT) for various approaches over 100 simulation realizations. RL-WT and RL-TT are reinforcement learning methods minimizing the waiting time and travel time, respectively. RL-WT-extended 1 and 2 allow the buses to learn to hold for a longer period of time. RL-WT-extended 2 allows the buses to hold even longer than RL-WT-extended 1, with a maximum holding duration of 200 s.

Approach	AWT (s)	ABT (s)	ATT (s)
Naive (bus bunching)	527 ± 24	540 ± 1	1073 ± 27
Holding strategy	171 ± 3	569 ± 3	741 ± 5
RL-WT	169 ± 3	569 ± 3	739 ± 4
RL-WT-extended 1	150 ± 4	693 ± 5	842 ± 4
RL-WT-extended 2	143 ± 4	755 ± 7	899 ± 7
RL-TT	174 ± 3	541 ± 2	717 ± 4

The averages are obtained from 100 simulation realizations with random initial condition. For each realization, the simulation is run for an initial period of 10,000 s first to smooth out any transient behavior before measurement is taken within the next 10,000 s. The different approaches considered include the naive approach (in which bus bunching occurs), the holding strategy, RL methods minimizing waiting time (RL-WT, RL-WT-extended 1 and 2), and RL method minimizing travel time (RL-TT). In RL-WT-extended 1 and 2, when the action to stay/hold is selected during the ε-greedy exploration, it also has a chance to execute this action multiple number of times consecutively (this number of times is uniformly distributed with a specified maximum number of times). This allows for the buses to learn to hold for a longer period of time. RL-WT-extended 2 allows the buses to hold even longer than RL-WT-extended 1, with a maximum holding duration of 200 s. It is observed that the RL-WT-extended methods result in shorter average waiting time compared to the holding strategy, but at the cost of much longer average travel time. On the other hand, RL-TT approach has similar performance to the holding strategy in terms of both the average waiting and travel times. The holding strategy and the RL methods clearly perform better than the naive approach.

The plots of the phase headway and position against time of the buses are shown in Fig. 1. As expected, the phase headway for the holding strategy is very close to the perfectly staggered phase of 120°. The RL-WT and RL-TT methods also exhibit phase headway near 120°. The plots for position for the RL-WT-extended methods show that the buses execute longer holding at the bus stops with higher passengers' arrival rate (located at 0° and 30°). By doing so, the buses are able to reduce the average waiting time for the passengers at these stops considerably and, as a consequence, reduce the overall global average waiting time. Another consequence of this behavior is that the phase headway can deviate significantly from the perfectly staggered phase, as shown in Fig. 1(d) and (e). However, we find that staying longer at particular stops does not help

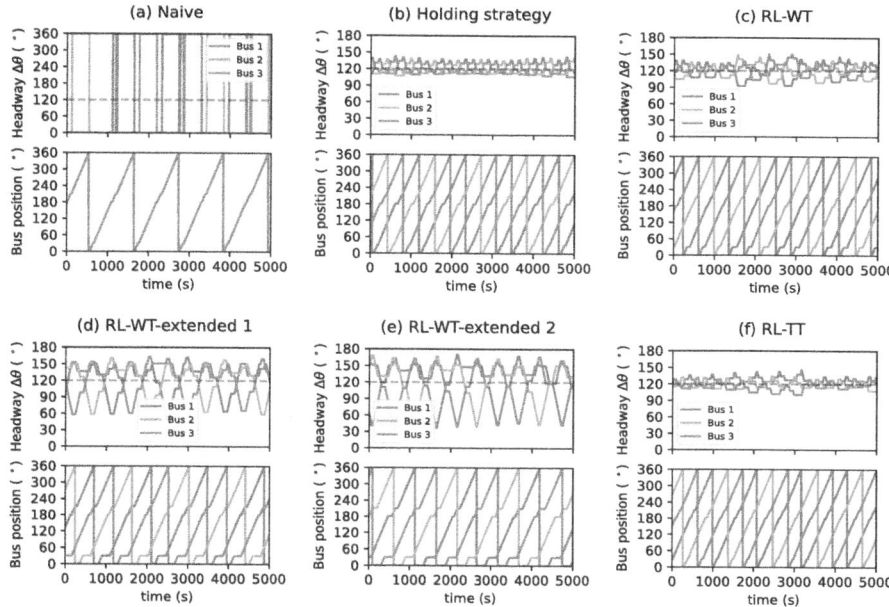

Fig. 1. Plots of phase headway (top) and position (bottom) vs time of the buses for the (a) naive approach, (b) holding strategy, (c) RL-WT, (d) RL-WT-extended 1, (e) RL-WT-extended 1, and (f) RL-TT. Dashed line on the top figures indicates the perfectly staggered phase headway of 120°. For bunched buses in (a), phase headway may 'jump' between 0° and 360° due to buses overtaking each other.

in reducing the average travel time and may in fact be detrimental to it. We provide an analytical explanation for this in the next section.

Analytical Results. Here, we consider a simple theoretical model of a bus loop system in which the buses have identical frequencies, the arrival of passengers is assumed to be continuous instead of discrete, and there is no fluctuation in the arrival rate. We also consider the staggered time-headway among the buses. At a bus stop, each bus drops all alighting passengers first and then picks up all waiting passengers until the bus stop is empty. Afterward, the bus can stay at the bus stop for an additional amount of time to pick up arriving passengers. However, the bus may only stay until the next bus arrives, such that only one bus can be at a bus stop at any point in time. Let the total time spent by a bus at stop i be τ_i and the time required for the bus from leaving stop i to reaching stop j in one loop be $T_{i,j}$. The period of a bus completing a loop is then

$$T = \sum_{i=1}^{N_{\text{stop}}} (\tau_i + T_{i,i+1}) = \left(\sum_{i=1}^{N_{\text{stop}}} \tau_i \right) + T^*, \tag{5}$$

where $T^* = \sum_i T_{i,i+1}$ is the time spent by a bus moving on the road and the index $i \in \{1, 2, \ldots, N_{\text{stop}}\}$ denoting the i-th bus-stop is periodically bounded (so that $i = N_{\text{stop}} + 1 = 1$).

If there are N_{bus} buses in the loop following the staggered time-headway strategy, then the time interval between a bus leaving a bus stop and the next bus leaving the same stop is T/N_{bus}. All passengers arriving within this time interval will be picked up by the next bus. Let the passenger arrival rate at stop i be r_i. Within the time interval from 0 to T/N_{bus}, a small passenger element, $dn_i = r_i \, dt$, arrives at stop i at time t and subsequently boards the next arriving bus. Since the bus leaves at time T/N_{bus}, the time interval between dn_i arriving and the bus leaving is given by $T/N_{\text{bus}} - t$. Therefore, the sum of elapsed time for all passengers for this stage is

$$\Delta t_1 = \sum_i \int_0^{\frac{T}{N_{\text{bus}}}} \left(\frac{T}{N_{\text{bus}}} - t \right) r_i \, dt = \frac{1}{2} \left(\frac{T}{N_{\text{bus}}} \right)^2 \sum_i r_i. \tag{6}$$

Next, the passengers travel in the bus to their destination stops. Let the passenger origin-destination probability from stop i to stop j be c_{ij} ($c_{ii} = 0$ and $\sum_j c_{ij} = 1$). Then the total number of passengers arriving at stop i within time interval T/N_{bus} who want to travel to stop j is $N_{ij} = r_i c_{ij} T/N_{\text{bus}}$. Noting that the travel time from origin stop i to destination stop j is precisely $T_{i,j}$ defined earlier, the sum of elapsed time for all passengers for this stage is therefore

$$\Delta t_2 = \sum_{ij} N_{ij} T_{i,j} = \frac{T}{N_{\text{bus}}} \sum_{ij} r_i c_{ij} T_{i,j}. \tag{7}$$

Lastly, at a destination stop j, the total number of passengers alighting is $\sum_i N_{ij}$. Assuming an alighting/boarding rate of l, the amount of time it takes to drop all alighting passengers is $\sum_i N_{ij}/l$ and the small alighting passenger element is $dn = l \, dt$. Therefore, the sum of elapsed time for all passengers for this stage is

$$\Delta t_3 = \sum_j \int_0^{\sum_i N_{ij}/l} t \, l \, dt = \sum_j \frac{l}{2} \left(\frac{\sum_i N_{ij}}{l} \right)^2 = \frac{1}{2l} \left(\frac{T}{N_{\text{bus}}} \right)^2 \sum_j \left(\sum_i r_i c_{ij} \right)^2. \tag{8}$$

Finally, the average travel time for all the passengers can be calculated from ATT $= (\Delta t_1 + \Delta t_2 + \Delta t_3)/N$, where $N = \sum_{ij} N_{ij} = (T/N_{\text{bus}}) \sum_i r_i$ is the total number of passengers picked up by a bus in one loop. After substituting Δt_1, Δt_2, and Δt_3, and some algebraic manipulation, ATT can be expressed as

$$\text{ATT} = \frac{T}{2N_{\text{bus}}} \left[1 + \frac{\sum_j \left(\sum_i r_i c_{ij} \right)^2}{l \sum_i r_i} \right] + \frac{\sum_{ij} r_i c_{ij} T_{i,j}}{\sum_i r_i}. \tag{9}$$

The task now is to find the appropriate values of T and $T_{i,j}$'s in Eq. (9) which minimize ATT. These quantities are linear functions of τ_i's, which are adjustable. The relation between T and τ_i is given in Eq. (5). As for $T_{i,j}$, since its definition is the travel time of a bus from stop i to stop j, it also includes the total time spent at bus stops in between i and j. For example, $\mathcal{T}_{1,4} = \mathcal{T}_{1,2} + \tau_2 + \mathcal{T}_{2,3} + \tau_3 + \mathcal{T}_{3,4}$. Here, $\mathcal{T}_{1,2}$, $\mathcal{T}_{2,3}$, and $\mathcal{T}_{3,4}$ cannot be decomposed any further as they are simply times spent purely on the road. So, the values of τ_i's need to be analyzed next.

The total time a bus spends at stop i, τ_i, can be broken down into three parts. First, once the bus arrives at the stop, it drops all $\sum_j N_{ji}$ alighting passengers with a rate of l. The total time spent for this part is

$$\tau_i^{(1)} = \frac{1}{l} \sum_j N_{ji} = \frac{T}{l N_{\text{bus}}} \sum_j r_j c_{ji}. \tag{10}$$

Next, the bus starts emptying the bus stop by picking up passengers. Recall that the time interval between a bus leaving the stop and the next bus leaving the same stop is T/N_{bus}. Therefore, the time interval between a bus leaving and the next bus arriving is given by $T/N_{\text{bus}} - \tau_i$. During this time interval plus an additional time $\tau_i^{(1)}$ defined above, passengers do not board yet and bus stop i keeps receiving passengers with a rate of r_i. The number of passengers accumulated at the stop after this time is then $r_i \times (T/N_{\text{bus}} - \tau_i + \tau_i^{(1)})$. Once the bus starts picking up passenger with the rate of l (with the stop still receiving passengers with a rate of r_i), the number of passengers at the stop reduces by a rate of $l - r_i$. The amount of time it takes to empty the bus stop is then

$$\tau_i^{(2)} = \frac{r_i}{l - r_i} \left(\frac{T}{N_{\text{bus}}} - \tau_i + \tau_i^{(1)} \right) = \frac{r_i}{l - r_i} \left[\frac{T}{N_{\text{bus}}} \left(1 + \frac{1}{l} \sum_j r_j c_{ji} \right) - \tau_i \right]. \tag{11}$$

After stop i is empty, the bus may stay at the stop for an additional amount of time s_i. We can now calculate the total time a bus spends at stop i as $\tau_i = \tau_i^{(1)} + \tau_i^{(2)} + s_i$ which, after substitution and some algebraic manipulation, yields

$$\tau_i = \left(1 - \frac{r_i}{l} \right) s_i + \frac{T}{l N_{\text{bus}}} \left(r_i + \sum_j r_j c_{ji} \right). \tag{12}$$

Taking the sum of τ_i over all stops and noting that $\sum_i \tau_i = T - T^*$ from Eq. (5) and $\sum_i c_{ji} = 1$ yields, after simplification, the expression

$$\frac{T}{l N_{\text{bus}}} = \frac{T^* + \sum_i \left(1 - \frac{r_i}{l} \right) s_i}{l N_{\text{bus}} - 2 \sum_i r_i}. \tag{13}$$

Substituting this back into Eq. (12) yields

$$\tau_i = \left(1 - \frac{r_i}{l} \right) s_i + \frac{T^* + \sum_j \left(1 - \frac{r_j}{l} \right) s_j}{l N_{\text{bus}} - 2 \sum_j r_j} \left(r_i + \sum_j r_j c_{ji} \right). \tag{14}$$

This is τ_i expressed in terms of s_i and other constants. More specifically, τ_i is a linear function of s_i's. Setting $s_i = 0$ for all i simultaneously minimizes τ_i for all i, which in turn minimizes T, $T_{i,j}$'s, and, consequently, ATT as well. From Eqs. (13) and (14), the minimum values of τ_i and T are

$$\tau_{i,\min} = \frac{T^*}{lN_{\text{bus}} - 2\sum_j r_j}\left(r_i + \sum_j r_j c_{ji}\right), \tag{15}$$

$$T_{\min} = \frac{lN_{\text{bus}}T^*}{lN_{\text{bus}} - 2\sum_i r_i}. \tag{16}$$

The optimal condition of $s_i = 0$ implies that the buses should not stay unnecessarily longer at the bus stops. In fact, staying longer can be detrimental to the average travel time, which also agrees with the results in the previous section.

We can use the parameters from the scenario of 3 buses serving 4 bus stops in the previous section to estimate the optimal average travel time. Inserting these parameters into Eqs. (15) and (16) yields $\tau_{1,\min} = \tau_{2,\min} = \tau_{3,\min} = \tau_{4,\min} \approx 21.74$ s and $T_{\min} \approx 1087$ s. Using Eq. (9), we obtain ATT ≈ 709 s. Despite the relative simplicity of this idealized model, its result is remarkably close to the holding strategy and the RL methods minimizing the travel time presented in Table 1 in the previous section. This is because in the scenario analyzed here, the time spent by each bus at the bus stops are much shorter than the time spent moving on the road. So, stopping at the bus stops should not appreciably change the phase difference between the buses. As a result, maintaining staggered time headway becomes similar to maintaining staggered phase headway.

5.2   Detuned Bus Frequency

We also applied our RL-TT method to a model based on a real world bus loop in Nanyang Technological University (NTU): the shuttle bus Blue route which consists of 12 reasonably staggered bus stops. Its measured parameters can be found in [4,14]. In particular, here we consider the lull period of the afternoon on the weekdays with two detuned buses serving the loop with frequencies of 0.93 mHz and 1.39 mHz. The passengers' arrival rates for the 12 stops are 0.001, 0.023, 0.015, 0.005, 0.016, 0.040, 0.018, 0.035, 0.024, 0.030, 0.007, and 0.010. All passengers are assumed to be traveling to the stop antipodal to their origin stop.

The results using the naive approach, holding strategy, and RL-TT method are shown in Table 2 and Fig. 2. The holding strategy yields a better average

Table 2. The average waiting time (AWT), average time spent traveling in a bus (ABT), and average travel time (ATT) for various approaches over 100 simulation realizations. RL-TT is reinforcement learning method minimizing the travel time.

Approach	AWT (s)	ABT (s)	ATT (s)
Naive (bus bunching)	449 ± 67	584 ± 5	1026 ± 67
Holding strategy	333 ± 5	704 ± 7	1041 ± 11
RL-TT	359 ± 10	581 ± 7	944 ± 15

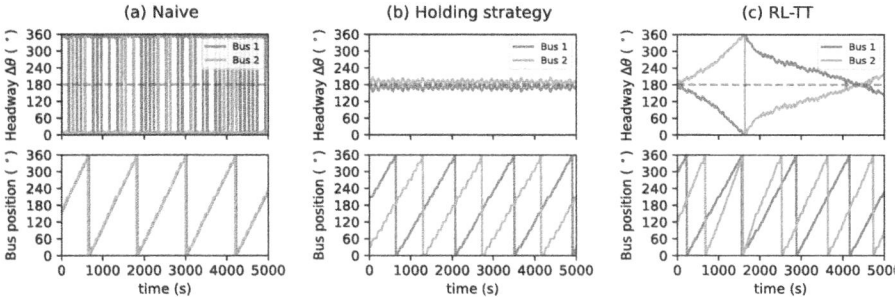

Fig. 2. Plots of phase headway (top) and position (bottom) vs time of the buses for the (a) naive approach, (b) holding strategy, and (c) RL-TT. Dashed line on the top figures indicates the perfectly staggered phase headway of $120°$. Bus 1 is the slower bus. For bunched buses in (a), phase headway may 'jump' between $0°$ and $360°$ due to buses overtaking each other.

waiting time but surprisingly no improvement in the average travel time compared to the naive approach. On the other hand, the RL-TT method improves average travel time by around 8% over the naive approach and around 10% over the holding strategy. It also improves the average waiting time by around 22% over the naive approach. Interestingly, using the RL-TT method, it can be observed from Fig. 2(c) that the buses may become temporarily bunched before they quickly unbunch. Obviously, in this case the buses do not maintain a perfectly staggered phase headway of $180°$. Perhaps unintuitively, allowing this temporary bunching turns out to yield a better average travel time.

6   Conclusion

We have explored the application of RL with the D3QN architecture to the bus loop system. For identical buses, waiting time optimization may result in better average waiting time compared to the holding strategy by allowing the buses to stay longer at stops with higher passengers' arrival rate, but at the cost of much longer average travel time. Using a theoretical model of a bus loop system, we also analytically proved that staying longer at any particular stops does not help in reducing the average travel time. This may not be a worthy trade-off considering the purpose of a transportation system is to move commuters to their destinations, preferably as quickly as possible. On the other hand, with travel time optimization, the buses managed to learn a close approximation of the holding strategy, maintaining headway near the perfectly staggered phase.

For detuned buses optimizing travel time, using the conditions of the NTU shuttle bus Blue route which consists of 12 reasonably staggered bus stops served by two buses, the buses surprisingly learned to allow temporary bunching to achieve better average travel time than both the holding and naive strategies, and also better average waiting time than the naive approach. The results highlights the potential benefit of RL methods to find novel strategies beyond just maintaining a target headway that are better than conventional approaches.

Finally, we found in our case that if the commuter inflow rate into the system is higher than the maximum delivery rate of the buses, unbounded growth of waiting commuters will occur which can only be solved by adding more buses. Therefore, we considered the scenario in which the Poissonian commuter inflow rate is comfortably lower than the maximum delivery rate. In this case, the buses may never be full and the bus capacity can effectively be treated as unlimited.

The weakness of our RL approach is the scaling problem. For a system with large number of buses and bus stops, we observe slow computation time and it becomes harder to converge to an optimal strategy. We intend to find ways to tackle this issue as our future work. In addition, since our approach is only based on the decision of each bus to stay or leave at bus stops, we believe our approach is general and thus also applicable for non-loop services. We will look into its application in bus line services and expanding to the city-scale bus network.

Acknowledgments. This work was supported by the Singapore Ministry of Education (MOE) Academic Research Fund (AcRF) Tier 2 Grant No. MOE-T2EP20222-0004.

Disclosure of Interests. The authors have no competing interests to declare that are relevant to the content of this article.

References

1. Silver, D., Hubert, T., Schrittwieser, J., et al.: A general reinforcement learning algorithm that masters chess, shogi, and go through self-play. Science **362**(6419), 1140–1144 (2018)
2. Silver, D., Huang, A., Maddison, C., et al.: Mastering the game of Go with deep neural networks and tree search. Nature **529**, 484–489 (2016)
3. Kasparov, G.: Chess, a drosophila of reasoning. Science **362**(6419), 1087 (2018)
4. Saw, V.-L., Chung, N.N., Quek, W.L., Pang, Y., Chew, L.Y.: Bus bunching as a synchronisation phenomenon. Sci. Rep. **9**, 6887 (2019)
5. Cats, O., Larijani, A.N., Koutsopoulos, H.N., Burghout, W.: Impacts of holding control strategies on transit performance. Transport. Res. Rec. J. Transport. Res. Board **2216**(1), 51–58 (2011)
6. Sun, A., Hickman, M.: The real-time stop-skipping problem. J. Intell. Transport. Syst. **9**(2), 91–109 (2005)
7. Furth, P.G.: Alternating deadheading in bus route operations. Transp. Sci. **19**(1), 13–28 (1985)
8. Delgado, F., Munoz, J.C., Giesen, R.: How much can holding and/or limiting boarding improve transit performance? Transport. Res. Part B: Methodol. **46**(9), 1202–1217 (2012)
9. Stewart, C., El-Geneidy, A.: All aboard at all Doors. Transport. Res. Rec. J. Transport. Res. Board **2418**(1), 39–48 (2014)
10. Quek, W.L., Chung, N.N., Saw, V.-L., Chew, L.Y.: Analysis and simulation of intervention strategies against bus bunching by means of an empirical agent-based model. Complexity **2021**, 2606191 (2021)

11. Saw, V.-L., Chew, L.Y.: No-boarding buses: synchronisation for efficiency. PLoS ONE **15**(3), e0230377 (2020)
12. Saw, V.-L., Chew, L.Y.: No-boarding buses: agents allowed to cooperate or defect. J. Phys. Complex. **1**(1), 015005 (2020)
13. Saw, V.-L., Vismara, L., Chew, L.Y.: Intelligent buses in a loop service: emergence of no-boarding and holding strategies. Complexity **2020**, 7274254 (2020)
14. Vismara, L., Chew, L.Y., Saw, V.-L.: Optimal assignment of buses to bus stops in a loop by reinforcement learning. XXPhys. A **583**, 126268 (2021)
15. Saw, V.-L., Vismara, L., Chew, L.Y.: Chaotic semi-express buses in a loop. Chaos **31**, 023122 (2021)
16. Ibarra-Rojas, O., Delgado, F., Giesen, R., Muñoz, J.: Planning, operation, and control of bus transport systems: a literature review. Transport. Res. Part B: Methodol. **77**, 38–75 (2015)
17. Saw, V.-L., Vismara, L., Suryadi, Y.B., Johansson, M., Chew, L.Y.: Inferring origin-destination distribution of agent transfer in a complex network using deep gated recurrent units. Sci. Rep. **13**, 8287 (2023)
18. Bahuleyan, H., Vanajakshi, L.D.: Arterial path-level travel time estimation using machine-learning techniques. J. Comput. Civ. Eng. **31**(3), 04016070 (2017)
19. Reddy, K.K., Kumar, B.A., Vanajakshi, L.: Bus travel time prediction under high variability conditions. Curr. Sci. **111**(4), 700 (2016)
20. Mendes-Moreira, J., Jorge, A.M., de Sousa, J.F., Soares, C.: Improving the accuracy of long-term travel time prediction using heterogeneous ensembles. Neurocomputing **150**, 428–439 (2015)
21. Panovski, D., Scurtu, V., Zaharia, T.: A neural network- based approach for public transportation prediction with traffic density matrix. In: Proceedings of 7th European Workshop on Visual Information Processing (EUVIP), pp. 1—6. IEEE, New York (2018)
22. Heghedus, C., Chakravorty, A., Rong, C.: Neural network frameworks: comparison on public transportation prediction. In: Proceedings of International Parallel and Distributed Processing Symposium Workshops (IPDPSW), pp. 842—849. IEEE, New York (2019)
23. Taparia, A., Brady, M.: Bus journey and arrival time prediction based on archived AVL/GPS data using machine learning. In: Proceedings of 7th International Conference on Models and Technologies for Intelligent Transportation Systems (MT-ITS), pp. 1—6. IEEE, New York (2021)
24. Xiao, M., Xiahou, J., Ge, M.: A reinforcement-learning-based bus scheduling model. 2022 IEEE 10th Joint International Information Technology and Artificial Intelligence Conference (ITAIC), pp. 923–927. IEEE, New York (2022)
25. Ai, G., Zuo, X., Chen, G., Wua, B.: Deep reinforcement learning based dynamic optimization of bus timetable. Appl. Soft Comput. **131**, 109752 (2022)
26. Liu, Y., Zuo, X., Ai, G., Liu, Y.: A reinforcement learning-based approach for online bus scheduling. Knowl.-Based Syst. **271**, 110584 (2023)
27. Chen, W., Zhou, K., Chen, C.: Real-time bus holding control on a transit corridorv based on multi-agent reinforcement learning. In: Proceedings of the IEEE 19th International Conference on Intelligent Transportation Systems (ITSC), pp. 100–106. IEEE, New Yor (2016)
28. Chen, C.X., Chen, W.Y., Chen, Z.Y.: A multi-agent reinforcement learning approach for bus holding control strategies. Adv. Transport. Stud., 41–54 (2015)
29. Alesiani, F., Gkiotsalitis, K.: Reinforcement learning-based bus holding for high-frequency services. In: BT - 21st International Conference on Intelligent Trans-

portation Systems, ITSC 2018, Maui, HI, USA, 4–7 November 2018, pp. 3162–3168 (2018)

30. Schulman, J., Wolski, F., Dhariwal, P., Radford, A., Klimov, O.: Proximal policy optimization algorithms. arXiv:1707.06347 (2017)
31. Wang, J., Sun, L.: Dynamic holding control to avoid bus bunching: a multi-agent deep reinforcement learning framework. Transport. Res. Part C: Emerg. Technol. **116**, 102661 (2020)
32. Wang, Z., Schaul, T., Hessel, M., Hasselt, H.V., Lanctot, M., Freitas, N.D.: Dueling Network Architectures for Deep Reinforcement Learning (2015). arXiv:1511.06581
33. Bradtke, S.J., Duff, M.O.: Reinforcement learning methods for continuous-time Markov decision problems. In: Tesauro, G., Touretzky, D., Leen, T. (eds.) Advances in Neural Information Processing Systems, NIPS 1994, vol. 7. MIT Press (1994)

Precise Language Deception: XAI Driven Targeted Adversarial Examples with Restricted Knowledge

Mateusz Gniewkowski[✉][ID], Paweł Walkowiak[ID], Marek Klonowski[ID], and Tomasz Walkowiak[ID]

Wrocław University of Science and Technology, Wrocław, Poland
mateusz.gniewkowski@pwr.edu.pl

Abstract. In this paper, we propose a novel approach for crafting targeted adversarial examples (attacks) using explainable artificial intelligence (XAI) techniques. Our method leverages XAI to identify key input elements that, when altered, can mislead NLP models, such as BERT and large language models (LLMs), into producing specific incorrect outputs. We demonstrate the effectiveness of our targeted attacks across a range of NLP tasks and models, even in scenarios where internal model access is restricted. Our approach is particularly effective in zero-shot learning settings, underscoring its adaptability and transferability to both traditional and conversational AI systems. In addition, we outline mitigation strategies, demonstrating that adversarial training and fine-tuning can enhance model defenses against such attacks. Although our work highlights the vulnerabilities of LLMs and BERT models to adversarial manipulation, it also lays the groundwork for developing more robust models, advancing the dual goal of understanding and securing black-box NLP systems. Through targeted adversarial examples and SHAP-based techniques, we not only expose the weaknesses of existing models but also propose strategies to enhance AI's resilience to deceptive linguistic input.

Keywords: LLM · BERT · Adversarial Example · XAI

1 Introduction

Adversarial Examples (AE), known also as adversarial attacks, are considered to be one of the most important obstacles to the further development of advanced deep learning methods and the implementation of trustworthy Artificial Intelligence (AI). Although tens of thousands of papers have been presented on this topic over the last decade, many AE issues in the field of Natural Language Processing (NLP) still seem underinvestigated. In this paper, we show that by using relatively simple methods of Explainable Artificial Intelligence (XAI), we are able not only to confuse the language model by constructing a wide class of various AEs but also to conduct a **targeted attack**. We can force the system to provide a pointed, specified (incorrect) answer. We refine attack strategies to control output classes across various NLP tasks. Beyond traditional

© The Author(s), under exclusive license to Springer Nature Switzerland AG 2025
M. H. Lees et al. (Eds.): ICCS 2025, LNCS 15903, pp. 49–60, 2025.
https://doi.org/10.1007/978-3-031-97626-1_4

transformer-based classifiers (BERT [6]), we also explore vulnerabilities in Large Language Models (LLMs) including ChatGPT and OpenChat [19], focusing on zero-shot learning scenarios.

Our main contributions are the following. We demonstrate a novel targeted attack method that requires only a small number of examples to mislead the model. The methods we propose are effective even with **restricted knowledge** of the model. That is, we do not require access to the model weights, making our approach applicable to models accessible through APIs, such as ChatGPT. We also examine mitigation strategies for BERT model, showing that standard fine-tuning and adversarial training can effectively defend against such attacks. Our findings contribute to both understanding and improving the security of black-box NLP models. In general, our result can be seen as an argument for the weakness of language models (in particular LLMs) facing AE. On the other hand, it seems that our results can also contribute to the building of more robust language models.

2   Our Approach and Related Work

Our work concerns the security of language models (esp. LLMs), explainable artificial intelligence (XAI), and so-called adversarial examples (AE). Each of these three areas has many thousands of relevant works and has already spawned numerous meta-surveys, so it is difficult to even list most important related works. Due to space limitations, in the following, we mention only the most important works and some results closest to our findings presented in this paper.

AE are carefully modified inputs to AI models that cause their incorrect/dangerous responses. In the current form they have been introduced in the seminal paper [16]. However, similar concepts have been explored earlier (e.g. [2]). Since AEs are considered to be one of the most serious threats to building trustworthy AI systems, tens of thousands of papers have been written on them in recent years. Most of them can be found in a constantly updated list [5] currently containing several thousand works. Nevertheless, few works have been written on AE for LLM and language models in general. The work [10] from a few years is the first paper to present AE in LLMs. An interesting approach to crafting AEs was proposed in [17]. This method allows generating a much richer class of AEs (compared to previous algorithms) in a semi-automatic model (with a human in the loop). An overview of security threats in LLMs, with a particular focus on AEs from a different perspective, can be found in [11]. Several works have shown that AEs are one of the main reasons for limiting LLMs in a number of applications where reliability is critical (i.e. [1]).

One of the key ideas behind the targeted attacks presented in this paper is to leverage XAI methods to generate effective adversarial examples. In our approach, explainability techniques are used to identify the most crucial elements of the input that influence the system's behavior. These elements are then modified in various (potentially subtle) ways to achieve the desired outcome with a sufficiently high probability in practice. To our knowledge, the connection between XAI and adversarial examples was first identified in [7], where the authors hypothesized *a deep relationship between model explainability and adversarial examples*. In [9], the authors demonstrated that the widely used game-theoretic SHAP method [13] can be used to mitigate adversarial attacks. On the other hand, the work [22] shows that AEs can be used to effectively manipulate SHAP scores and extract sensitive data. Similar results for manipulating LIME-based explanations [15] of text classification can be found in [4].

In the context of LLMs, significant research has focused on "jailbreaking" [20], i.e., techniques for extracting restricted content from a model, such as instructions for potentially harmful actions public safety. The study in [23] shows that jailbreak attacks can be effectively automated and, more surprisingly, exhibit a degree of transferability between models. Other important works include [21], where general mechanisms of protection against jailbreak attacks are considered. Although jailbreak can be seen as a form of adversarial attack, it remains unclear how these methods could be applied to the LLM classification problem explored in our work. Recent efforts have specifically targeted adversarial prompt generation to induce harmful behaviors in LLMs. Much of this research highlights potential risks or proposes mitigation strategies, such as output filtering [18] or modifying training data to reduce vulnerabilities [12]. Only a few studies have attempted to explain the underlying mechanisms that enable these attacks. A notable finding is presented in [20], where the authors identify competing objectives and mismatched generalization as key factors contributing to the existence of adversarial examples in LLMs.

Our work is most closely related to [8], which explores methods to effectively find adversarial examples using SHAP functions. In particular, in the current paper we also make extensive use of SHAP. However, the key distinction is that our focus is on targeted attacks, specifically constructing adversarial examples where the incorrect output follows predefined properties (for example, ensuring that an element from class A is always misclassified as B). Naturally, this constraint limits the generation of adversarial examples and requires a different methodological approach. Moreover, we have implemented the proposed methods in state-of-the-art NLP solutions, specifically LLMs.

3 Methods

The goal of the proposed method is to execute targeted attacks, forcing the AI system to produce a specific, predefined (incorrect) response. To achieve this, we utilize differentiated ranking lists to steer the attack toward designated classes. By manipulating importance-based rankings (obtained by the SHAP method), we can precisely influence the attack trajectory, enhancing control over the final output of the model. This refined approach enables the generation of more precise adversarial examples, making it particularly effective for tasks that require class-specific misclassifications.

Let $R_A = \{(w, S_A(w))\}$ and $R_B = \{(w, S_B(w))\}$ be the ranked lists of tokens w with their corresponding SHAP importance scores $S_A(w)$ and $S_B(w)$ for classes A and B, respectively. To execute a targeted adversarial attack aimed at shifting the prediction from class A to class B, we compute the differential importance score for a token w as $\Delta S(w) = S_A(w) - S_B(w)$. The resulting list is defined as $R_{A \to B} = ((w, \Delta S(w)) \mid w \in R_A \cup R_B)$ sorted in descending order with respect to $\Delta S(w)$. This ranking highlights tokens that are highly influential for class A while minimally supporting class B, making them ideal candidates for modification. Intuitively, if $S_A(w) >= S_B(w)$, the token w supports class A and is prioritized for alteration. If $S_B(w) > S_A(w)$, modifying w is less favorable as it already aligns with the class B. Using tokens with the highest positive $\Delta S(w)$, the attack effectively reduces the influence of class A while steering the model toward class B.

3.1 Computing Importance for LLM

Prompt: Classify sentence into one of the following classes 0 - cat, 1 - dog.
Return only a single digit related to class:
<Text>

Text: A dog is barking, a cat is meowing, a cow does muu.

Result:
Class cat: A dog is barking, a cat is meowing, a cow does muu

Class dog: A dog is barking, a cat is meowing, a cow does muu

Fig. 1. Classification example. SHAP results for GPT-4o-mini, shows the local importance of each token for the respective classes. Red indicates higher importance, while blue represents lower importance. (Color figure online)

To compute token-level importance scores for LLMs, particularly those accessible via API, we employ a modified SHAP-based approach tailored for black-box settings. We begin with a text sample and a corresponding prompt crafted to guide the LLM's response. Consider the example from Fig. 1.

To compute SHAP values for individual tokens, we modify only the input text (using any tokenizer) while keeping the prompt constant. Each modified version is then sent to the remote LLM using OpenAI API. The model needs to return two things: a list of tokens and log probabilities (`logprobs`) for its generated outputs. These log probabilities can be converted into standard probabilities, allowing us to assess how changes in the text influence the likelihood of each class. In binary classification tasks, we concentrate on the top-$n = 2$ most relevant tokens. This approach enables us to calculate Shapley values by observing how the removal or alteration of specific tokens affects the model's predictions. If the returned token does not correspond to any predefined class label, it is simply excluded from the explanation process.

3.2 Attack Methodology

To investigate the behavior of large language models (LLMs) under adversarial conditions, we adapted the attack methodology from [8], incorporating explainable AI (XAI) techniques to test the models. The original approach leveraged SHAPley value-based global explanations (computed on a separate dataset split) to identify the most influential parts of the victim text for classification. Additionally, we employed the following text-disrupting methods:

- **WordNet-XAI**, which replaces selected words in the text with their synonyms retrieved from plWordNet. Candidate synonyms are further filtered based on cosine similarity, computed using FastText [3] word embeddings. Only candidates with a similarity score that exceeds the threshold ϵ_w (set to 65%) are considered for substitution.
- **WordNet-XAI-CharDiscard** (WordNet-XAI-ChD), which introduces perturbations by randomly deleting letters from a word w_i with a given probability p (set to 0.4%).

The candidate attack sentences, after applying the substitutions, are filtered based on a cosine similarity threshold ϵ (set to 95%), using sentence embeddings generated by the Sentence Transformer [14]. For LLM evaluation, a zero-shot prompting approach was chosen. Furthermore, the original class labels were represented as digits (see Fig. 1), allowing the extraction of SHAP values from the models. The results presented were obtained using the same algorithmic parameters as those in [8].

4   Results

Table 1. The characteristics of the dataset used: language of the dataset, number of labels, sizes of dataset parts, average length of the texts in words. All datasets and their parts are balanced.

Dataset	Lang	No. of classes	Train size	Test size	XAI size	Aver. len
AG_News	EN	4	120,000	6,840	100	38
Wiki_PL	PL	4	801	358	40	186

In reported experiments, we tested the BERT model, which we trained for straightforward text classification. Furthermore, we compared its performance with classification results from the OpenChat and GPT-4o-mini models using a zero-shot learning approach. The LLM prompts followed the structure illustrated in Fig. 2. This figure demonstrates that even minor modifications to a sample can sometimes be enough to mislead these classifiers. Moreover, it highlights that the constraints imposed on the generation methods ensure that the modified samples remain highly similar to the original ones.

Table 1 provides a summary of the datasets used in our study, namely AG_News[1] and Wiki_PL[2]. Each dataset contains four distinct classes and has been divided into three subsets. The first subset is a training set, used primarily for fine-tuning the BERT and OpenChat models. The second is a test set, which serves both for evaluating classifier accuracy and for executing adversarial attacks. Finally, the third subset, referred to

---

[1] http://groups.di.unipi.it/~gulli/AG_corpus_of_news_articles.html.

[2] http://hdl.handle.net/11321/216.

Prompt: Classify sentence into one of the following classes:
0: Articles related to international events, global news, and world affairs. This category includes stories on political events, international conflicts, diplomacy, and relations between countries.
1: This category includes news related to sports events, athletes, match outcomes, developments across various sports, as well as updates on teams and sporting events.
2: Articles in this category cover financial, economic, and market-related news. It includes content on companies, market trends, investments, financial matters, and economic topics.
3: Articles focused on science and technology news. Topics include new technologies, scientific research, discoveries, and trends or innovations in fields such as medicine, IT, computers, and beyond.
Return only a single digit related to class:
\<Text>

Text: Panel Urges N.Y. to Pay \$14 Billion More for City Schools Court appointed referees recommended the state → commonwealth pay an additional \$14 billion over four years to improve New York City schools.

Result: 0 → 2

Fig. 2. Example of successful directed attacks for AG_News classification (Test set) using GPT-4o-mini model, achieved by altering just a single word in the sentence. The green indicates original form, the red one change after the attack. The importance score of the word 'state' is 0.021 for class "0" and -0.019 for class "2". The difference between these values (0.04) makes it the best candidate for replacement in class "0" to class "2" attack scenario. (Color figure online)

Table 2. Classification accuracy (ACC) for Wiki_PL and AG_News datasets. All datasets are balanced across classes to ensure fair evaluation. The BERT model is fine-tuned on the respective training sets, while OpenChat and GPT-4o-mini perform zero-shot classification using prompts that include natural language descriptions of each class.

Dataset	BERT ACC [%]	OpenChat ACC [%]	GPT-4o-mini ACC [%]
Wiki_PL	99.00	99.00	99.00
AG_News	95.00	85.00	85.00

as the XAI set, comprises a smaller sample of the data and is used to generate word-level importance rankings for modification purposes. The complete data processing pipeline is depicted in Fig. 3.

All datasets are approximately balanced, with only minor deviations in the number of examples per class. Table 2 presents preliminary classification results for each of the examined methods. As can be observed, all approaches achieve satisfactory accuracy scores, confirming their general effectiveness in tackling the classification task.

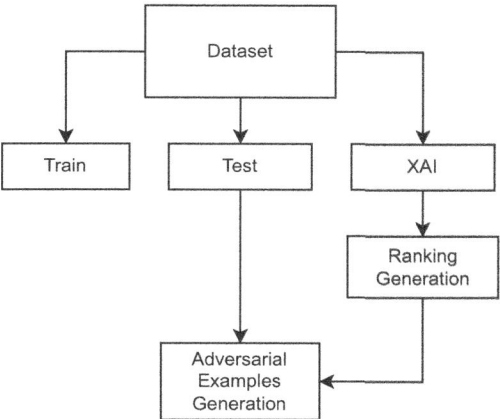

Fig. 3. General data flow used in the experimental setup. The original dataset is split into three parts: training data for fine-tuning BERT classifier, test data for evaluation (used also for Open-Chat and GPT-4o-mini), and data used for generating explanations via an XAI method. The XAI component produces feature importance rankings, which guide the *Adversarial Examples Generation* module. This module uses the ranked features to create targeted perturbations on the test set, resulting in adversarial examples used for robustness evaluation of the models.

We began with a standard adversarial attack, where the goal was to alter the classification to any other class. This is illustrated in Table 3. In particular, attention should be paid to the results of methods applied to the XAI subset. These results demonstrate how an attack would perform on samples where we had prior information (the ranking remains global, averaged across the samples in the XAI subset). It is likely that slightly better results could be achieved by using the importance scores computed for individual samples. Although the data sets are not very large, they reveal that at least some of the samples are vulnerable to the attack (1% for the AG_News data set and 2.5% for Wiki_PL). Using the same rankings, we performed attacks on the test subsets. The results show a high effectiveness of the attacks, with the most successful method being the simple removal of characters from individual (most important) words. A slightly more sophisticated approach, replacing words with synonyms, also yields satisfactory results. An interesting observation is that methods based on LLMs seem to be effectively resistant to attacks for the Polish language. However, this is associated with the very high confidence of these models in basic classification tasks. Tables 4 and 5 illustrate the effectiveness of both targeted and untargeted attacks. Comparing these values allows us to evaluate how well the proposed method directs the attacks. The only case where the method did not perform as expected was AG_News A ⇸ B, likely due to the strong separation between these classes. For the Wiki_PL dataset, we expected that targeted attacks, designed to highlight key modifications, would be more effective in misleading the model. However, our results indicate that the small changes introduced in

Table 3. Results of adversarial attacks. The **WordNet-XAI** method replaces important words with their synonyms, while **WordNet-XAI-ChD** introduces noise by randomly deleting characters from relevant words. BERT results are taken from [8]. The OpenChat and GPT-4o-mini results are based on zero-shot classification. An attack is considered successful if the model's prediction changes from a true positive to any other class.

Attack type	Dataset	Part	Success % BERT	Success % OpenChat	Success % GPT-4o-mini
WordNet-XAI	AG_News	XAI	4.00	10.00	11.00
WordNet-XAI-ChD	AG_News	XAI	2.00	10.00	13.00
WordNet-XAI	AG_News	Test	2.27	7.44	7.65
WordNet-XAI-ChD	AG_News	Test	2.99	7.85	7.69
WordNet-XAI	Wiki_PL	XAI	5.00	7.50	5.00
WordNet-XAI-ChD	Wiki_PL	XAI	20.00	12.50	12.50
WordNet-XAI	Wiki_PL	Test	1.68	0.00	0.28
WordNet-XAI-Chd	Wiki_PL	Test	10.89	0.28	0.84

the samples were insufficient, as successfully deceiving the model would require significantly larger alterations, making the modifications more noticeable to the reader. Additionally, it is important to consider that LLM models have probably encountered articles from this dataset in multiple languages, given that the data originate from Wikipedia.

Fine-tuning models on adversarially altered samples is a promising strategy for mitigating the impact of attacks. We successfully fine-tuned BERT, which led to improved results, although its performance declined slightly when using the character removal method. For OpenChat, Supervised Fine-Tuning (SFT) can be employed, and when executed properly, it should yield better results by adapting the model to a specific task. However, training such models is highly resource intensive, and performing SFT correctly presents challenges, particularly since task-specific specialization can degrade performance in other areas. A more efficient and cost-effective approach may involve detecting modifications before inputting the data into the LLM, thereby reducing the need for extensive fine-tuning. The classification stability of BERT remained unchanged after fine-tuning on the perturbed Wiki_PL dataset. Regarding adversarial attacks, the model showed increased resistance to WordNet-XAI attacks (0.28% success rate), likely due to its exposure to synonym substitutions during training. However, it struggled more with character removal attacks (7.54% success rate). This is because once a word is altered, it undergoes different tokenization and the modified tokens in the test data may not align with those seen during training.

Table 4. Success rates of directed adversarial attacks on a BERT-based classifier. The **WordNet-XAI** method generates adversarial examples by replacing semantically important words with their synonyms using the WordNet. The **WordNet-XAI-ChD** method applies additional perturbations by randomly deleting characters from those important words. The notation $A \rightarrow B$ indicates an attempt to intentionally change a sample originally and correctly classified as class A into being misclassified as class B. Mean success rates are reported for each scenario.

Method	Attack type	Dataset	A → B	A → C	A → D	Mean
Targeted	WordNet-XAI	AG_News	2.28	2.34	2.4	2.34
	WordNet-XAI-ChD		2.75	3.45	2.75	2.98
Untargeted	WordNet-XAI		0.94	0.76	0.53	0.74
	WordNet-XAI-ChD		1.46	1.58	0.7	1.25
Targeted	WordNet-XAI	Wiki_PL	4.44	2.22	2.22	2.96
	WordNet-XAI-ChD		6.67	6.67	8.89	7.41
Untargeted	WordNet-XAI		0.00	0.00	1.11	0.37
	WordNet-XAI-ChD		1.11	0.00	1.11	0.74

Table 5. Success rates of directed adversarial attacks on large language models (LLMs): Open-Chat and GPT-4o-mini. The **WordNet-XAI** method generates adversarial examples by replacing semantically important words with their synonyms using the WordNet. The **WordNet-XAI-ChD** method applies additional perturbations by randomly deleting characters from those important words. The notation $A \rightarrow B$ indicates an attempt to intentionally change a sample originally and correctly classified as class A into being misclassified as class B.

Method	Attack type	Dataset	Openchat				GPT-4o-mini			
			A → B	A → C	A → D	Mean	A → B	A → C	A → D	Mean
Targeted	WordNet-XAI	AG_News	2.75	2.57	2.51	2.61	3.10	3.04	2.98	3.04
	WordNet-XAI-ChD		3.10	3.10	3.27	3.16	2.92	3.39	3.27	3.19
Untargeted	WordNet-XAI		3.74	2.87	1.35	2.65	3.16	3.22	1.17	2.52
	WordNet-XAI-ChD		4.33	3.04	1.64	3.00	4.09	3.10	1.46	2.88
Targeted	WordNet-XAI	Wiki_PL	0.00	1.11	0.00	0.37	0.00	0.00	0.00	0.00
	WordNet-XAI-ChD		1.11	0.00	1.11	0.74	0.00	0.00	0.00	0.00
Untargeted	WordNet-XAI		0.00	0.00	0.00	0.00	0.00	0.00	0.00	0.00
	WordNet-XAI-ChD		0.00	0.00	0.00	0.00	0.00	0.00	0.00	0.00

5   Conclusion

In this study, we demonstrate that insights derived from explainable artificial intelligence (XAI) can be leveraged to craft targeted adversarial attacks on natural language processing models operating under black-box constraints. By observing the attribution scores returned by local explanation methods, we are able to focus on the input fragments that matter most to the classifier and subsequently adjust them with minimal effort.

Our experiments show that inconspicuous edits can alter predictions even for LLMs such as GPT-4o-mini. Furthermore, a brute-force search can produce up to 500 times more candidates to be tested, and an XAI-driven strategy achieves similar success with far fewer queries. Targeted attacks are shown to be more effective than untargeted ones, highlighting the practical value of explanation-guided evaluation in real-world settings.

The results indicate that models such as BERT and OpenChat, although demonstrating promising performance in classification tasks, remain susceptible to adversarial attacks. In the context of targeted attacks, we found that all the analyzed models can be manipulated with subtle modifications. The effectiveness of these attacks varies; the removal of random characters has proven to be more effective, whereas other methods, such as synonym substitution, demonstrate greater robustness. This suggests that while targeted attacks are possible, they depend heavily on specific circumstances. Furthermore, using local explanation methods, we can identify key features that contribute to these vulnerabilities and potentially reduce the impact of such attacks.

Importantly, local explanation techniques expose tokens that contribute most strongly to model decisions. Knowledge of these fragile anchors can be used to both intensify attacks and design defenses. Future research should therefore explore mitigation strategies such as adversarial training, input sanitization, or confidence calibration that reduce vulnerability without eroding predictive accuracy.

Acknowledgments. Financed by: (1) CLARIN ERIC (2024–2026), funded by the Polish Minister of Science (agreement no. 2024/WK/01); (2) CLARIN-PL, the European Regional Development Fund, FENG program (FENG.02.04-IP.040004/24); (3) statutory funds of the Department of Artificial Intelligence, Wroclaw Tech; (4) the EU project 'DARIAH-PL', under investment A2.4.1 of the National Recovery and Resilience Plan. (5) the European Regional Development Fund as part of the 2014–2020 Smart Growth Operational Program (POIR.04.02.00-00C002/19); and (6) by the National Science Center, Poland, grant number 2018/29/B/HS2/02919.

References

1. Albrecht, J., Kitanidis, E., Fetterman, A.J.: Despite "super-human" performance, current LLMs are unsuited for decisions about ethics and safety (2022). https://arxiv.org/abs/2212.06295
2. Biggio, B., et al.: Evasion attacks against machine learning at test time. In: Blockeel, H., Kersting, K., Nijssen, S., Železný, F. (eds.) ECML PKDD 2013. LNCS (LNAI), vol. 8190, pp. 387–402. Springer, Heidelberg (2013). https://doi.org/10.1007/978-3-642-40994-3_25
3. Bojanowski, P., Grave, E., Joulin, A., Mikolov, T.: Enriching word vectors with subword information. Trans. Assoc. Comput. Linguist. **5**, 135–146 (2017)
4. Burger, C., Chen, L., Le, T.: Are your explanations reliable? Investigating the stability of lime in explaining text classifiers by marrying XAI and adversarial attack (2023). https://arxiv.org/abs/2305.12351
5. Carlini, N.: A complete list of all (arxiv) adversarial example papers (2019–2025). https://nicholas.carlini.com/writing/2019/all-adversarial-example-papers.html
6. Devlin, J., Chang, M.W., Lee, K., Toutanova, K.: BERT: pre-training of deep bidirectional transformers for language understanding. In: Burstein, J., Doran, C., Solorio, T. (eds.) Proceedings of the 2019 Conference of the North American Chapter of the Association for Computational Linguistics: Human Language Technologies, Volume 1 (Long and Short Papers),

pp. 4171–4186. Association for Computational Linguistics, Minneapolis, Minnesota (2019). https://doi.org/10.18653/v1/N19-1423

7. Fidel, G., Bitton, R., Shabtai, A.: When explainability meets adversarial learning: detecting adversarial examples using SHAP signatures. In: 2020 International Joint Conference on Neural Networks, IJCNN 2020, Glasgow, United Kingdom, 19–24 July 2020, pp. 1–8. IEEE (2020). https://doi.org/10.1109/IJCNN48605.2020.9207637

8. Gniewkowski, M., et al.: Do not trust me: explainability against text classification. In: ECAI 2023 - 26th European Conference on Artificial Intelligence, 30 September–4 October 2023, Kraków, Poland - Including 12th Conference on Prestigious Applications of Intelligent Systems (PAIS 2023). Frontiers in Artificial Intelligence and Applications, vol. 372, pp. 875–882. IOS Press (2023). https://doi.org/10.3233/FAIA230356

9. Hickling, T., Aouf, N., Spencer, P.: Robust adversarial attacks detection based on explainable deep reinforcement learning for UAV guidance and planning. IEEE Trans. Intell. Veh. 8(10), 4381–4394 (2023)

10. Jia, R., Liang, P.: Adversarial examples for evaluating reading comprehension systems. CoRR abs/1707.07328 (2017). http://arxiv.org/abs/1707.07328

11. Jia, X., et al.: Global challenge for safe and secure LLMs track 1. arXiv preprint arXiv:2411.14502 (2024)

12. Lukas, N., et al.: Analyzing leakage of personally identifiable information in language models. In: 44th IEEE Symposium on Security and Privacy, SP 2023, San Francisco, CA, USA, 21–25 May 2023, pp. 346–363. IEEE (2023). https://doi.org/10.1109/SP46215.2023.10179300

13. Lundberg, S.M., Lee, S.I.: A unified approach to interpreting model predictions. In: Advances in Neural Information Processing Systems, vol. 30 (2017)

14. Reimers, N., Gurevych, I.: Sentence-BERT: sentence embeddings using Siamese BERT-networks. In: Proceedings of the 2019 Conference on Empirical Methods in Natural Language Processing. Association for Computational Linguistics (2019). http://arxiv.org/abs/1908.10084

15. Ribeiro, M.T., Singh, S., Guestrin, C.: "Why should I trust you?": explaining the predictions of any classifier. In: Krishnapuram, B., Shah, M., Smola, A.J., Aggarwal, C.C., Shen, D., Rastogi, R. (eds.) Proceedings of the 22nd ACM SIGKDD International Conference on Knowledge Discovery and Data Mining, San Francisco, CA, USA, 13–17 August 2016, pp. 1135–1144. ACM (2016). https://doi.org/10.1145/2939672.2939778

16. Szegedy, C., et al.: Intriguing properties of neural networks. arXiv preprint arXiv:1312.6199 (2013)

17. Wallace, E., Rodriguez, P., Feng, S., Yamada, I., Boyd-Graber, J.L.: Trick me if you can: human-in-the-loop generation of adversarial question answering examples. Trans. Assoc. Comput. Linguist. 7, 387–401 (2019). https://doi.org/10.1162/TACL_A_00279

18. Wang, B., et al.: Exploring the limits of domain-adaptive training for detoxifying large-scale language models. In: Advances in Neural Information Processing Systems 35: Annual Conference on Neural Information Processing Systems 2022, NeurIPS 2022, New Orleans, LA, USA, 28 November–9 December 2022 (2022). http://papers.nips.cc/paper_files/paper/2022/hash/e8c20cafe841cba3e31a17488dc9c3f1-Abstract-Conference.html

19. Wang, G., Cheng, S., Zhan, X., Li, X., Song, S., Liu, Y.: OpenChat: advancing open-source language models with mixed-quality data. In: The Twelfth International Conference on Learning Representations (2024). https://openreview.net/forum?id=AOJyfhWYHf

20. Wei, A., Haghtalab, N., Steinhardt, J.: Jailbroken: how does LLM safety training fail? In: Advances in Neural Information Processing Systems 36: Annual Conference on Neural Information Processing Systems 2023, NeurIPS 2023, New Orleans, LA, USA, 10–16 December 2023 (2023). http://papers.nips.cc/paper_files/paper/2023/hash/fd6613131889a4b656206c50a8bd7790-Abstract-Conference.html

21. Wei, A., Haghtalab, N., Steinhardt, J.: Jailbroken: how does LLM safety training fail? (2023). https://arxiv.org/abs/2307.02483
22. Yeghiazaryan, M., et al.: Texture- and shape-based adversarial attacks for vehicle detection in synthetic overhead imagery (2024). https://arxiv.org/abs/2412.16358
23. Zou, A., Wang, Z., Carlini, N., Nasr, M., Kolter, J.Z., Fredrikson, M.: Universal and transferable adversarial attacks on aligned language models. arXiv preprint arXiv:2307.15043 (2023)

Tensorial Implementation for Robust Variational Physics-Informed Neural Networks

Askold Vilkha[1] , Carlos Uriarte[2] , Paweł Maczuga[1] ,
Tomasz Służalec[1] , and Maciej Paszyński[1(✉)]

[1] AGH University of Krakow, Kraków, Poland
maciej.paszynski@agh.edu.pl
[2] Basque Center for Applied Mathematics, Bilbao, Spain

Abstract. Variational Physics-Informed Neural Networks (VPINN) train the parameters of neural networks (NN) to solve partial differential equations (PDEs). They perform unsupervised training based on the physical laws described by the weak-form residuals of the PDE over an underlying discretized variational setting; thus defining a loss function in the form of a weighted sum of multiple definite integrals representing a testing scheme. However, this classical VPINN loss function is not robust. To overcome this, we employ Robust Variational Physics-Informed Neural Networks (RVPINN), which modifies the original VPINN loss into a robust counterpart that produces both lower and upper bounds of the true error. The robust loss modifies the original VPINN loss by using the inverse of the Gram matrix computed with the inner product of the energy norm. The drawback of this robust loss is the computational cost related to the need to compute several integrals of residuals, one for each test function, multiplied by the inverse of the proper Gram matrix. In this work, we show how to perform efficient generation of the loss and training of RVPINN method on GPGPU using a sequence of einsum tensor operations. As a result, we can solve our 2D model problem within 350 s on A100 GPGPU card from Google Colab Pro. We advocate using the RVPINN with proper tensor operations to solve PDEs efficiently and robustly. Our tensorial implementation allows for 18 times speed up in comparison to *for*-loop type implementation on the A100 GPGPU card.

Keywords: Robust Variational Physics Informed Neural Network · GPGPU · Parallelization of tensor operations

1 Introduction

Recently, there has been a growing interest in designing and training Deep Neural Networks (DNN) for solving challenging Partial Differential Equations (PDEs). The most popular methods for training the DNN solutions of PDEs are Physics Informed Neural Networks (PINN) [1–3], and Variational Physics Informed Neural Networks (VPINN) [4]. Since their introduction in 2019, they have gained

© The Author(s), under exclusive license to Springer Nature Switzerland AG 2025
M. H. Lees et al. (Eds.): ICCS 2025, LNCS 15903, pp. 61–75, 2025.
https://doi.org/10.1007/978-3-031-97626-1_5

exponential growth in the number of papers and citations. It is an attractive alternative for solving PDEs, in comparison with traditional solvers such as the Finite Element Method. With the introduction of modern stochastic optimizers such as ADAM [5] they easily find high-quality minimizers of the loss functions employed.

Physics-Informed Neural Network, proposed in 2019 by Prof. Karniadakis, revolutionized the way in which neural networks find solutions to boundary-value problems described by means of PDEs [1]. In the PINN method, the neural network is treated as a function approximating the solution of a PDE. After computing the necessary differential operators, the neural network and its appropriate differential operators are inserted into the PDE. The residuum of the PDE and the boundary conditions are assumed as the loss function. The learning process consists of sampling the loss function at different points by calculating the PDE residuum and the boundary conditions. PINNs have been successfully applied to solve a wide range of problems, from fluid mechanics [2,3], in particular, Navier-Stokes equations [6], wave propagation [7,8], phase-field modeling [9], biomechanics [10], quantum mechanics [11], electrical engineering [12], problems with point singularities [13], uncertainty qualification [14], dynamic systems [15,16], or inverse problems [17,18], among many others.

Prof. Karniadakis has also proposed Variational Physics Informed Neural Networks VPINN [4]. VPINN uses the idea of a variational formulation in which the PDE is averaged using the integration over a given domain with prescribed distributions, called the test functions. The relation between PINN and VPINN is similar to the relation between finite difference and finite element methods (FDM/FEM). In the first class of methods, the continuous PDE is considered in the strong form at the set of selected points. In the second class of methods, the PDE is considered in the weak form, averaged using a family of distributions called the test functions. The VPINN method has found several applications, from Poisson and advection-diffusion equations [19], non-equilibrium evolution equations [20], solid mechanics [21], fluid flow [22], and inverse problems [23,24], among others.

In this paper, we focus on the VPINN method. We show that the loss functions employed by the VPINN method are not robust. The loss function of VPINN can significantly differ from the true error. Thus, we employ the robust loss proposed in the RVPINN method [25]. The authors in [25] show that the robust loss proposed there is a lower bound for the true error. It is also an upper bound up to some oscillatory term.

The drawback of this robust loss is the computational cost related to the need to compute several integrals of residuals, one for each test function, multiplied by the inverse of the proper Gram matrix. In this paper, we focus on model Laplace problems. As it is shown in [25], for this kind of problem, the Gram matrix has to be computed in the weighted H_0^1 inner product. In this paper, we select the trigonometric test functions defined over the entire domain. These test functions result in the diagonal Gram matrix, as well as its inverse. We show

how to perform efficient generation of the loss on GPGPU using a sequence of einsum tensor operations.

Using our parallel tensor operations designed for RVPINN, we can solve our model 2D PDEs within 350 s on A100 GPGPU card from Google Colab Pro. Our parallel implementation allows for 18 times speed up in comparison to *for*-loop type implementation on the A100 GPGPU card executed using Google Colab Pro.

2 Neural-Network Framework

To numerically approximate PDEs, we consider a DNN function with input $\mathbf{x} = (x_1, \ldots, x_d)$ and output $u_{NN}(\mathbf{x}; \theta)$, where $\theta \in \mathbb{R}^S$ represents the trainable parameters. We employ a fully-connected Feed-Forward Neural Network (NN) composed of L layers. Each layer l in the NN consists of a set of neurons. The output of layer l, with $l = 1, \ldots, L-1$, is given by:

$$\mathbf{z}^{(l)} = \sigma(\mathbf{w}^{(l)}\mathbf{z}^{(l-1)} + \mathbf{b}^{(l)}), \tag{1}$$

where σ is a tanh activation function, $\mathbf{w}^{(l)}$, $\mathbf{b}^{(l)}$ are the weights and biases, respectively, associated with the layer l, and $\mathbf{z}^{(0)} = \mathbf{x}$ is the input to the first layer. The final layer L is innactivated:

$$u_{NN}(\theta) = \mathbf{w}^{(L)}\mathbf{z}^{(L-1)} + \mathbf{b}^{(L)}. \tag{2}$$

Using ADAM optimization algorithm [5], the NN weights and biases are learned. We denote the manifold of different realizations of the neural network functions as U_{NN}.

3 Numerical Results for VPINNs

In this section we solve two model two-dimensional problems by using VPINN [4] method. The goal of this section is to show the lack of robustness of the VPINN loss. For that, we illustrate the discrepancy between the VPINN loss function and the true error. By the true error we mean the relative error in the energy-norm defined by

$$\frac{\|u_{NN} - u_{EXACT}\|_{H_0^1}}{\|u_{EXACT}\|_{H_0^1}}, \tag{3}$$

where $\|u\|_{H_0^1} = \int_\Omega \nabla u(\mathbf{x}) \cdot \nabla u(\mathbf{x}) \, d\mathbf{x}$ is the norm of the underlying trial space $H_0^1 = H_0^1(\Omega)$. Besides its theoretical foundations described in [25], this energy norm gives a good estimate how the derivatives of the NN solution approximate the derivatives of the exact solution.

3.1   Laplace Model Problem with Sin-Sin Solution

Given $\Omega = (0,1)^2 \subset \mathbb{R}^2$ we seek the solution of the model problem with manufactured solution

$$- \Delta u = f, \tag{4}$$

with homogeneous Dirchlet boundary conditions that we enforce on the NN in a strong way, following the ideas presented in [26]. In this subsection, we select the solution

$$u(x_1, x_2) = sin(2\pi x_1)sin(2\pi x_2), \tag{5}$$

presented in Fig. 1. In order to obtain this solution, we consider the source

$$f(x_1, x_2) = -\Delta u(x_1, x_2) = -8\pi^2 sin(2\pi x_1)sin(2\pi x_2). \tag{6}$$

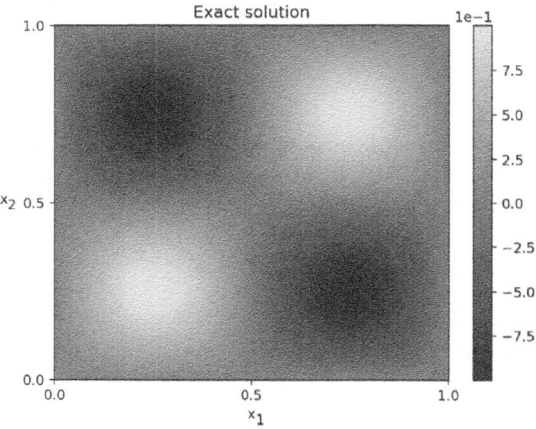

Fig. 1. Solution to the first problem.

We consider the weak form of the PDE, obtained by integration by parts with selected test functions $v \in V = H_0^1(\Omega)$. We also assume, that the solution u is approximated by the neural network $u \approx u_{NN}(\theta) \in U_{NN}$. Namely, find $u_{NN}(\theta) \in U_{NN}$ such that

$$b(u_{NN}(\theta), v) := \int_\Omega \nabla u_{NN}(\mathbf{x}) \cdot \nabla v(\mathbf{x}) \, d\mathbf{x} \tag{7}$$

$$= \int_\Omega f(\mathbf{x}) \, v(\mathbf{x}) \, d\mathbf{x} =: l(v), \, \forall v \in V, \tag{8}$$

where $\mathbf{x}$ is an abbreviation for (x_1, x_2).

This weak formulation can equivalently be read as vanishing the following residual form:

$$r(u_{NN}(\theta), v) := b(u_{NN}(\theta), v) - l(v) = 0, \, \forall v \in V. \tag{9}$$

For the test discretization setting, we define the finite-dimensional space $V_M \subset V$

$$V_M = \mathrm{span}(\{v_m\}_{m=1}^M) \qquad (10)$$

of trigonometric test functions $v_m = \sin(m_1 \pi x) \sin(m_2 \pi y)$ with $m = (m_1, m_2)$ and $1 \leq m_1 \leq M_1$ and $1 \leq m_2 \leq M_2$. Thus, $M = M_1 M_2$.

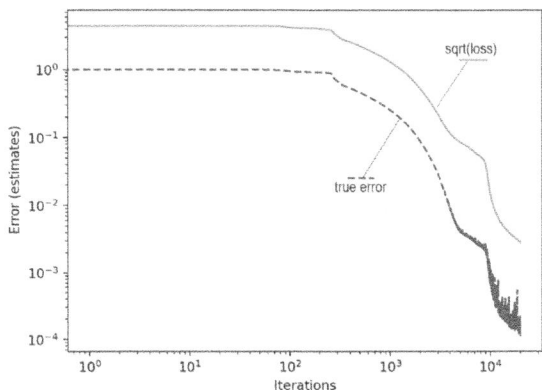

Fig. 2. First problem. VPINNs approximation using strong imposition of boundary conditions, 30×30 trigonometric test functions, 400×400 integration points for training, and 400×400 integration points for the true error. Convergence of the loss function and the convergence of the true relative error as measured in the energy norm.

In this way, the original VPINN loss function (see [4,19]) is defined as the result of adding up all the squared residual contributions for each test basis function as follows:

$$LOSS(\theta) = \sum_{m=1}^{M} \left\{ r\left(u_{NN}(\theta), v_m\right)\right\}^2. \qquad (11)$$

We perform numerical integration to approximate each residual contribution employing Monte Carlo integration, i.e.,

$$r(u_{NN}(\theta), v_m) \approx$$
$$\frac{1}{K} \sum_{k=1}^{K} \nabla u_{NN}(\mathbf{x}_k) \cdot \nabla v_m(\mathbf{x}_k) - f(\mathbf{x}_k)\, s_m(\mathbf{x}_k), \qquad (12)$$

where K is the total number of integration points.

For the VPINN approximation we use strong imposition of boundary conditions, 30×30 trigonometric test functions, 400×400 integration points for training, and 400×400 integration points for computing the true error for the convergence plot. Figure 2 show the convergence of the VPINN minimization.

We can see from this figure that the plot of the true error $\|u_{NN} - u_{EXACT}\|$ is far from the square root of the loss (as well as from the loss itself). Here u_{NN} is the neural network solution, u_{EXACT} is the exact solution (that is usually not known in the real problems). Ideally, we would like these two plots coincide.

3.2 Laplace Model Problem with Sin-Exp Solution

Following model problem (4), we consider the solution

$$u(x_1, x_2) = -exp(\pi(x_1 - 2x_2)) \sin(2\pi x_1) \sin(\pi x_2), \qquad (13)$$

whose source term is

$$f(x_1, x_2) = -\pi^2 exp(\pi(x_1 - 2x_2)) sin(2\pi x_1)(4 cos(\pi x_2). \qquad (14)$$

This exact solution is presented in Fig. 3.

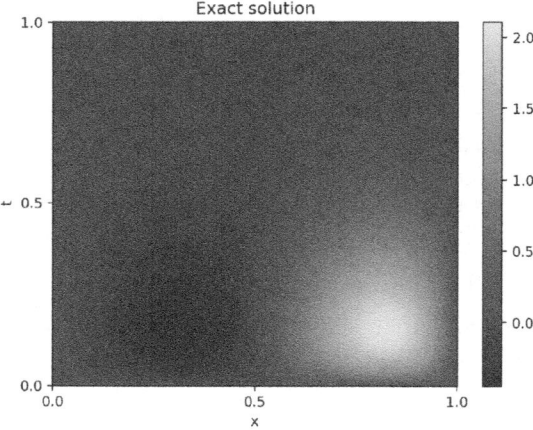

Fig. 3. Solution to the second problem.

We employ the same weak formulation as above but with different right-hand side, same test discretization setting, and same VPINN loss function. For the VPINNs approximation we use strong BCs imposition, 30×30 spectral test functions, 400×400 integration points for training, 400×400 integration points to compute the truth error for the convergence plot. The convergence of the VPINN is presented in Fig. 4. We can read from this figure, that the plot of the true error $\|u_{NN} - u_{EXACT}\|$ does not coincide at all with the square root of the loss (or with the loss itself). Here u_{NN} is the neural network solution, u_{EXACT} is the exact solution (that is usually not known in the real problems). Ideally, we would like these two plots to coincide.

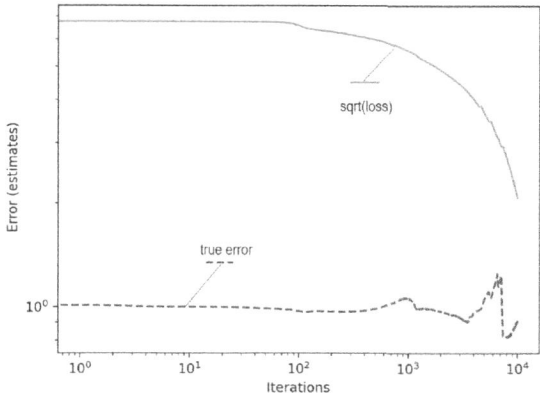

Fig. 4. Second problem. VPINNs approximation: strong BCs imposition, 30×30 spectral test functions, 400×400 integration points for training, 400×400 integration points for the true error. Convergence of the loss function and the convergence of the true relative error as measured in the energy norm.

3.3   Summary of VPINN Results

The convergence of training with ADAM optimizer [5] is presented in Figs. 2 and 4. We can see from these Figures that the loss functions are far from the true error computed between the approximated solution u_{NN} and the known exact solutions. Thus, the VPINN loss is not robust. Changing the number of neurons or layers, improving the quadrature as in [27], or changing the training rate, does not help to make this loss robust. Still, the VPINN loss allows us to obtain the correct solutions presented in Figs. 1 and 3, but looking at the loss function convergence in Figs. 2 and 4 we cannot see what is the true error of the trained solution. Thus, we do not know what is the quality of the trained solution. This is especially true if we do not know the exact solution, which is the case in practical problems.

4   Numerical Results for RVPINNs

In this section, we present how to modify the originally proposed VPINN loss function in [4,19] to obtain its robust counterpart proposed in [25]: instead of considering a single residual contribution for each selected test basis function, we have to test all the residual contributions against each other. Moreover, such testing has to be consistently weighted via the corresponding Gram matrix.

We emphasize that simply avoiding crossed multiplications of residual terms and Gram-matrix coefficients does not guarantee robustness during training, i.e., lower and upper bounds for the true error during loss minimization, as seen in the VPINN experiments of Figs. 2 and 4. We refer to [25] for details.

Following a general test-space discretization V_M spanned by basis functions $\{v_m\}_{m=1}^M$, the robust version of the VPINN loss function is as follows:

$$LOSS(\theta) = \sum_{m,n=1}^M \{r(u_{NN}(\theta), v_m)\} \, G_{m,n}^{-1} \, \{r(u_{NN}(\theta), v_n)\} \, .$$

One might argue that original VPINNs is a simplification of RVPINNs when the Gram matrix is the identity.

In our discretization setting, it is easy to check that our trigonometric basis functions are orthogonal with respect to the inner product in $H_0^1(\Omega)$, producing a corresponding diagonal inverse of the Gram matrix as follows:

$$G_{m_1 m_2, n_1 n_2}^{-1} = \begin{cases} \frac{4}{(m_1^2 + m_2^2)\pi^2}, & \text{if } m_1 = n_1, m_2 = n_2, \\ 0, & \text{otherwise,} \end{cases} \tag{15}$$

Here, $G_{m_1 m_2, n_1 n_2} = (v_{m_1 m_2}, v_{n_1 n_2})_{H_0^1}$ denotes the Gram matrix. This reduces our additive complexity of the RVPINN loss function to

$$LOSS(\theta) = \sum_{m_1, m_2} G_{m_1 m_2}^{-1} \, \{r(u_{NN}(\theta), v_{m_1, m_2})\}^2 \, , \tag{16}$$

where, by abuse of notation, $G_{m_1 m_2}^{-1}$ denotes the diagonal of the inverse of the Gram matrix given by the coefficients in (15).

4.1   Laplace Model Problem with Sin-Sin Solution

For the first model problem, the loss function (16) is robust. This time, the square root of the loss function is equal to the true error, as it is presented in Fig. 5. This indicates that the robust loss accurately reflects the error between the neural network solution u_{NN} and the exact solution u_{EXACT}, namely $\|u_{NN} - u_{EXACT}\|$. This is true even when the exact solution is unknown. Thus, we know when to stop the training to get a good quality solution.

For the RVPINNs approximation we use strong BCs imposition, 20×20 spectral test functions, 200×200 integration points for training, 1000×1000 integration points to compute the truth error for the convergence plot. The convergence of the VPINN is presented in Fig. 4.

4.2   Laplace Model Problem with Sin-Exp Solution

For the second model problem, the loss function (16) is also robust. In contrary to the VPINN, illustrated in Fig. 4, in the RVPINN, the square root of the loss function is a good estimate of the true error, as it is presented in Fig. 6. For the VPINNs approximation we use strong BCs imposition, 20×20 spectral test functions, 200×200 integration points for training, 1000×1000 integration points to compute the truth error for the convergence plot.

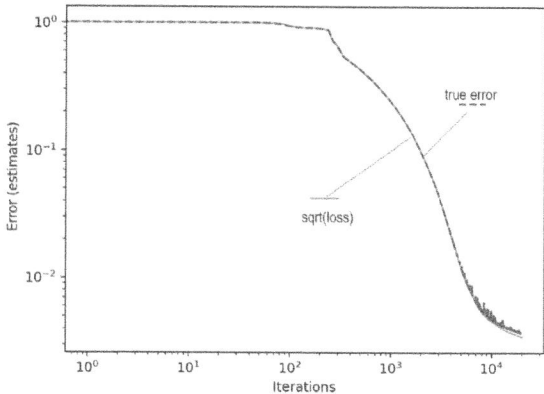

Fig. 5. First problem. RVPINNs approximation: strong BCs imposition, 20×20 spectral test functions, 200×200 integration points for training, 1000×1000 integration points for truth error. Convergence of the loss function and the convergence of the true relative error as measured in the energy norm.

5   Tensor Implementation for (R)VPINN

The RVPINN method with trigonometric test functions requires the addition of several residual terms and multiplied by the coefficients of the inverse of the Gram matrix in the loss function.

Although we can easily implement this scheme using component-by-component operations over a few number of *for*-type loops, it should be noted that such an approach is highly inefficient in interpreted programming languages like Python, which is where neural-network-based models are nowadays majorly developed. In this way, an efficient tensor workflow implementation consists of properly organizing operation functions from GPGPU-developed libraries like TensorFlow, PyTorch, or JAX. Trying to design a flowchart with operations outside these libraries typically produces disproportionately inefficient execution times. We considered PyTorch as our coding platform.

5.1   Linear Algebra with Einstein Summation

Our loss function consists of a combination of numerical integration, Eq. (12), and the residual summation of Eq. (16) as follows:

$$\sum_m G_m^{-1} \left\{ \frac{1}{K} \sum_{k=1}^{K} \nabla u_{NN}(\mathbf{x}_k) \cdot \nabla v_m(\mathbf{x}_k) - f(\mathbf{x}_k)\, v_m(\mathbf{x}_k) \right\}^2 .$$

where $m = (m_1, m_2)$ and $\mathbf{x}_k = (x_{1k}, x_{2k})$.

This summation expression involves the appropriate combination of tensors. These operations have a user-friendly implementation on tensor-oriented platforms that follow the Einstein summation convention.

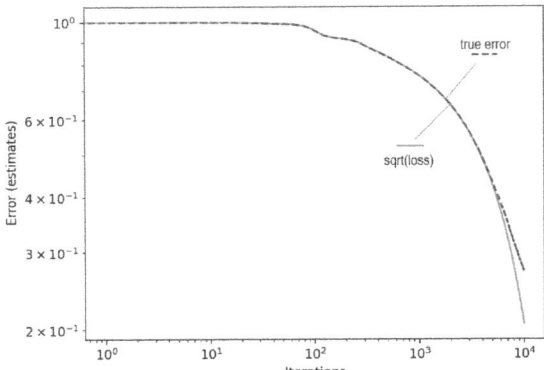

Fig. 6. Second problem. RVPINNs approximation: strong BCs imposition, 20×20 spectral test functions, 200×200 integration points for training, 1000×1000 integration points for truth error. Convergence of the loss function and the convergence of the true relative error as measured in the energy norm.

In the following, we describe and support with graphics the implementation of our loss evaluation.

We start from the operation constructing 3D tensors. From now on, x, t, n and m replace x_1, x_2, n, and m, respectively. The einsum function prevents representing the indexes of each axis by more than one character during codification.

```
x_times_n = torch.einsum("xt,n->xtn",
    x.reshape(n_x, n_t), n)
t_times_m = torch.einsum("xt,m->xtm",
    t.reshape(n_x, n_t), m)
```

Now, we construct a 4D tensor with values of two-dimensional test functions, from two 3D tensors.

```
test_x = torch.sin(math.pi*x_times_n)
test_t = torch.sin(math.pi * t_times_m)
test = torch.einsum("xtn,xtm->xtnm",
test_x, test_t)
```

Next, we construct the derivatives of the test functions along x.

```
test_x_dx = torch.pi *
    torch.einsum("n,xtn->xtn", n,
    torch.cos(torch.pi*x_times_n))
```

We also construct a 4D tensor with values of derivatives of two-dimensional test functions with respect to x, out of two 3D tensors.

```
test_dx = torch.einsum("xtn,xtm->xtnm",
    test_x_dx, test_t)
```

Next we compute derivatives of the test functions with respect to t.

```
test_t_dt = torch.pi *
```

```
torch.einsum("m,xtm->xtm", m,
torch.cos(torch.pi*t_times_m))
```

We also construct a 4D tensor with values of derivatives of two-dimensional test functions with respect to t, out of two 3D tensors.

```
test_dt = torch.einsum("xtn,xtm->xtnm",
test_x, test_t_dt)
```

Finally, we construct the first part of the loss function, a 2D tensor, out of the Neural Network and the derivatives of test functions with respect to x.

```
loss1 = dx * dt * epsilon *
torch.einsum("xt,xtnm->nm",
dpinn_dx, test_dx)
```

We also construct in the analogous way the second part of the loss function, a 2D tensor, out of the Neural Network and the derivatives of test functions with respect to t.

```
loss2 = dx * dt * epsilon *
torch.einsum("xt,xtnm->nm",
dpinn_dt, test_dt)
```

The last part of the loss function is constructed out of the right-hand side 2D tensor and the 4D tensor representing the test functions.

```
loss3 = dx * dt *
torch.einsum("xt,xtnm->nm", rhs, test)
```

We sum up all the loss contributions

```
loss = loss1 + loss2 - loss3
```

we take the second power and multiply by the Gram matrix.

```
loss = loss**2 * self.G
```

and we return the sum of all the obtained loss values.

```
return loss.sum()
```

6   Numerical Experiments

6.1   Google ColabPro Comparison

We perform 20,000 iterations with 100×100 integration points, 20×20 basis functions, 2 layers of the neural network with 200 neurons each. Using our tensorial implementation, we can solve the experiments conducted in Sects. 3 and 4 within 350 s on a A100 GPGPU card from Google Colab. Execution times in a implementation using *for*-type loops takes 108 min. We have 18 times speed up the training process using tensor magic.

6.2   CYFRONET Supercomputing Center Experiment

To investigate further the scalability of RVPINN we have executed our model problems on two GPUs, both NVIDIA A100-SXM4-40GB with a total Memory of 39.56 GB each, 108 multiprocessors each, with CUDA Capability: 8.0. The timing for growing number of training points is presented in Fig. 7.

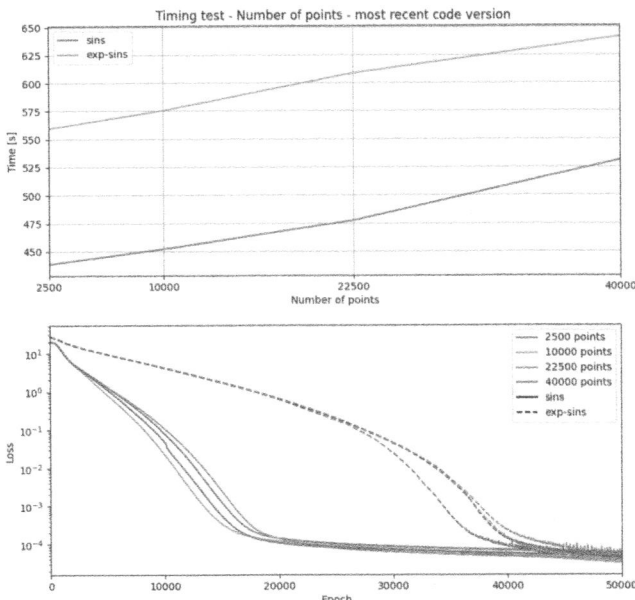

Fig. 7. RVPINNs approximation of the first and second problem, for 20×20 test functions, for growing number of integration points for training (from $50 \times 50 = 2500$, $100 \times 100 = 10000$, $150 \times 150 = 22500$, and $200 \times 200 = 400000$.

We can conclude that the 50×50 points are enough for both problems. For the first problem we need 20,000 iterations, which takes less than 450 s, for the second problem we need 40,000 iterations which takes less than 650 s.

7   Conclusions

Robust Variational Physics-Informed Neural Networks (RVPINNs) are a pivotal advancement in addressing the inherent unrobust nature of Variational Physics-Informed Neural Networks (VPINN). By recalibrating the VPINN loss function to provide a good estimation of the true error, RVPINN offers a more reliable and comprehensive estimation of solution accuracy. In addition, it is critical to implement in terms of tensor algebra in order to exploit GPGPU power during training. A typical implementation in terms of *for*-type loops is inefficient in a tensor workflow, such as in (R)VPINNs. Our tensorial implementation allows for 18 times speed up in comparison to *for*-loop type implementation on a A100 GPGPU card from Google Colab. The future work may involve extension of the method to other classical problems solved by finite element method [28,29], and including adaptive algorithms [30–33] for the test space.

Acknowledgements. This Project has received funding from the European Union's Horizon Europe research and innovation programme under the Marie Sklodowska-Curie grant agreement No. 101119556.

Carlos Uriarte is supported by: PID2023-146678OB-I00 funded by MICIU/AEI/ 10.13039/501100011033 and by FEDER, EU; PID2023-146668OA-I00 funded by MICIU /AEI / 10.13039/501100011033 and by FEDER, EU; the BCAM Severo Ochoa accreditation of excellence CEX2021-001142-S funded by MICIU/AEI/10.13039/ 501100011033; and the BERC 2022–2025 program funded by the Basque Government.

References

1. Raissi, M., Perdikaris, P., Karniadakis, G.E.: Physics-informed neural networks: a deep learning framework for solving forward and inverse problems involving non-linear partial differential equations. J. Comput. Phys. **378**, 686–707 (2019)
2. Cai, S., Mao, Z., Wang, Z., Yin, M., Karniadakis, G.E.: Physics-informed neural networks (PINNs) for fluid mechanics: a review. Acta. Mech. Sin. **37**(12), 1727–1738 (2021)
3. Mao, Z., Jagtap, A.D., Karniadakis, G.E.: Physics-informed neural networks for high-speed flows. Comput. Methods Appl. Mech. Eng. **360**, 112789 (2020)
4. Kharazmi, E., Zhang, Z., Karniadakis, G.E.: Variational physics-informed neural networks for solving partial differential equations, arXiv preprint arXiv:1912.00873 (2019)
5. Kingma, D.P., Ba, J.: Adam: a method for stochastic optimization, arXiv preprint arXiv:1412.6980 (2014)
6. Ling, J., Kurzawski, A., Templeton, J.: Reynolds averaged turbulence modelling using deep neural networks with embedded invariance. J. Fuild Mech. **807**, 155–166 (2016). https://doi.org/10.1017/jfm.2016.615
7. Rasht-Behesht, M., Huber, C., Shukla, K., Karniadakis, G.E.: Physics-informed neural networks (pinns) for wave propagation and full waveform inversions. J. Geophys. Res.: Solid Earth **127**(5), e2021JB023120 (2022)
8. Maczuga, P., Paszyński, M.: Influence of activation functions on the convergence of physics-informed neural networks for 1D wave equation. In: Mikyška, J., de Mulatier, C., Paszynski, M., Krzhizhanovskaya, V.V., Dongarra, J.J., Sloot, P.M. (eds.) Computational Science - ICCS 2023, pp. 74–88. Springer, Cham (2023)
9. Goswami, S., Anitescu, C., Chakraborty, S., Rabczuk, T.: Transfer learning enhanced physics informed neural network for phase-field modeling of fracture. Theor. Appl. Fracture Mach. **106** (2020). https://doi.org/10.1016/j.tafmec.2019. 102447
10. Alber, M., et al.: Integrating machine learning and multiscale modeling-perspectives, challenges, and opportunities in the biologica biomedical, and behavioral sciences. NPJ Digit. Med. **2** (2019). https://doi.org/10.1038/s41746-019-0193-y
11. Jin, H., Mattheakis, M., Protopapas, P.: Physics-informed neural networks for quantum eigenvalue problems. In: International Joint Conference on Neural Networks (IJCNN), pp. 1–8 (2022). https://doi.org/10.1109/IJCNN55064.2022. 9891944
12. Nellikkath, R., Chatzivasileiadis, S.: Physics-informed neural networks for minimising worst-case violations in DC optimal power flow. In: 2021 IEEE International Conference on Communications, Control, and Computing Technologies

for Smart Grids (SmartGridComm), pp. 419–424 (2021). https://doi.org/10.1109/SmartGridComm51999.2021.9632308

13. Huang, X., et al.: A universal pinns method for solving partial differential equations with a point source. In: Proceedings of the Fourteen International Joint Conference on Artificial Intelligence (IJCAI 2022), pp. 3839–3846 (2022)

14. Yang, Y., Perdikaris, P.: Adversarial uncertainty quantification in physics-informed neural networks. J. Comput. Phys. **394**, 136–152 (2019). https://doi.org/10.1016/j.jcp.2019.05.027

15. Sun, F., Liu, Y., Sun, H.: Physics-informed spline learning for nonlinear dynamics discovery. In: Proceedings of the Thirtieth International Joint Conference on Artificial Intelligence (IJCAI 2021), pp. 2054–2061 (2021)

16. Kim, J., Lee, K., Lee, D., Jhin, S.Y., Park, N.: DPM: a novel training method for physics-informed neural networks in extrapolation. In: Proceedings of the AAAI Conference on Artificial Intelligence, vol. 35, no. 9, pp. 8146–8154 (2021). https://doi.org/10.1609/aaai.v35i9.16992

17. Mishra, S., Molinaro, R.: Estimates on the generalization error of physics-informed neural networks for approximating a class of inverse problems for PDEs. IMA J. Numer. Anal. **42**(2), 981–1022 (2022)

18. Lu, L., Pestourie, R., Yao, W., Wang, Z., Verdugo, F., Johnson, S.G.: Physics-informed neural networks with hard constraints for inverse design. SIAM J. Sci. Comput. **43**(6), B1105–B1132 (2021). https://doi.org/10.1137/21M1397908

19. Kharazmi, E., Zhang, Z., Karniadakis, G.E.: hp-VPINNs: variational physics-informed neural networks with domain decomposition. Comput. Methods Appl. Mech. Eng. **374**, 113547 (2021). https://doi.org/10.1016/j.cma.2020.113547

20. Huang, S., He, Z., Chem, B., Reina, C.: Variational onsager neural networks (VONNs): a thermodynamics-based variational learning strategy for non-equilibrium PDEs. J. Mech. Phys. Solids **163** (2022). https://doi.org/10.1016/j.jmps.2022.104856

21. Liu, C., Wu, H.A.: A variational formulation of physics-informed neural network for the applications of homogeneous and heterogeneous material properties identification. Int. J. Appl. Mech. **15**(08) (2023). https://doi.org/10.1142/S1758825123500655

22. Kim, Y., Kwak, H., Nam, J.: Physics-informed neural networks for learning fluid flows with symmetry. Korean J. Chem. Eng. **40**(9), 2119–2127 (2023). https://doi.org/10.1007/s11814-023-1420-4

23. Liu, C., Wu, H.: cv-PINN: efficient learning of variational physics-informed neural network with domain decomposition. Extreme Mech. Lett. **63** (2023). https://doi.org/10.1016/j.eml.2023.102051

24. Badia, S., Li, W., Martin, A.F.: Finite element interpolated neural networks for solving forward and inverse problems. Comput. Methods Appl. Mech. Eng. **418**(A) (2024). https://doi.org/10.1016/j.cma.2023.116505

25. Rojas, S., Maczuga, P., Muñoz-Matute, J., Pardo, D., Paszyński, M.: Robust variational physics-informed neural networks. Comput. Methods Appl. Mech. Eng. **425**, 116904 (2024). https://doi.org/10.1016/j.cma.2024.116904

26. Sun, L., Gao, H., Pan, S., Wang, J.-X.: Surrogate modeling for fluid flows based on physics-constrained deep learning without simulation data. Comput. Methods Appl. Mech. Eng. **361**, 112732 (2020). https://doi.org/10.1016/j.cma.2019.112732

27. Berrone, S., Canuto, C., Pintore, M.: Variational physics informed neural networks: the role of quadratures and test functions. J. Sci. Comput. **92**(3) (2022). https://doi.org/10.1007/s10915-022-01950-4

28. Demkowicz, L.: Computing with hp-Adaptive Finite Elements, vol. 1, Wiley (2006)
29. Demkowicz, L., Kurtz, J., Pardo, D., Paszynski, M., Rachowicz, W., Zdunek, A.: Computing with hp-Adaptive Finite Elements: Volume II Frontiers: Three Dimensional Elliptic and Maxwell Problems with Applications, 1st edn. Chapman and Hall/CRC (2007)
30. Paszyńska, A., Paszyński, M., Grabska, E.: Graph transformations for modeling hp-adaptive finite element method with triangular elements. In: Bubak, M., van Albada, G.D., Dongarra, J., Sloot, P. (eds.) ICCS 2008. LNCS, vol. 5103, pp. 604–613. Springer, Heidelberg (2008). https://doi.org/10.1007/978-3-540-69389-5_68
31. Paszyński, M., Paszyńska, A.: Graph transformations for modeling parallel hp-adaptive finite element method. In: Wyrzykowski, R., Dongarra, J., Karczewski, K., Wasniewski, J. (eds.) PPAM 2007. LNCS, vol. 4967, pp. 1313–1322. Springer, Heidelberg (2008). https://doi.org/10.1007/978-3-540-68111-3_139
32. Paszyński, M., Grzeszczuk, R., Pardo, D., Demkowicz, L.: Deep learning driven self-adaptive hp finite element method. In: Paszynski, M., Kranzlmüller, D., Krzhizhanovskaya, V.V., Dongarra, J.J., Sloot, P. (eds.) ICCS 2021. LNCS, vol. 12742, pp. 114–121. Springer, Cham (2021). https://doi.org/10.1007/978-3-030-77961-0_11
33. Paszyńska, A., et al.: Quasi-optimal elimination trees for 2D grids with singularities. Sci. Program. (1), 303024 (2015)

Neural Parabolic Wave Equation for Refractivity Estimation

Mikhail S. Lytaev[✉] 🆔

St. Petersburg Federal Research Center of the Russian Academy of Sciences,
14-th Linia, V.I., No. 39, Saint Petersburg 199178, Russia
mikelytaev@gmail.com

Abstract. The inverse problem of estimating the refractive index in a waveguide based on wave field measurement data is studied. A differentiable finite-difference scheme for the parabolic wave equation is constructed. The desired function of spatial coordinates, corresponding to the refractive index, is represented as a deep neural network. Optimization problem with respect to unknown refractive index function is formulated and solved. Automatic differentiation of the numerical scheme is used for efficient gradient computation. Numerical examples confirm that the proposed method outperforms the existing approaches to solving underwater and tropospheric tomography problems.

Keywords: inverse problem · physics informed machine learning · ill-posed problem · radiowave propagation · underwater acoustics · JAX

1 Introduction

Refraction has a decisive impact on wave propagation in large unbounded domains such as troposphere [14] or underwater environments [9]. Tropospheric refractive index may form waveguides that transmit radio signals for hundreds of kilometers near the Earth's surface. Similarly, acoustic signals propagate in the sea for hundreds and thousands of kilometers under the influence of the underwater sound speed profile. Despite this, reliable methods for real-time measurement or estimation of atmospheric refractivity parameters [20,21] or oceanic parameters [19] have not yet been developed. The size of the region is too large for real-time direct measurements, so inversion based on indirect measurements seems the most promising. Mathematically, the complexity of the inversion problem lies in its nonlinearity and ill-posedness in the sense of Hadamard [6].

From the point of view of classical theory, nonlinear ill-posed problems are rather hopeless for a reliable solution [25]. Even if a solution can be found, it takes hours or days of extensive computations, i.e., the results become irrelevant [7]. On the other hand, problems solved by modern machine learning (ML), including scientific ML [24], are also ill-posed but are often successfully and quickly solved by modern neural network architectures and optimization methods. This suggests the use of ML tools in the problem of refractive index inversion.

M. H. Lees et al. (Eds.): ICCS 2025, LNCS 15903, pp. 76–90, 2025.
https://doi.org/10.1007/978-3-031-97626-1_6

Physics-informed machine learning models often suffer from a lack of interpretability. The ML approach usually relies on data rather than laws and equations. However, high-quality data in physical problems is a rarity. A quite successful attempt to overcome this issue is the method of physics-informed neural networks (PINN) [15]. PINN allows incorporating physical laws into the objective function, thereby increasing the accuracy and interpretability of the results. PINN is suitable for solving both direct and inverse problems. One of disadvantages of PINN is that it does not take into account the specifics of numerical modeling of the processes it works with. In particular, this is evident in wave propagation modeling in waveguides, where the main difficulty lies in the numerical solution, as the computational domain is very large.

To account for the specifics of numerical implementation, one can substitute numerical scheme for the original physical laws into the objective function. This allows taking into account numerical features but requires differentiating the numerical schemes. Differentiable numerical schemes have already shown their effectiveness in problems of hydrodynamics [1,3], mechanics, thermodynamics [26], and underwater acoustics [17].

In this work, for the first time, the unknown profile is sought in the form of a deep neural network. The parabolic equation method is used as the numerical scheme for the corresponding forward problem, which is equally well suited for solving underwater acoustics problems [5] and tropospheric radio wave propagation [14]. This explains the title of the present paper. The idea of this research is largely inspired by the works [4,12] on neural differential equations, which proposed building models that simultaneously include differential equations and neural networks. This approach allows taking into account wave dynamics using strict wave equations, while poorly interpretable features such as refractive index inhomogeneities are modeled and estimated using neural networks.

2 Mathematical Formulation of the Problem

This section discloses the relationship between the direct and inverse wave propagation problems.

2.1 Direct Problem and Its Solution

The wave process is modeled by the two-dimensional Helmholtz equation [14]

$$\frac{\partial^2 \psi}{\partial x^2} + \frac{\partial^2 \psi}{\partial z^2} + k^2 n^2 (z) \psi = 0, \tag{1}$$

where $\psi(x, z)$ is the complex-valued two-dimensional distribution of the wave field, $n(z)$ is the refractive index of the medium, $k = 2\pi/\lambda$ is the wavenumber, λ is the wavelength. Depending on the specifics of a particular problem, function ψ satisfies certain initial and boundary conditions.

It is assumed that the length (along the x coordinate) of the computational domain significantly exceeds the height (along the z coordinate), i.e., propagation occurs in an elongated waveguide. Under these conditions, refractive index $n(z)$ has a decisive influence on the long range wave propagation.

The problem of finding wave field $\psi(x, z)$ given the refractive index $n(z)$, initial and boundary conditions is called the direct one. It is generally well-posed, i.e., has a unique solution. There are several methods for solving the direct problem for the Helmholtz equation, but the parabolic equation method and its generalization, called the one-way Helmholtz equation, best suit the specifics of the problem being solved [5, 14].

Ignoring backscattering, Eq. (1) can be formally rewritten in the one-way form [8, 18]

$$\frac{\partial \psi}{\partial x} = i\sqrt{\frac{\partial^2}{\partial z^2} + k^2 n^2(z)}\psi.$$

Using the operator exponential, the step-by-step solution can be written as

$$u(x + \Delta x, z) = \mathcal{P}(n)u(x, z),$$

$$u(x, z) = \psi(x, z)\exp(-ikx),$$

$$\mathcal{P}(n)u = \exp\left(ik\Delta x\left(\sqrt{\frac{1}{k^2}\frac{\partial^2}{\partial z^2} + n^2(z)} - 1\right)\right)u. \tag{2}$$

Thus, the direct problem reduces to the numerical approximation of the operator exponential (2). In this work, we use the finite-difference rational approximation method [18]. Within the present research, it is essential that the numerical approximation of (2) is implemented in a finite number of sequential steps. Indeed, as it is shown in [18], the entire step-by-step solution process essentially consists of sequentially solving one-dimensional differential equations of the form

$$\left[1 + b_i\left(\frac{1}{k^2}\frac{\partial^2}{\partial z^2} + n^2(z) - 1\right)\right]u_{i+1}(z) = \left[1 + a_i\left(\frac{1}{k^2}\frac{\partial^2}{\partial z^2} + n^2(z) - 1\right)\right]u_i(z). \tag{3}$$

After discretization along the z variable, this equation reduces to a tridiagonal system of linear algebraic equations, which is solved in linear time using the tridiagonal matrix algorithm.

2.2   Inverse Problem Formulation

In the inverse problem, the refractive index $n(z)$ is unknown. However, the values of the wave field ψ at some points in space (x_i, z_i), $i = 1..N$ are known. We denote the vector of these measurements as v.

The complexity of the inverse problem lies in its ill-posedness in the sense of Hadamard. It is unknown whether a solution exists and whether it is unique. Indeed, there may be too few measurements, or they may be too noisy.

Let $\mathcal{G}(n)$ denote operator that solves the direct problem at points (x_i, z_i), $i = 1..N$ for the refractive index n. Then the inverse operator $\mathcal{G}^{-1}(v)$, mapping the wave field measurement data to the refractive index, will be the solution to the inverse problem. We express the inverse operator $\mathcal{G}^{-1}(\boldsymbol{v})$ in terms of $\mathcal{G}(n)$ and the functional minimization problem

$$\mathcal{G}^{-1}(\boldsymbol{v}) = \arg\min_{n} Loss\left(\mathcal{G}(n), \boldsymbol{v}\right), \tag{4}$$

where

$$Loss\left(\mathcal{G}(n), \boldsymbol{v}\right) = \left\|\mathcal{G}(n) - \boldsymbol{v}\right\|^2 + \gamma \left\|\frac{\partial n}{\partial z}\right\|^2. \tag{5}$$

Indeed, the inverse problem can be viewed as finding such a refractive index $n(z)$ that minimizes the difference between the measured field and the field predicted by the direct model. This is what the first term of the objective functional (5) is responsible for. The second term is responsible for regularization [25]. It eliminates strongly oscillating solutions that formally minimize the functional but have no physical meaning.

Note that this functional does not have any special properties such as convexity or linearity, so its minimization is a highly non-trivial problem.

3   Refractive Index Inversion

As we saw in the previous section, the inverse problem of refractive index inversion is formulated as a minimization problem of a functional that includes solution to the direct problem. The space of functions is infinite-dimensional, so the first thing to do for the numerical minimization of functional (5) is to determine the search space. Usually, vertical refractive index profile is sought in the form of a finite set of values on a given grid [10, 27]. In this work, it is proposed to estimate the refractive index in the form of a multilayer perceptron with one input (height z), one output (real-valued refractive index), and several hidden layers. Thus, the minimization problem reduces to finding the optimal value of a finite number of neural network weights (θ).

There are essentially two large classes of solution methods: stochastic global methods [23] and local methods [7] based on gradient descent. Global methods are convenient because they do not require any additional information about the minimized functional. A black box that outputs the value of the functional at any point in the search space is sufficient. Unfortunately, even with the most successful parameterization, the number of parameters to be determined is one or several tens. Global methods converge extremely slowly, given that it is not any but a specific global minimum that is being sought.

Local optimization methods, which use the gradient of the minimized function with respect to the unknown parameters, are significantly more efficient. Following the gradient direction significantly increases the convergence rate, at least to a local minimum. This is the basis of all existing methods for training neural networks. They are successfully trained, although may have millions of unknown parameters. The difficulty here is precisely the requirement of having a gradient. The use of the finite-difference approach is inefficient due to the catastrophic cancellation problem [2].

For the neural networks training, the automatic differentiation method [2] is used. Its essence lies in representing the network as a computational acyclic graph consisting of elementary operations. Although there may be quite a few of them, sequential automatic application of differentiation rules allows efficient analytical computation of the gradient.

For a long time, this approach was limited to neural networks. A certain revolution here was made by the JAX framework [22], which allowed representing functions and algorithms of a very arbitrary form as a computational graph and, accordingly, automatically differentiating them. At the same time, the programming interface, as much as possible, repeats the widely used numerical modeling libraries numpy and scipy.

Functional (5) is significantly more complex than those usually used in machine learning, as it contains the operator of the direct problem solution. One can, following the adjoint equation method [10, 19], try to differentiate operator $\mathcal{G}$ analytically. This leads to the need to derive and numerically solve a new adjoint equation. There is low flexibility in choosing the representation of the function to be determined within the adjoint method. A much more efficient approach seems to be representing the numerical implementation of the operator $\mathcal{G}$ as a computational graph using the JAX framework.

Note that numerical scheme (3) consists of a known sequence of elementary operations. Therefore, it can be represented as a computational graph. The method is implemented within the PyWaveProp library and is freely available [16].

The proposed solution is schematically presented in Fig. 1. $n_\theta(z)$ is sought in the form of a multilayer perceptron with a finite unknown set of real parameters θ (the nodes of the network). At the same time, $n_\theta(z)$ is an argument of the operator $\mathcal{G}$, the numerical implementation of which is an automatically differentiable computational graph with respect to θ. In addition, the vector of field measurements v is fed as input. This configuration allows automatic computation of the required gradient $\nabla_\theta Loss$.

The resulting gradient computation algorithm is used by one or another local optimization algorithm to search for the optimal parameters θ. Following most works on deep learning, including PINN, in this research we use the Adam method [13] for optimization.

4   Numerical Results and Discussion

The general scheme of computational experiments is based on the inversion of synthetic data. A typical refractive index profile is selected. Using the direct problem solution method, the value at points corresponding to the location of the receivers is computed. Random noise is added to these values. The resulting noisy synthetic measurements are fed to the inversion algorithm, which estimates the refractive index. At the end, the original and inverted profiles are compared.

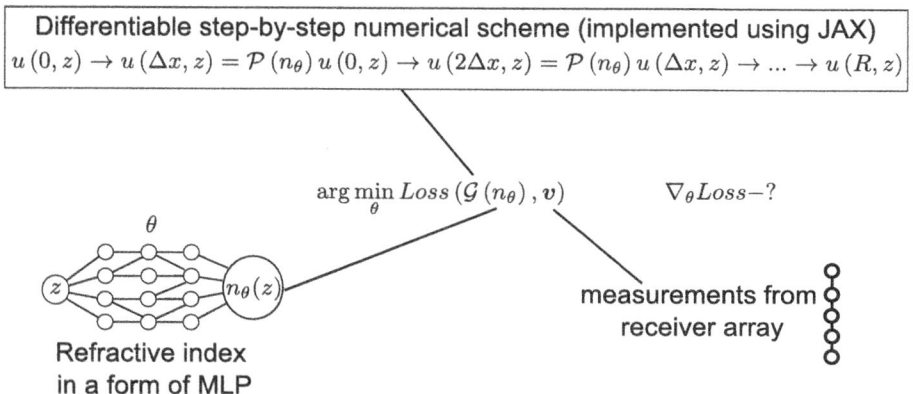

Fig. 1. Schematic description of the inversion algorithm.

4.1   Tropospheric Refractive Index Estimation

The schematic description of the tropospheric refractive index inversion problem is shown in Fig. 2. A source with known parameters emits a radio signal received by a vertical array of receivers. As the signal propagates between the source and the receiver, it is influenced by the inhomogeneities of the tropospheric refractive index. By analyzing the received signal, it is required to determine the tropospheric refractive index.

In this work, only a monochromatic source emitting at a frequency of 3 GHz is considered. The receiver array is located at a distance of 5 km from the source. The array consists of 17 point receivers uniformly located at heights of 5–170 m. The signal-to-noise ratio at the receivers is assumed to be 30 dB.

Unless otherwise specified, refractive index profile is sought in the form of a multilayer perceptron with 4 hidden layers of width 50. The Adam method with a learning rate of 0.05 and regularization parameter $\gamma = 10^{-3}$ is used.

Let us check the fundamental possibility of inversion for typical tropospheric waveguides [14]: surface duct, surface-based duct, and elevated duct. It can be seen from Fig. 3 that the proposed method successfully inverted four different typical tropospheric refractive index profiles. At the same time, the method does not require any prior information about the profile distribution.

Figure 4 depicts the two-dimensional distribution of the electromagnetic field computed for the original and inverted profiles. The influence of the waveguide effects on propagation near the Earth's surface is clearly observable. The elevated waveguide focuses the field near the Earth's surface (up to 100 m), with zones of strong signal and shadow alternating with each other. It can be seen that the patterns for the original and inverted profiles differ slightly. For clarity, Fig. 5 shows a pointwise comparison of the amplitudes. Although the overall qualitative and quantitative picture for the true and inverted profiles is the same, there are some local deviations that can exceed 20 dB. This should be taken into account when processing the results of real experiments.

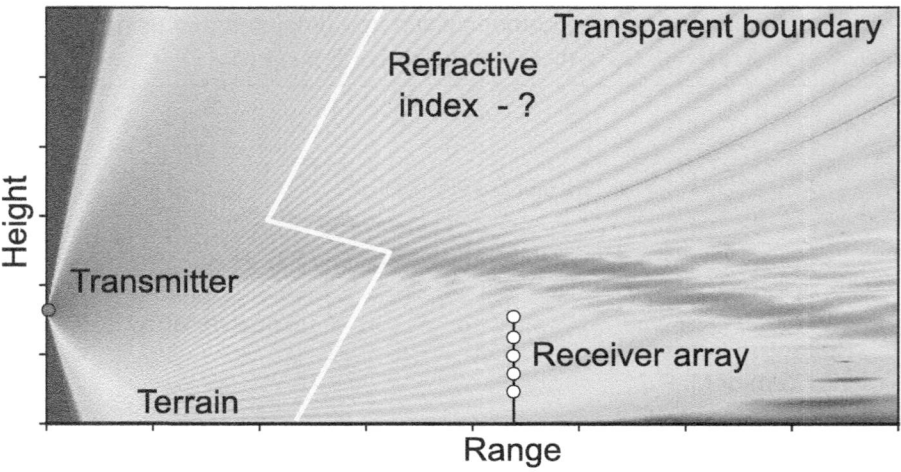

Fig. 2. Tropospheric refractive index inversion. The white line indicates the (unknown) refractive index profile.

Now let us analyze the speed and dynamics of the optimization algorithm convergence. Figure 6 (left) shows the dependence of the minimized functional value on the iteration number of the optimization algorithm. It can be seen that the convergence does not depend significantly on the type of waveguide. The loss function decreases most rapidly up to about 100 iterations and then decreases much more slowly. Figure 6 (right) demonstrates the dependence of the relative error between the true and inverted profiles on the iteration number. Interestingly, for the surface-based duct, the relative error continues to decrease even when the loss function values practically stop decreasing.

Next, let us analyze how the width and depth of the multilayer perceptron affect the efficiency of the proposed algorithm. Figure 7 shows the dependence of the relative error between the true and inverted mixed profile for several different values of the depth and width of the network. It can be seen that in the single-layer case, a lower accuracy was achieved in a reasonable number of iterations than with multilayer networks. Figure 8 demonstrates this visually. The inversion errors of the single-layer network are clearly visible, and deviations reach 10 M-units, while the multilayer network allows almost perfect inversion, even despite the noise.

A completely logical question may arise: why use neural networks if one could search for the unknown profile in the form of a simple piecewise linear or piecewise constant function, as was done in all previous works [10, 27]? Figure 9 shows a comparison of the convergence dynamics of the multilayer perceptron and the piecewise linear function with 50 nodes. It can be seen that the multilayer perceptron converged to the solution two orders of magnitude faster. At the same time, the found solution turned out to be closer to the true one. It should also be noted that the proposed approach allows easy use of any neural network

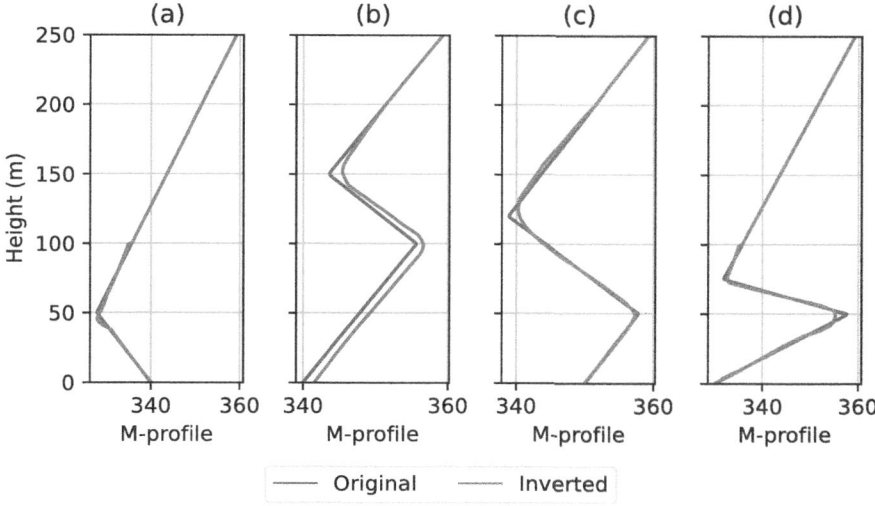

Fig. 3. Original profile and inversion result. (a) Surface duct (b) Elevated duct (c) Surface-based duct (d) Combination of elevated and surface-based profiles (mixed profile).

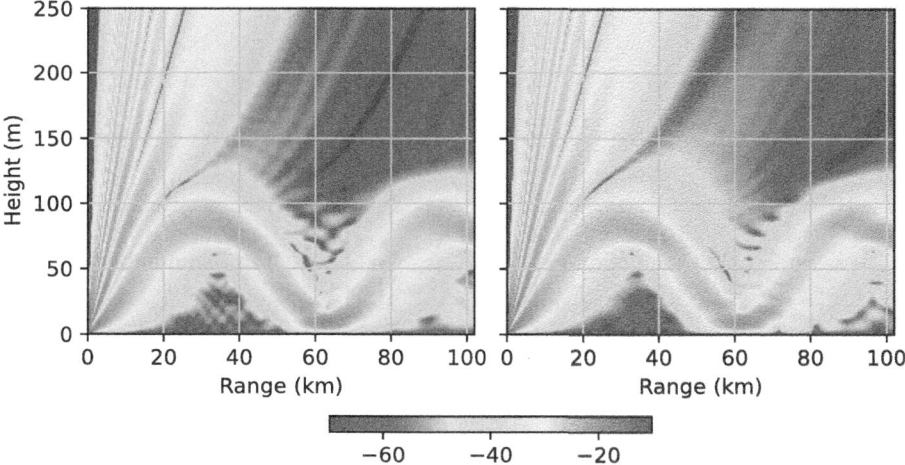

Fig. 4. Distribution of the electromagnetic wave amplitude ($20\log|\psi(x,z)|$) in the true (left) and inverted (right) surface-based ducts.

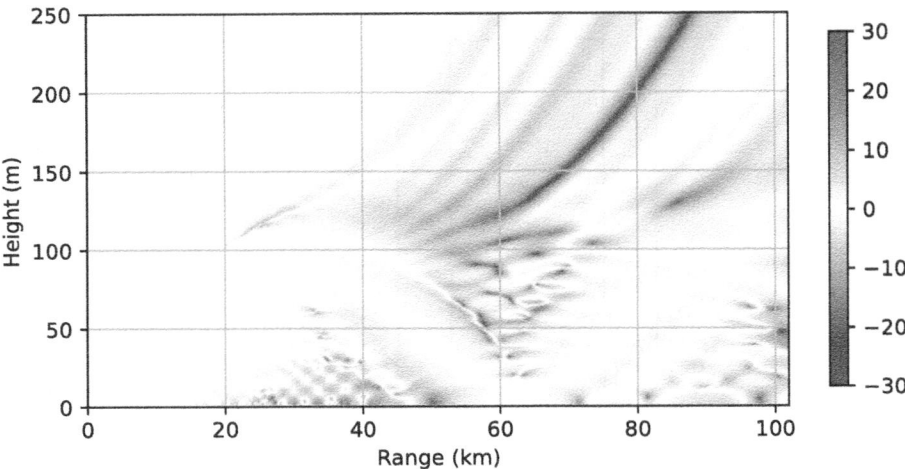

Fig. 5. Difference in amplitudes between the true and inverted surface-based ducts.

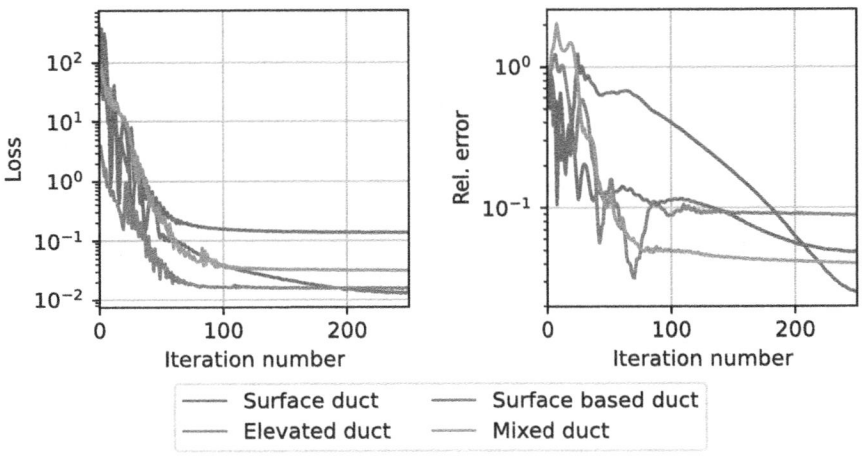

Fig. 6. Dependence of the loss function value (left) and relative error (right) on the iteration number of the optimization algorithm.

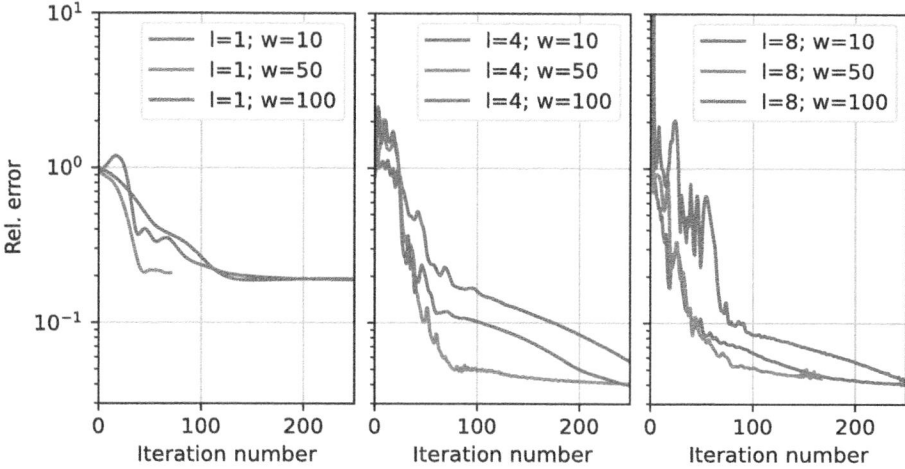

Fig. 7. Dependence of the relative error between the true and inverted mixed profile for depth (l) 1, 4, and 8, and width (w) 10, 50, and 100.

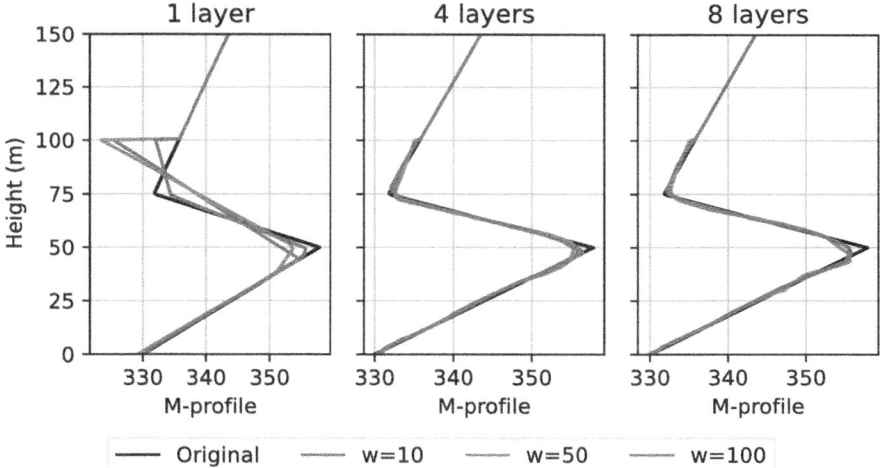

Fig. 8. Original mixed profile and inversion result for various sizes of the multilayer perceptron.

configurations and other representations of the desired function. At the same time, changes in the method and its software implementation are minimal.

Table 1 provides a summary of the comparison of various configurations of the desired function. It can be seen that the most preferred option for the considered mixed profile was the use of a 4-layer perceptron with a layer width of 10. Single-layer models either take too long to optimize or are unable to achieve adequate accuracy. As the number of layers increases, the expressive power of the model increases, which contributes to faster and more accurate convergence to the exact solution. At the same time, excessive increase in the number of layers and network width leads only to the complication of the model and its convergence time but no longer leads to an increase in accuracy. Thus, there is some optimal network topology in terms of accuracy and convergence speed, but the question of how to quickly find it remains open.

It should be noted that previously proposed inversion methods [10, 27] relied heavily on the initial approximation. In fact, they work well only when the initial approximation is close to the true value of the refractive index. The proposed method, on the other hand, does not require any prior information about the refractive index distribution, which makes this method significantly more universal.

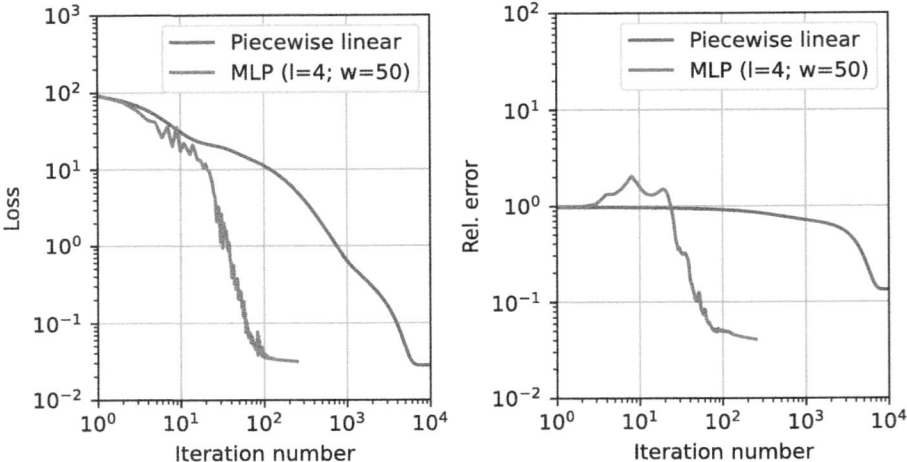

Fig. 9. Dependence of the loss function value (left) and relative error (right) on the iteration number for the multilayer perceptron and the piecewise linear function.

Table 1. Convergence parameters of various refractive index models.

Method	Number of iterations	Inversion time (s)	Error
MLP ($l=1$, $w=10$)	312	59	0.188
MLP ($l=1$, $w=50$)	71	23	0.207
MLP ($l=1$, $w=100$)	1477	400	0.068
MLP ($l=4$, $w=10$)	**548**	**168**	**0.022**
MLP ($l=4$, $w=50$)	304	376	0.038
MLP ($l=4$, $w=100$)	384	863	0.032
MLP ($l=8$, $w=10$)	385	165	0.0373
MLP ($l=8$, $w=50$)	168	519	0.044
MLP ($l=8$, $w=100$)	456	1885	0.0298
Piecewise linear (50 points)	10000	2169	0.13

4.2   Underwater Sound Speed Determination

A similar problem in essence and importance arises in underwater acoustics. It is required to determine the dependence of the sound speed in water on depth. The schematic description of the problem is shown in Fig. 10. An acoustic wave source is submerged underwater and emits an acoustic signal at a certain frequency. As the signal propagates from the source to the hydrophone array, it is distorted under the influence of sound speed inhomogeneities. The task is to determine the vertical sound speed profile based on the acoustic pressure measurements at the hydrophones. Although this problem has a completely different physical nature [18], it corresponds to the same mathematical model as the tropospheric inversion problem. In both problems, there is a waveguide formed by the refractive index. In both cases, wave propagation satisfies the Helmholtz equation (1).

Acoustic waveguides caused by inhomogeneous vertical stratification of sound speed can facilitate the propagation of acoustic signals over hundreds and thousands of kilometers or, conversely, contribute to their attenuation near the source. Direct measurements of sound speed are often difficult, so the problem of inversion based on indirect data is very relevant. Figure 11 shows the inversion results for four typical sound speed profiles in shallow water. An array of 15 hydrophones was located at a distance of 3 km from the source. The signal frequency was 200 Hz. It can be seen that the proposed method successfully inverted all four considered profiles. As before, the method does not require any prior information about the desired profiles.

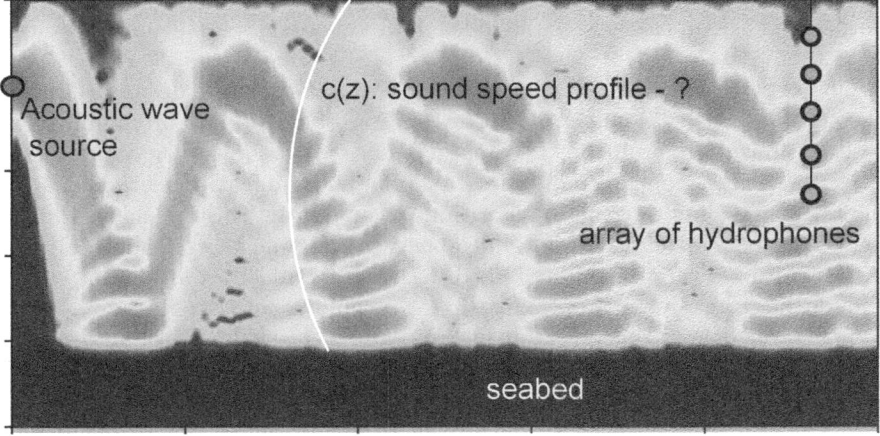

Fig. 10. Schematic description of underwater sound speed inversion.

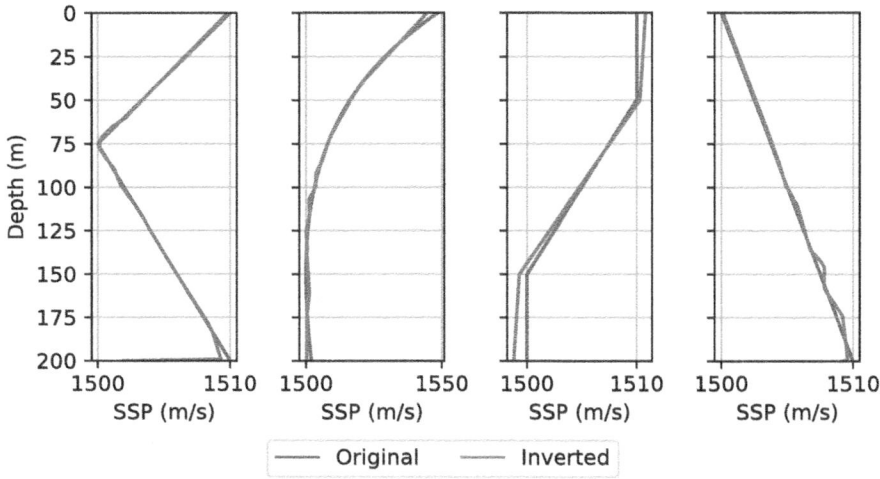

Fig. 11. Original sound speed profile and inversion result.

5 Conclusion

Unlike PINN, the proposed approach uses the existing well established numerical method for solving the direct problem, as it more reliable and efficient. That is, the method uses not only physical laws («physics informed») but also the specifics of their numerical implementation, making it «numerical method informed». This approach seems significantly more interpretable and numerically efficient.

The use of a deep neural network as representation of the desired refraction index profile allows significantly better finding of the global minimum of the posed optimization problem. The method does not require prior information and works orders of magnitude faster than global optimization methods such as simulated annealing or genetic algorithms. Apparently, this is achieved due to the greater expressive power of the deep networks. It may seem counterintuitive that a neural network with hundreds of parameters finds a solution better than a simple piecewise linear approximation. It would seem that it should simply overfit. Nevertheless, this does not happen, and a deeper network shows better results. The theoretical aspects of this curious and useful result remain to be clarified.

Meta-parameters sometimes have to be manually tuned, which complicates the use of this method in applied problems. It will be burdensome and time-consuming for a hydroacoustic engineer or radio physicist to select the artificial parameters. Therefore, an actual direction for further research is the application of autoML [11] and the selection of the optimal neural network automatically.

Acknowledgments. This study was supported by the State research (Topic No. FFZF-2025-0006).

References

1. Ataei, M., Salehipour, H.: XLB: a differentiable massively parallel lattice Boltz-mann library in python. Comput. Phys. Commun. **300**, 109187 (2024)
2. Baydin, A.G., Pearlmutter, B.A., Radul, A.A., Siskind, J.M.: Automatic differen-tiation in machine learning: a survey. J. Mach. Learn. Res. **18**(153), 1–43 (2018)
3. Bezgin, D.A., Buhendwa, A.B., Adams, N.A.: Jax-fluids: a fully-differentiable high-order computational fluid dynamics solver for compressible two-phase flows. Comput. Phys. Commun. **282**, 108527 (2023)
4. Chen, R.T., Rubanova, Y., Bettencourt, J., Duvenaud, D.K.: Neural ordinary dif-ferential equations. In: Advances in Neural Information Processing Systems, vol. 31 (2018)
5. Collins, M.D., Siegmann, W.L.: Parabolic Wave Equations with Applications. Springer, Cham (2019)
6. Colton, D., Kress, R.: Inverse Acoustic and Electromagnetic Scattering Theory. Springer, Cham (2012)
7. Fallat, M.R., Dosso, S.E.: Geoacoustic inversion via local, global, and hybrid algo-rithms. J. Acoust. Soc. Am. **105**(6), 3219–3230 (1999)
8. Fishman, L., McCoy, J.J.: Derivation and application of extended parabolic wave theories. I. The factorized Helmholtz equation. J. Math. Phys. **25**(2), 285–296 (1984)
9. Jensen, F.B., Kuperman, W.A., Porter, M.B., Schmidt, H.: Computational Ocean Acoustics. Springer, Cham (2014)
10. Karabaş, U., Diouane, Y., Douvenot, R.: A variational adjoint approach on wide-angle parabolic equation for refractivity inversion. IEEE Trans. Antennas Propag. **69**(8), 4861–4870 (2021)
11. Karmaker, S.K., Hassan, M.M., Smith, M.J., Xu, L., Zhai, C., Veeramachaneni, K.: Automl to date and beyond: challenges and opportunities. ACM Comput. Surv. (CSUR) **54**(8), 1–36 (2021)
12. Kidger, P., Foster, J., Li, X.C., Lyons, T.: Efficient and accurate gradients for neural SDEs. Adv. Neural. Inf. Process. Syst. **34**, 18747–18761 (2021)
13. Kingma, D.P.: Adam: a method for stochastic optimization. arXiv preprint arXiv:1412.6980 (2014)
14. Levy, M.F.: Parabolic Equation Methods for Electromagnetic Wave Propagation. The Institution of Electrical Engineers, UK (2000)
15. Lu, L., Meng, X., Mao, Z., Karniadakis, G.E.: DeepXDE: a deep learning library for solving differential equations. SIAM Rev. **63**(1), 208–228 (2021)
16. Lytaev, M.S.: PyWaveProp (2024). https://github.com/mikelytaev/wave-propagation
17. Lytaev, M.: Automatically differentiable higher-order parabolic equation for real-time underwater sound speed profile sensing. J. Mar. Sci. Eng. **12**(11), 1925 (2024)
18. Lytaev, M.S.: Rational interpolation of the one-way Helmholtz propagator. J. Comput. Sci. 101536 (2022)
19. Papadakis, J.S., Karasmani, E.: Gradient of the cost function via the adjoint method for underwater acoustic inversion. J. Theor. Comput. Acoust. **28**(01), 1950010 (2020)
20. Pastore, D.M., et al.: Comparison of atmospheric refractivity estimation methods and their influence on radar propagation predictions. Radio Sci. **56**(9), 1–17 (2021)
21. Pastore, D.M., et al.: Refractivity inversions from point-to-point x-band radar propagation measurements. Radio Sci. **57**(2), 1–16 (2022)

22. Sapunov, G.: Deep Learning with JAX. Manning (2024)
23. Sen, M.K., Stoffa, P.L.: Global Optimization Methods in Geophysical Inversion. Cambridge University Press, Cambridge (2013)
24. Thiyagalingam, J., Shankar, M., Fox, G., Hey, T.: Scientific machine learning benchmarks. Nat. Rev. Phys. **4**(6), 413–420 (2022)
25. Tichonov, A.N., Leonov, A.S., Jagola, A.G.: Nonlinear Ill-Posed Problems, vol. 1. Chapman & Hall London (1998)
26. Xue, T., et al.: Jax-fem: a differentiable GPU-accelerated 3D finite element solver for automatic inverse design and mechanistic data science. Comput. Phys. Commun. **291**, 108802 (2023)
27. Zhao, X., Wang, D.: Ocean acoustic tomography from different receiver geometries using the adjoint method. J. Acoust. Soc. Am. **138**(6), 3733–3741 (2015)

Discover the Tractable Latent Space of Floating Offshore Wind Turbine Based on a Novel GNN-Encoder-Decoder-LSTM Deep Learning Architecture

Kobe Hoi-Yin Yung[1,2](✉) ⃝, Qing Xiao[1] ⃝, Xiuqing Xing[2] ⃝, Atilla Incecik[1] ⃝, and Chang Wei Kang[2] ⃝

[1] University of Strathclyde Glasgow, Glasgow, UK
kobehy.yung@strath.ac.uk
[2] Institute of High Performance Computing (IHPC) Agency for Science, Technology and Research (A*STAR), Singapore, Singapore

Abstract. Floating Offshore Wind Turbines (FOWT) provided new potential in harvesting wind energy in far offshore deep-sea regions and contributed to the world decarbonization Net-Zero target. Providing structural health monitoring (SHM) is crucial for ensuring the structural integrity of FOWT in lifecycle. However, the SHM is technically challenging with high Operational and Maintenance Expenditure (OPEX). Recently, Digital Twin (DT) and advanced sensor technologies offer alternative solutions to provide effective strategy in SHM remotely. Data-driven DT with deep learning models can formulate highly nonlinear dynamics systems. Yet, these existing models only perform the "black box" prediction without explicitly modeling the spatial-temporal relationship and consider only homogenous loading exerted in contrast to the complicated loading combination of FOWT with wind, wave and sea current.

To address the existing modelling limitations, a new Graph Neural Network (GNN)-Encoder-Decoder-Long Short-Term Memory (LSTM) surrogate model of FOWT is presented in this work, which can perform 50 times faster than the real-time of simulation data set with accurate prediction of wind turbine tower bottom forces in the dominant dynamic modes force-aft and side-side directions. The training data is based on the software QBlade simulation and focuses on the OC4 5MW DeepCwind FOWT structure. A holistic quantitative analysis is carried out to validate the tractable latent space vectors for this complex FOWT system.

Keywords: Floating Offshore Wind Turbine · Artificial Intelligence · Deep Learning · Digital Twin · Tractable Latent Space · Surrogate model

1 Introduction

Offshore wind energy has been demonstrated promising renewable energy over the past decade, which was implemented all over the globe, including China, UK and Germany [1]. Especially, FOWT are considered to have potential for harvesting wind energy from

© The Author(s), under exclusive license to Springer Nature Switzerland AG 2025
M. H. Lees et al. (Eds.): ICCS 2025, LNCS 15903, pp. 91–107, 2025.
https://doi.org/10.1007/978-3-031-97626-1_7

far offshore deep-sea regions compared to fix-bottom offshore wind. When implementing construction projects, there are huge technical challenges in providing maintenance of large-scale wind farm in which pose huge OPEX. With the advancement of state-of-the-art sensor technology and cloud technology, it allows the development of DT technologies [2], which is a virtual representation of the physical asset. Finite Element Modeling (FEM) is widely used to model multibody dynamics, but it is computationally expensive to run in real-time for DT. Surrogate models allow DT running instantaneously but keeping the accuracy in highly nonlinear dynamics.

In this work, we presented a novel GNN-Encoder-Decoder-LSTM for solving mentioned above. In deep learning architecture, there are autoencoder and encoder-decoder used for dimension compression, which are the nonlinear reduced order model approaches. However, in literature, there is a lack of organized analysis explaining the behavior of latent space characteristics. Therefore, we provide a systematic analysis for the relationship between the higher dimensions graph embedding and the latent space vector and provide the first application of Modal Assurance Criterion (MAC) [3] to demonstrate the tractability of the latent representation of performing this temporal dynamic prediction task of FOWT.

2   Related Work

Nowadays mid-fidelity engineering tools e.g. OpenFAST [4] and QBlade [5] can simulate complex coupling in between the aerodynamic, hydrodynamic, structural dynamic, servo-dynamic, and provide efficient computation time comparing to high fidelity tool e.g. Computation Fluid Dynamics. However, mid-fidelity engineering tools still have limitations in running real-time in a desktop computer and typically for time series dynamic simulation that require the numerical transient period for the iterating the convergence in between different dynamic modules. For example, QBlade employed Hilber-Hughes-Taylor formulation [6] integrator for solving the Differential-Algebraic Equations. To address the inherent computational time problem, Reduced Order Model (ROM) is commonly adopted to speed up the simulation. In this section, we review recent developments in ROM for structural dynamics.

2.1   Physics-Based and Conventional ROM

A reduced ordered state space model with linearized approach can be used to run the interested parameters simulation of Ordinary Differential Equation. Which requires to linearize the nonlinear function at certain operating points e.g. sea state for FOWT. For instance, [7] formulated the time series ROM with N4SID system identification technique and created a bank of state space models at several sea states for Kalman Filtering and then merged the estimation result with Bayesian Multiple-Model Adaptive Estimation algorithm. This provided the mooring forces estimation for FOWT for unseen sea states with 20 times faster than real-time. There are also effective compression approaches for high dimensional data, such as the method of Proportional Orthogonal Decomposition (POD) [8] and Dynamic Mode Decomposition (DMD) [9] are used to truncate the high order component to provide the low rank matrix for faster computation.

2.2 Deep Learning ROM

Recently, machine learning and neural network provide new alternatives with the concept of "Universal approximation theorem", and deeper layer network "Deep Learning" can provide better prediction. Especially Autoencoder [10] forms the architecture that learns the lower dimension representation by training the neural network with means square error loss. For instance, [11] used AE-LSTM to extract the low-dimension and pass to LSTM. However, this does not explicitly incorporate the physical properties of the structure e.g. mass, stiffness and geometrical space. Furthermore, the existing deep learning applications in FEM only considered simple structures e.g. a single beam element or a rectangular multistorey frame with homogeneous loading input e.g. one horizontal force or one acceleration. [12] used the Variational Autoencoder (VAE) and POD for time series force prediction and homogeneous loading or ground motion acceleration, and the real-time factor was not reported. On the contrary, FOWT experiences a complicated combination of wind, wave and current with different amplitudes and turbulence, which are heterogenous loading properties problem to the structural system. The above-mentioned deep learning methods have not addressed the highly nonlinear problem for FOWT.

Recently, a Physics-Guided Spatial Temporal Graph Neural Network (GNN) [13] was presented that overcome the limitations mentioned above and able to handle heterogeneous loadings input. The GNN encoded the aero-hydrodynamic loadings with geometry and material properties, and combined with LSTM for predicting the tower forces 14 times faster than real-time. With the "baseline" model GATv2LSTM in [13], we presented a new architecture that compose of GNN-Encoder-Decoder-LSTM which can further accelerate the prediction by reducing the LSTM hidden layer parameter size substantially with the additional Encoder-Decoder and present the latent representation.

3 Simulation in QBlade

This study focuses on the FOWT OC4 5MW DeepCwind semisubmersible [14] as shown in Fig. 1. The model file is openly available on QBlade website [5]. The metocean data is based on the West of Barra Scotland from LIFES50 + project [15].

This study refers to the Design Load Case DLC 1.2 in the technical specification IEC 61400-3-2 [16]. The sea state follows Pierson-Moskowitz spectrum with significant wave height 3.5 m and period 10.68 s. The hydrodynamic modeling is Potential Flow with Morison Drag. To capture the higher order hydrodynamics effect, vertical stretching for wave and the second-order full Quadratic Transfer Functions (QTF) are used. For sea current, the near surface current 0.88m/s with reference depth of 30 m and subsurface 0.84 m/s with power 1/7 are adopted.

TurbSim [17] is used for modeling the turbulence wind with a mean wind speed of 13.63 m/s at hub height with turbulence Class IC and Kaimal model. The aerodynamic loading from the rotor is solved by the Unsteady Blade Element Momentum (UBEM) [18]. Øye dynamic stall and the tower shadow models are activated. A tower drag coefficient of 0.5 is used.

Figure 2(a) shows the simulation setting in QBlade and Fig. 2(b) shows the discretized nodes and the critical tower bottom for the focus of this study. Detailed GNN is described in Sect. 4 with Table 1 and 2. Figure 3 refers to the wind velocity at hub height and Fig. 4 refers to the wave elevation time series that the FOWT system experiences.

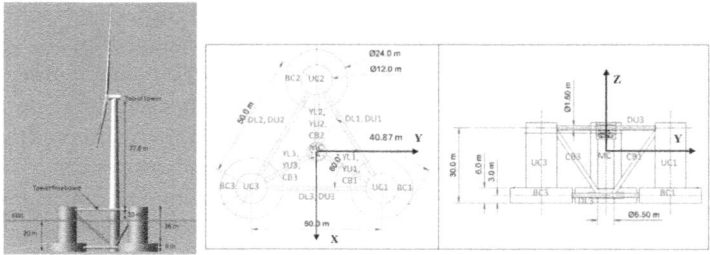

Fig. 1. Geometric specification of OC4 5MW DeepCwind FOWT [14]

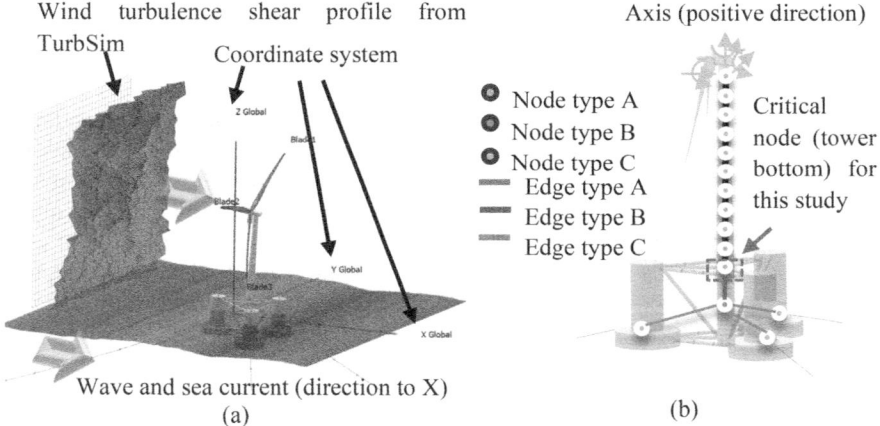

Wind turbulence shear profile from TurbSim

Coordinate system

Axis (positive direction)

● Node type A
● Node type B
● Node type C
━ Edge type A
━ Edge type B
━ Edge type C

Critical node (tower bottom) for this study

Wave and sea current (direction to X)

(a)   (b)

Fig. 2. (a) QBlade simulation and (b) GNN discretized of OC4 5MW DeepCwind FOWT

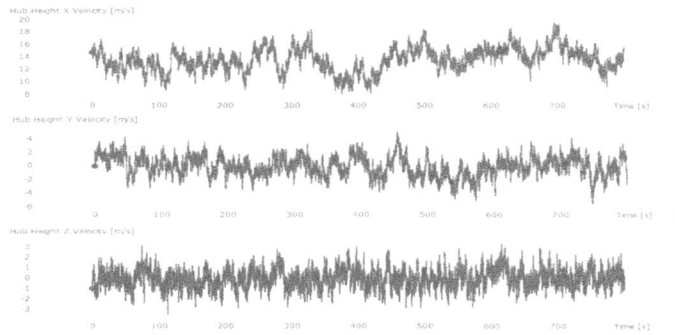

Fig. 3. Wind velocity generated from TurbSim

Fig. 4. Wave elevation profile time series data in QBlade

4   Novel GNN-Encoder-Decoder-LSTM

Table 1 and 2 summarize the node features and edge attribute of the GNN respectively as illustrated in Fig. 2(b).

Table 1. Node features inputs and node target outputs of the GNN

	Node features and inputs	Node target outputs
Node type A (11 nodes)	• Nodal mass • Node coordinates • Aerodynamic load	• Internal forces and moment in X, Y and Z directions
Node type B (1 node)	• Lumped mass of hull structure • 6DOF displacement, velocity and acceleration • Wave elevation • Water kinematic velocity	• 6DOF displacement, velocity and acceleration • Wave elevation • Water kinematic velocity
Node type C (3 nodes)	• Node coordinates • Wave elevation • Water kinematic velocity	• Support reaction forces from mooring fairlead

Figure 5 illustrates the whole proposed architecture: firstly the node embedding from GNN is concatenated to the graph embedding of size 240 and compressed to the latent space vector with the encoder; then the latent vector is passed to LSTM and the predicted latent space vector is mapped to the node feature prediction with the decoder and Fully

Table 2. Edge attributes of the GNN

	Edge attributes
Edge type A	• Length • Stiffness (bending, axial and torsion)
Edge type B and C	• Length

Connected (FC) layer. The activation functions of Rectified Linear Unit (ReLU) and Exponential Linear Unit (ELU) are tested with different combinations. Figure 5 refers to the optimum architecture of LSTM192ED128-128relu-elu32res. The procedure of Hyperparameter identification will be elaborated in Sect. 5.

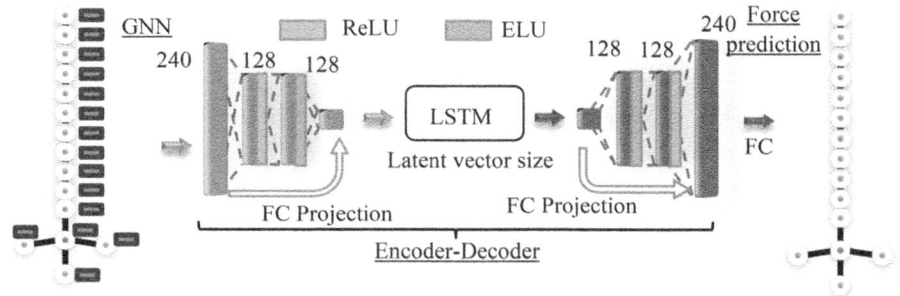

Fig. 5. GNN-Encoder-Decoder-LSTM architecture

The Graph Attention Network GATv2Conv [19, 20] is adopted and the node embeddings are calculated in the following:

$$x_i' = \alpha_{i,i}\Theta_s x_i + \sum_{j\in\mathcal{N}(i)} \alpha_{i,j}\Theta_t x_j \tag{1}$$

The attention weight $\alpha_{i,j}$ is calculated as

$$\alpha_{i,j} = \frac{\exp(a^T LeakyReLu(\Theta_s x_i + \Theta_t x_j + \Theta_e e_{i,j})}{\sum_{k\in\mathcal{N}(i)\cup\{i\}}\exp(a^T LeakyReLu(\Theta_s x_i + \Theta_t x_k + \Theta_e e_{i,k})} \tag{2}$$

The node features inputs x_i are defined in matrix [number of nodes, number of features] and edge attribute $e_{i,j}$ in matrix [number of edges, number of features]. The inputs to GAT include node type A feature $x_A \in \mathbb{R}^{11\times11}$, node type B feature $x_A \in \mathbb{R}^{1\times20}$, node type C feature $x_C \in \mathbb{R}^{3\times2}$, edge type A attributes $e_B \in \mathbb{R}^{10\times4}$, edge type B attributes $e_B \in \mathbb{R}^{1\times1}$ and edge attributes type C $e_C \in \mathbb{R}^{3\times1}$ as listed in Table 1 and 2. After the node embeddings are calculated with 1-hop neighbors for each edge type messaging passing, they are concatenated to form the whole graph embedding vector $z \in \mathbb{R}^{240}$.

Long Short-Term Memory (LSTM) model is used for handling temporal dynamics:

$$i_t = \sigma(W_{ii}\tilde{z}_t + b_{ii} + W_{hi}h_{t-1} + b_{hi}) \tag{3a}$$

$$f_t = \sigma(W_{if}\tilde{z}_t + b_{if} + W_{hf}h_{t-1} + b_{hf}) \tag{3b}$$

$$g_t = \tanh(W_{ig}\tilde{z}_t + b_{ig} + W_{hg}h_{t-1} + b_{hg}) \tag{3c}$$

$$o_t = \tanh(W_{io}\tilde{z}_t + b_{io} + W_{ho}h_{t-1} + b_{ho}) \tag{3d}$$

$$c_t = f_t \odot c_{t-1} + i_t \odot g_t \tag{3e}$$

$$h_t = o_t \odot \tanh(c_t) \tag{3f}$$

where $\tilde{z}_t$ is the input latent vector at time t, h is the hidden state, W are the weight matrix, b are the bias, i_t, f_t, g_t, o_t are the input, forget, cell, and output gates, respectively. σ is the sigmoid function, $\odot$ is the Hadamard product. A time window size 100s is used for the sliding window.

The FC layer is used to map the output of LSTM to the node target prediction matrix y in the format of [number of nodes, number of target features] for $y_A \in \mathbb{R}^{11 \times 6}$, $y_B \in \mathbb{R}^{1 \times 20}$ and $y_C \in \mathbb{R}^{3 \times 3}$. Additionally, the residual skip connection, FC projection, is used to improve gradient flow for the deeper layers. The deep learning model is implemented in PyTorch [21] and PyTorch Geometric [22] and trained with Backpropagation ADAM optimizer [23], weight decay 0.0001, learning rate 0.001 and epoch 100 and mean squared error (squared L2 norm). 731.5 s of time series data is used for training and another unseen 731.5 s of time series data for prediction and testing. The training data is preprocessed with z-score normalization.

5   Results

5.1   Validation of Simulation Training Data

A pitching decay test is used to benchmark the present QBlade simulation model for preparing Deep Learning model training data is aligned with the reported [24] QBlade model (QB) and the OpenFAST (OF) result as shown in Fig. 4 (Fig. 6).

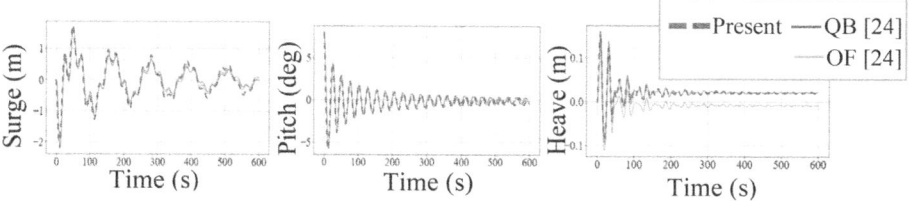

Fig. 6. Pitch decay test for benchmarking

5.2 Deep Learning Model Prediction Results

This section shows the prediction experiments on the neural network Hyperparameters identification procedure e.g. hidden layer sizes and activation function of the proposed GNN-Encoder-Decoder-LSTM and compares with the baseline model GATv2LSTM (without the Encoder-Decoder), against the "truth value" QBlade simulation. This study will only reveal the critical tower bottom forces and moments as stated in Fig. 7, 8, 9, 10, 11, 12, 13, 14 and 15. The time series prediction of forces are shown in detail to illustrate the trend difference. The forces and moments for each architecture results are summarized in Table 3. An example of legend name LSTM192ED128-128relu-elu32res means LSTM192 [hidden size] ED128-128 [Encoder-Decoder structure] relu-elu [activation function in the Encoder-Decoder] 32 [latent vector size] res [residual skip connection].

Study on Latent Space Vector Size. Based on the LSTM hidden size 100, the latent space vector size is varied from 10 to 32 with all fully connected layers attached with ReLU activation function.

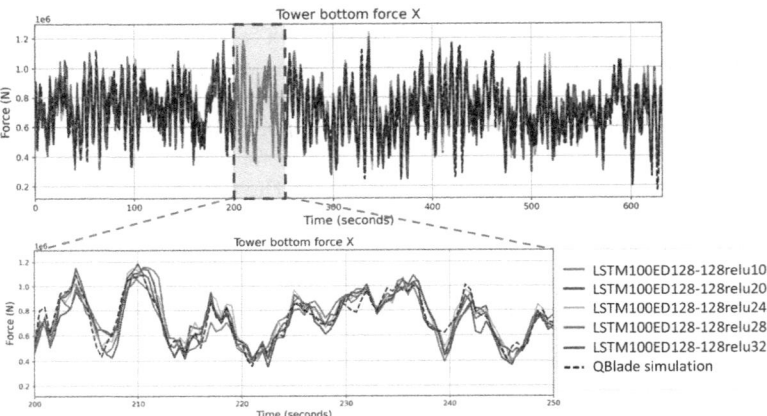

Fig. 7. Tower bottom force X prediction with varying latent vector size.

Higher latent space size can preserve more information from the high dimensional vector. The latent size 32 balances the capacity for model complexity and prevents the noisy details of larger size so it provides better prediction to the QBlade simulation.

Study on LSTM Hidden Size. Based on the optimal value latent space vector 32, the LSTM hidden size is varied from 100 to 192 with all fully connected layers attached with ReLU activation function.

Increasing the LSTM hidden size can capture more nonlinearity and intricate temporal pattern, especially the mixture of low-frequency and high-frequency dynamics. The optimum number of 192, multiple of 8, also fits with the tensor core operation. Hence, it matches better with the QBlade simulation.

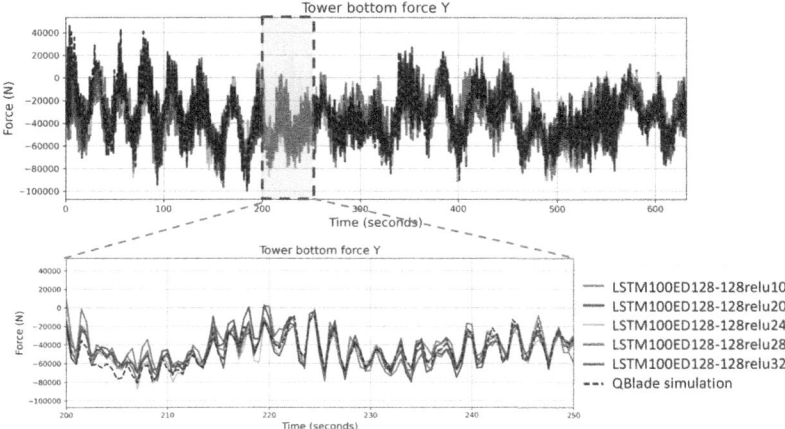

Fig. 8. Tower bottom force Y prediction with varying latent vector size.

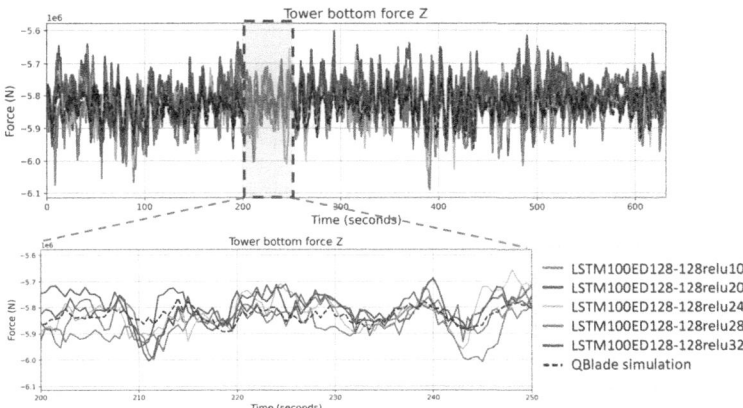

Fig. 9. Tower bottom force Z prediction with varying latent vector size.

Study on Residual Skip Connection. Based on the optimal value latent space vector 32 and LSTM hidden size 192, residual skip connection is applied for testing the effect of gradient flow. ELU consists of a small negative slope which can perform a better gradient flow than ReLU in the inner layer.

Figure 13, 14 and 15 reveal the residual skip connection can improve the prediction in finer detailed variation with better flow of information and gradient, and LSTM192ED128-128relu-elu32 can produce precise result as the baseline GATv2LSTM against the "truth value" QBlade simulation.

The prediction stage runs on the CPU 11th Gen Intel(R) Core(TM) i5-1145G7 @ 2.60 GHz 2.61 GHz. The first 100 s data (time window) is excluded from the calculation of Real-Time factor so 631.5 s is used for the reference value. A common performance

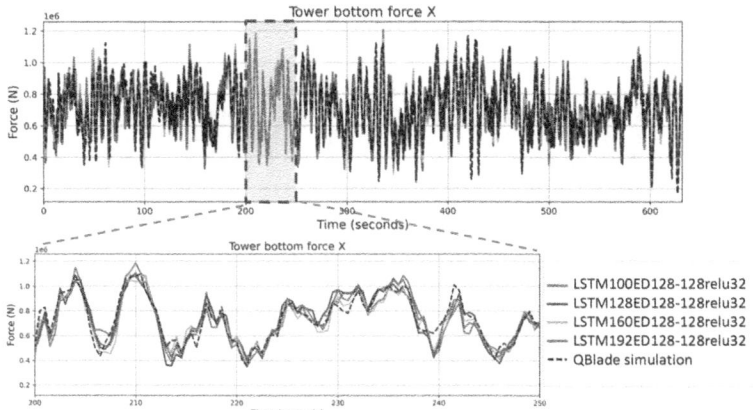

Fig. 10. Tower bottom force X prediction with varying LSTM hidden size.

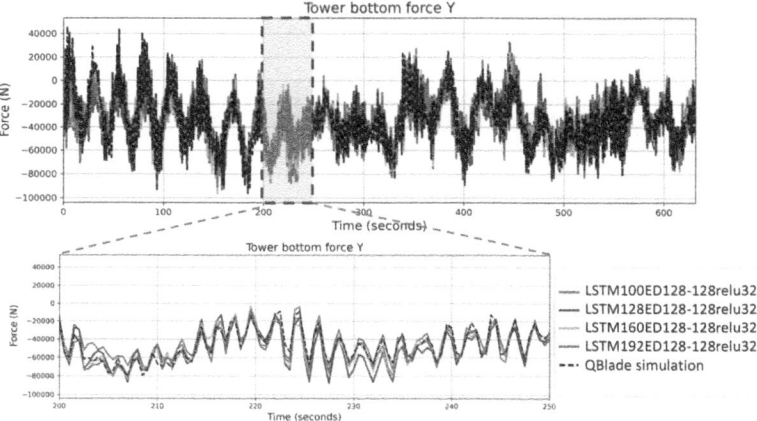

Fig. 11. Tower bottom force Y prediction with varying LSTM hidden size.

metric for machine learning prediction is Percent Bias (PBIAS) defined as

$$PBIAS = 100\% \times \frac{\sum(Q_{pred} - Q_{sim})}{\sum Q_{sim}} \qquad (4)$$

where Q_{pred} is the prediction and Q_{sim} is the reference from QBlade simulation.

To conclude, most models shown in Table 3 can provide accurate prediction for the dominant modes force-aft (force X and moment Y) and side-side (force Y and moment X) forces and moments with ±5%. With about 50 times faster than the real-time data. The large differences are evaluated for moment Z (torsion mode) for sorting the suitable candidates for further study, and they are LSTM100ED128-128relu24, LSTM100ED128-128relu32, LSTM192ED128-128relu32, LSTM192ED128-128relu32res and LSTM192ED128-128relu-elu32res.

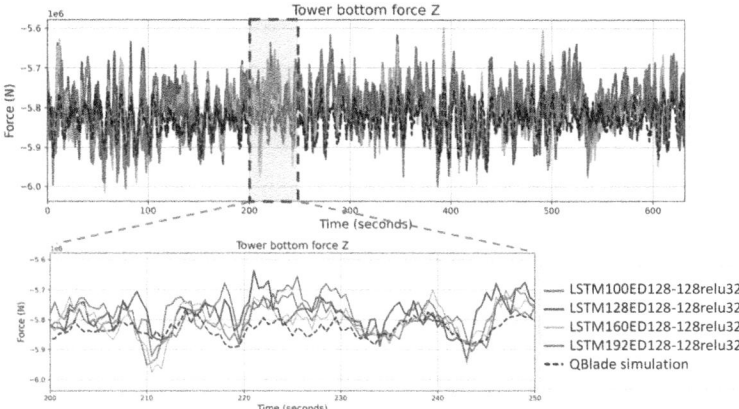

Fig. 12. Tower bottom force Z prediction with varying LSTM hidden size.

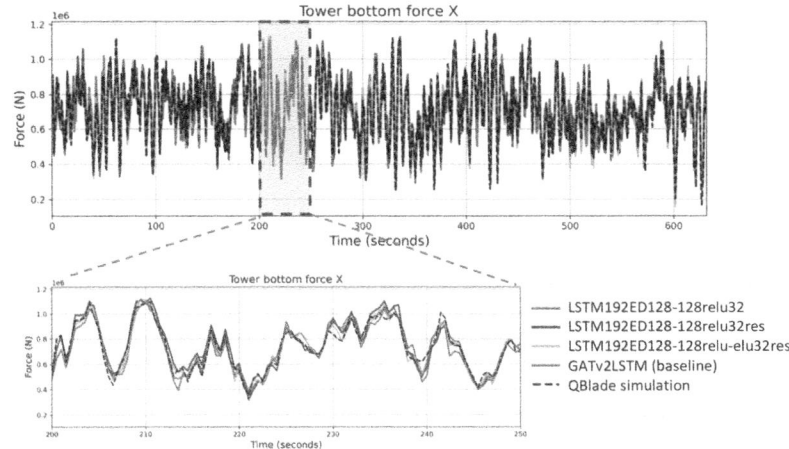

Fig. 13. Tower bottom force X prediction with the effect of residual skip connection.

The following section reveals the tractability of latent space vector comparing to the baseline model in high dimension, the detailed Singular Value Decomposition (SVD) analysis is used to demonstrate how the variance (structural energy) is preserved, and the MAC analysis is used to demonstrate how the mode shape is preserved.

Figure 16 shows the LSTM192ED128-128relu-elu32res can preserve the most singular values up to mode 18 as the baseline model. MAC require two set of data in the same dimension. Therefore, Principal Component Analysis is applied to the baseline model vector to reduce the dimension to the same as the latent vector. As the basis of reduced dimension vector of the baseline model can be different from the latent vector produced from the encoder, Procrustes analysis (function scipy.spatial.procrustes [25]) is used to align both set of vectors without changing the mode shape, and then followed

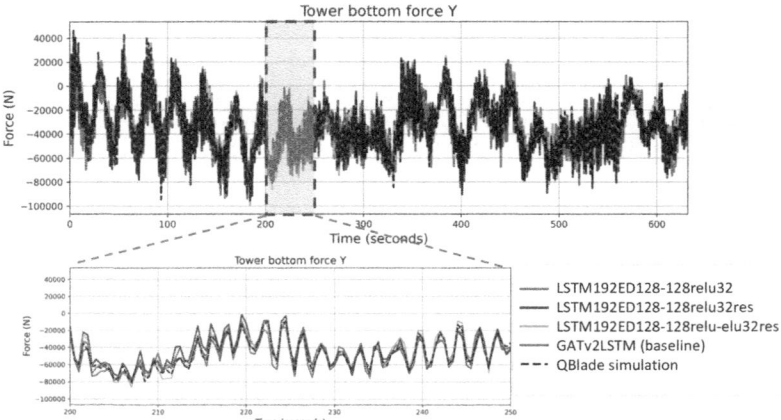

Fig. 14. Tower bottom force Y prediction with the effect of residual skip connection.

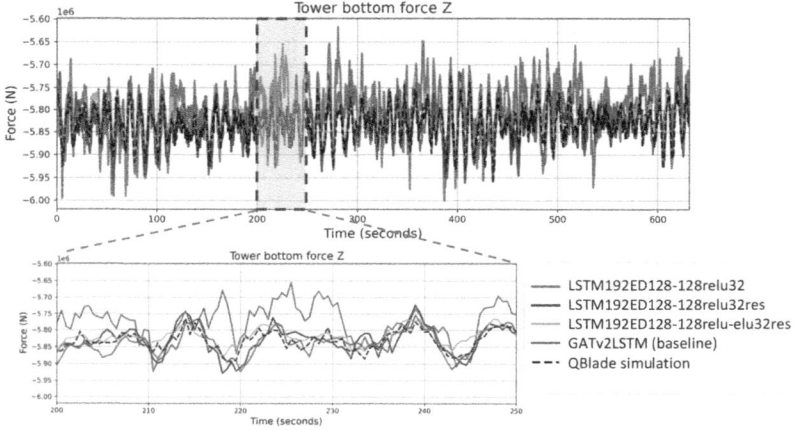

Fig. 15. Tower bottom force Z prediction with the effect of residual skip connection.

by MAC calculated as

$$MAC(\phi, \psi) = \frac{\left(\phi^T \psi\right)^2}{\left(\phi^T \phi\right)\left(\psi^T \psi\right)} \tag{5}$$

where ϕ is the eigenvector of latent space vector and ψ is the eigenvector of baseline model vector after Procrustes analysis.

Figure 17, 18, 19, 20 and 21 show the improvement of capturing mode shapes from model LSTM100ED128-128relu24 to LSTM192ED128-128relu-elu32res with MAC = 0.9 indicate highly correlated results. Particularly Fig. 20 and 21 show the significant improvement of skip residual connection from capturing 2 modes to at least 9 modes.

Table 3. Percent Bias and CPU execution wall time for the testing data set (631.5s)

| Prediction model | Percent Bias (%) | | | | | | | Real-Time Factor |
	force X (N/N)	force Y (N/N)	force Z (N/N)	moment X (Nm/Nm)	moment Y(Nm/Nm)	moment Z(Nm/Nm)	Wall time (s)	631.5/Wall time
GATv2LSTM (baseline)	0.01	1.03	−0.11	0.27	−0.09	−9.79	51.5	12.3
LSTM100ED128-128relu10	−0.08	2.01	0.15	0.60	−0.62	−53.13	11.6	54.4
LSTM100 ED128-128relu20	−0.20	1.77	−0.57	0.48	−1.11	−66.08	12.3	51.3
LSTM100 ED128-128relu24	1.31	2.30	−0.12	0.64	1.07	6.14	11	57.4
LSTM100 ED 128-128relu32	−1.23	−3.31	−0.42	−1.25	−2.25	−10.86	10.4	60.7
LSTM128ED128-128rel32	−0.33	2.76	−0.55	0.51	−1.20	32.67	14.5	43.6
LSTM160ED128-128relu32	−0.85	4.80	−0.42	1.10	−1.58	−28.33	12.2	51.8
LSTM192ED128-128relu32	−0.22	1.61	−0.46	0.50	−0.85	−26.46	13.3	47.5
LSTM192ED128-128relu32res	0.62	−0.88	−0.03	−0.42	0.54	−26.99	13.2	47.8
LSTM192ED128-128relu−elu32res	−0.49	−3.12	−0.26	−1.15	−1.14	−11.60	12.8	49.3

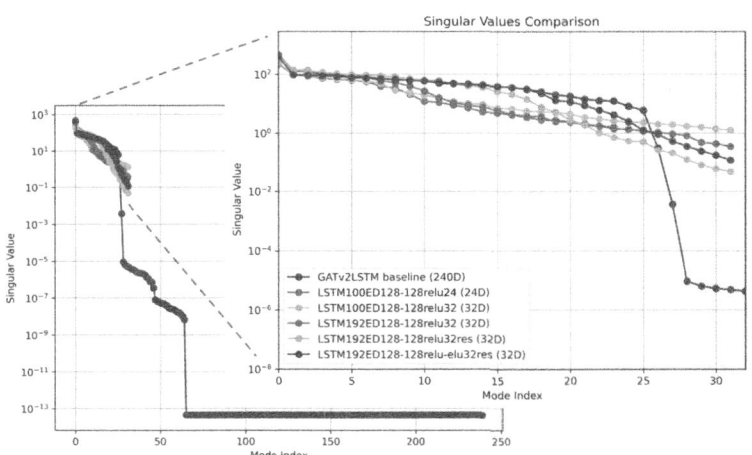

Fig. 16. Singular value comparison

This is because the high dimensional vector supplementing the information to the latent vector.

The disparity value of Procrustes analysis indicates similarity of two sets of mode shape data and smaller disparity means better fit. Figure 21 shows that LSTM192ED128-128relu-elu32res preserves the most system dynamics up to mode 13 with the smallest disparity of 0.112. In summary, LSTM192ED128-128relu-elu32res preserves the most dynamic properties of the high dimension vector of baseline model.

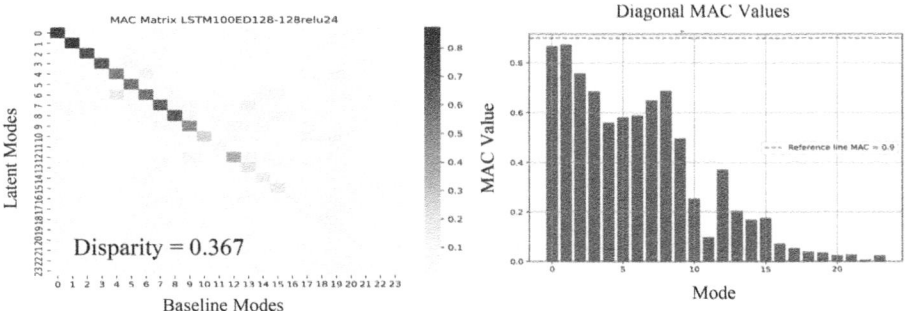

Fig. 17. MAC of LSTM100ED128-128relu24 model

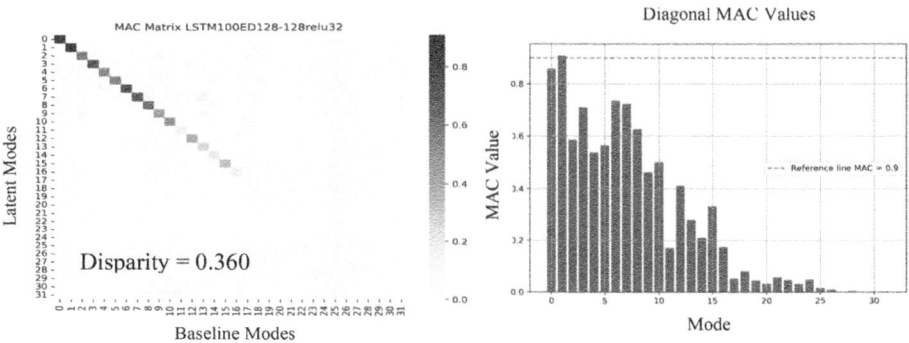

Fig. 18. MAC of LSTM100ED128-128relu32 model

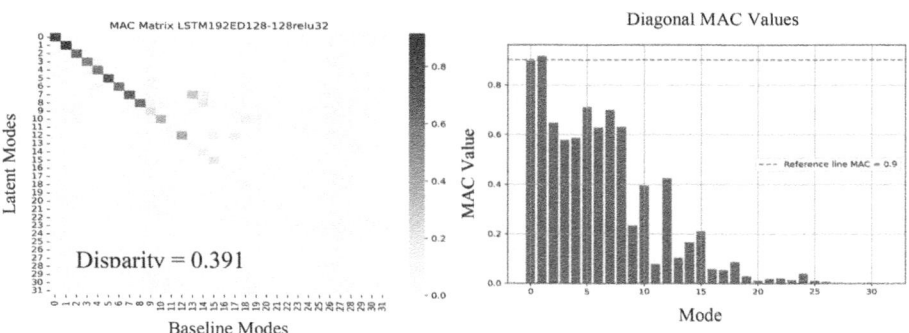

Fig. 19. MAC of LSTM192ED128-128relu32 model

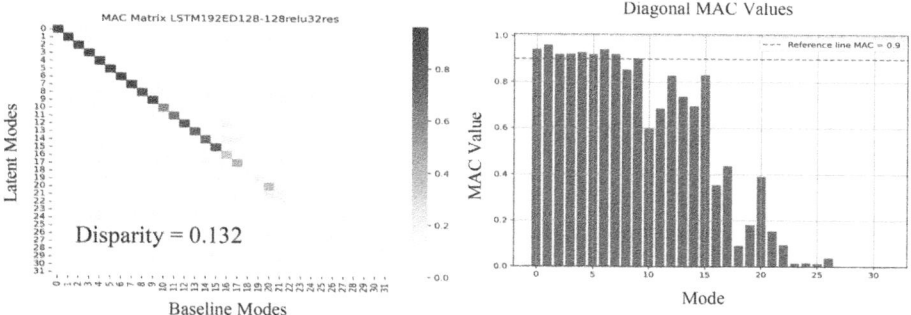

Fig. 20. MAC of LSTM192ED128-128relu32res model

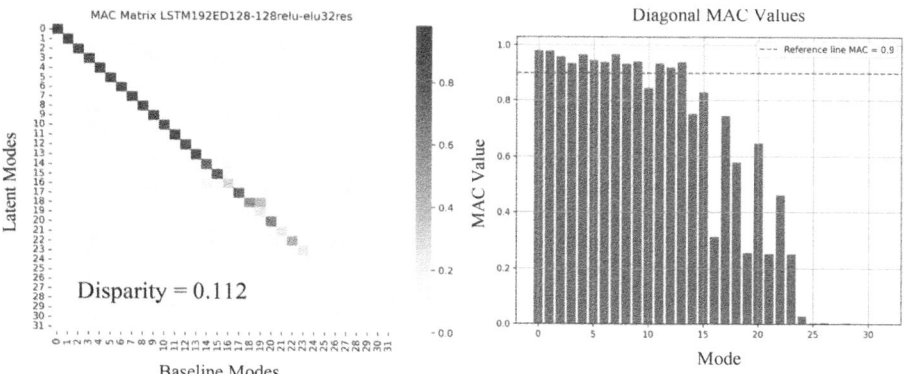

Fig. 21. MAC of LSTM192ED128-128relu-elu32res model

6 Conclusion

In this work, we present a novel GNN-Encoder-Decoder-LSTM for FOWT time series internal force prediction and provide a holistic analysis of the latent space vector characteristic comparing to the high dimensional vector. The addition of Encoder-Decoder can accelerate the prediction time by 4 times compared with the baseline model with only minimal difference in accuracy, and it is also about 50 times faster than real-time data set. Which is excellent for Digital Twin implementation. In terms of the quality of the latent vector from the Encoder-Decoder, the SVD and MAC analysis demonstrate the preservation of variance energy and the mode shapes up to mode 13 with the optimum model LSTM192ED128-128relu-elu32res. This result aligns with the neural network hyperparameter experiment prediction results of time series data. Which means the latent vector is tractable as the high dimensional graph embedding vector.

In future work, more loading combinations and scenarios will be included to extend the generalizability of the prediction model.

Acknowledgments. The first author would like to acknowledge the research grant of Research Attachment Programme from the Agency for Science, Technology and Research (A*STAR) Singapore.

References

1. Global Offshore Wind Report: Global Offshore Wind Report (2024)
2. Glaessgen, E., Stargel, D.: The digital twin paradigm for future NASA and US air force vehicles. In: Proceedings of the 53rd AIAA/ASME/ASCE/AHS/ASC Structures, Structural Dynamics and Materials Conference 20th AIAA/ASME/AHS Adaptive Structures Conference 14th AIAA (2012)
3. Lein, C., Beitelschmidt, M.: Comparative study of model correlation methods with application to model order reduction. In: Proceedings 26th ISMA (International Conference on Noise and Vibration Engineering) (2014)
4. Jonkman, Jason, et al.: OpenFAST v3.0.0. https://github.com/OpenFAST/openfast/releases/tag/v3.0.0
5. Marten, D.: Qblade. https://qblade.org/
6. Tasora, A.: Time integration in CHRONO::engine, project::chrono technical documentation (2018). http://www.projectchrono.org/assets/white_papers/ChronoCore/integrator.pdf
7. Yung, K. H.-Y., Xiao, Q., Incecik, A., Thompson, P.: Mooring force estimation for floating offshore wind turbines with augmented Kalman filter: a step towards digital twin, in ASME 2023 5th International Offshore Wind Technical Conference (2023)
8. Berkooz, G., Holmes, P., Lumley, J.L.: The proper orthogonal decomposition in the analysis of turbulent flows. Annu. Rev. Fluid Mech. **25**(1), 539–575 (1993)
9. Schmid, P.J.: Dynamic mode decomposition of numerical and experimental data. J. Fluid Mech. **656**, 5–28 (2010)
10. Hinton, G.E., Salakhutdinov, R.R.: Reducing the dimensionality of data with neural networks. Science **313**(5786), 504–507 (2006)
11. Simpson, T., Dervilis, N., Couturier, P., Maljaars, N., Chatzi, E.: Reduced order modeling of non-linear monopile dynamics via an AE-LSTM scheme. Front. Energy Res. **11** (2023)
12. Simpson, T., Vlachas, K., Garland A., Dervilis N., Chatzi, E.: VpROM: a novel variational autoencoder-boosted reduced order model for the treatment of parametric dependencies in nonlinear systems. Sci. Rep. **14**(6091) (2024)
13. Yung, K. H.-Y., Xiao Q., Incecik A., Xing, X., Kang, C. W.: Structural surrogate modelling of a floating offshore wind turbine with physics-guided spatial-temporal graph neural network. In: EERA DeepWind Conference 2025, Trondheim, Norway (2025)
14. Robertson, A., Jonkman, J., Masciola, M., Song, H., Goupee, A., Coulling, A., Luan, C.: Definition of the Semisubmersible Floating System for Phase II of OC4.. National Renewable Energy Laboratory (2014)
15. European Union: Deliverable 1.1 Oceanographic and meteorological conditions for the design. LIFES50+ (2015)
16. International Electrotechnical Commision: Wind energy generation systems–Part 3-2: Design requirements for floating offshore wind turbines (2021)
17. Jonkman, J.: TurbSim User's Guide v2.00.00. National Renewable Energy Laboratory (2014)
18. Madsen, H., Larsen, T., Pirrung, G., Li, A., Zahle, F.: Implementation of the blade element momentum model on a polar grid and its aeroelastic load impact. Wind Energy Sci. **5**(1–27) (2020)
19. Brody, S., Alon, U., Yahav, E.: How attentive are graph attention networks?. In: International Conference on Learning Representations (2022)

20. PyTorch Geometric: conv.GATv2Conv. https://pytorch-geometric.readthedocs.io/en/2.5.1/generated/torch_geometric.nn.conv.GATv2Conv.html

21. Paszke, A., et al.: Pytorch: an imperative style, high-performance deep learning library. Adv. Neural Inf. **32**, 8024–8035 (2019)

22. Fey, M. and Lenssen, J.E.: Fast graph representation learning with PyTorch Geometric, arXiv preprint arXiv:1903.02428

23. Kingma, D.P., Ba, J.L.: Adam: a method for stochastic optimization. arXiv preprint arXiv:1412.6980 (2014)

24. Saverin, J., Perez-Becker, S., Luna, R.B.D., Marten, D., Gilloteaux, J.-C., Kurnia, R.: D1.2. Higher order hydroelastic module (2021)

25. SciPy: scipy.spatial.procrustes. https://docs.scipy.org/doc/scipy/reference/generated/scipy.spatial.procrustes.html

Towards Weight-Space Interpretation of Low-Rank Adapters for Diffusion Models

Jacek Duszenko$^{(\boxtimes)}$ and Piotr Bielak

Department of Artificial Intelligence, Wrocław University of Science and Technology,
Wrocław, Poland
jacek.duszenko@gmail.com, piotr.bielak@pwr.edu.pl

Abstract. Low-rank adapters (LoRAs) have emerged as an efficient
method for customizing large-scale diffusion models, but their internal
representations remain poorly understood. We present a comprehensive
investigation of the interpretability of adapter weight-spaces for image
diffusion models. To that end, we open-source a dataset of 100,000 Stable Diffusion adapters fine-tuned across a hierarchy of image concepts
amounting to 264 leaf classes, complete with training metadata. Through
systematic analysis, we demonstrate that adapter weights encode meaningful semantic information about their training data, enabling direct
interpretation without image generation. We evaluate multiple weight-
space representations, including raw parameters, statistical summaries,
and learned embeddings, to determine their effectiveness in predicting training data characteristics. To demonstrate real-world impact, we
apply our findings to the critical task of detecting potentially harmful
content on newly introduced NSFW (Not Safe For Work) toy dataset
of Stable Diffusion LoRAs fine-tuned on harmful content. This work
advances the interpretability of adapter-based fine-tuning and provides
practical tools for understanding and auditing adapted diffusion models.

Keywords: Weight-space models · Metanetworks · Low-rank
adapters · Image diffusion · Interpretability

1 Introduction

Text-to-image diffusion models, particularly Stable Diffusion [1], have revolutionized the field of AI-generated imagery by enabling widespread access to high-
quality image synthesis. Although these models offer impressive general-purpose
capabilities, many applications require customized image generation aligned with
specific artistic styles, concepts, or domain requirements. This customization
typically requires fine-tuning the model on specialized datasets.

However, full model fine-tuning presents significant computational challenges,
often requiring extensive GPU resources and training time. Low-Rank Adaptation (LoRA) [2] has emerged as an efficient alternative, allowing for a lightweight

M. H. Lees et al. (Eds.): ICCS 2025, LNCS 15903, pp. 108–121, 2025.
https://doi.org/10.1007/978-3-031-97626-1_8

model adaptation through training small adapter modules while keeping the base model frozen. This approach has led to a proliferation of publicly shared LoRA adaptations across various communities.

A critical challenge has emerged with the widespread adoption of these adapter modules: the lack of transparency regarding their encoded content and behaviors. Users who incorporate third-party LoRA modules into their generation pipeline have no reliable way to verify the nature of the adaptations without trial-and-error image generation. This raises concerns about potential inappropriate, biased, or harmful content being inadvertently introduced into the generation process.

Current approaches to understanding LoRA adaptations rely primarily on empirical testing through image generation, which is both time-consuming and computationally expensive. Moreover, this black-box testing approach may fail to reveal the full scope of encoded behaviors. There is a clear need for methods to analyze and interpret LoRA adapters directly without requiring the execution of the full diffusion model.

In this work, we experiment with various weight-space representations and predictive models to interpret LoRA adapters. Our approach enables direct examination of adapter weights eliminating the need for image generation or model execution, which are both resource and time intensive. This methodology has immediate practical applications for popular model repositories such as Civit.ai [3] and Hugging Face [4], where thousands of community-contributed LoRA adapters are shared. Such platforms could leverage the weight-space approach to automatically categorize and tag adapters based on their encoded content, significantly improving content organization and enabling more effective content moderation - all without the computational overhead of running the adapters themselves.

We summarize our contributions as follows:

1. We prepare and open-source the largest dataset of LoRA adapters for Stable Diffusion fine-tuned across a variety of image categories. We publish multiple versions differing in size (number of LoRA adapters), i.e. 1K, 10K, 50K and 100K adapters. Our dataset includes not only the final adapter weights, but also complete training metadata.
2. We provide an experimental evaluation of various weight-space representations and report their performance in classification of the fine-tuning set class.
3. Additionally, we contribute a small real-world use case dataset for predicting whether LoRA adapters were trained on harmful content and empirically prove that it's possible to do solely by looking at adapter's weights.
4. We make our models and code publicly available, ensuring reproducibility of our experiments, to accelerate weight-space research for LoRA adapters.

2   Related Work

Image Synthesis. Image synthesis has advanced significantly with Generative Adversarial Networks (GANs) [5], which enabled realistic image generation through a generator-discriminator framework. Diffusion models [6,7] offered more stable training and better image quality by iteratively denoising images, though at the cost of slower inference. CLIP [8] enabled text-conditioned generation in both GANs [9–11] and diffusion models [12–14]. Stable Diffusion [1] improved efficiency by operating in latent space, reducing computational costs while preserving quality, and became widely adopted due to its open-source nature, spurring innovations such as parameter-efficient fine-tuning.

Model Adaptation. While fine-tuning entire models has been the traditional approach to customization, parameter-efficient adaptation methods have recently gained prominence. Low-Rank Adaptation (LoRA) [2] has emerged as a particularly effective approach. For a given weight matrix $\mathbf{W} \in \mathbb{R}^{d \times k}$, LoRA introduces a low-rank decomposition: $\Delta\mathbf{W} = \mathbf{BA}$ where $\mathbf{B} \in \mathbb{R}^{d \times r}$ and $\mathbf{A} \in \mathbb{R}^{r \times k}$ with rank $r \ll \min(d, k)$. The final weight matrix becomes: $\mathbf{W}' = \mathbf{W} + \alpha\Delta\mathbf{W}$ where α is a scaling factor. This decomposition typically reduces trainable parameters by several orders of magnitude. Variations include AdaLoRA [15], which dynamically adjusts rank during training, and adapter layers [16], which insert trainable modules between existing layers. In diffusion models, LoRA has become the standard for community adaptations due to its efficiency and ease of implementation. To prevent overfitting of the original base pipeline on the training set, the standard practice is to combine LoRA adaptation with DreamBooth [17] due to its ability to preserve prior class features while enabling personalized subject adaptation with a small dataset. We employed DreamBooth while creating the dataset.

Weight Space Models for LoRA Adapters. Several recent works have focused on using LoRA weights for various tasks. Salama et al. (2024) [18] propose a task of predicting the number of training samples used to fine-tune a model based on LoRA weights. Their method exploits the relationship between the singular value spectrum of LoRA matrices and dataset size and leverages a simple K-NN approach for training set size prediction. They evaluate their approach on a dataset of 2 thousand fine-tuned LoRA adapters with snapshots of their weights amounting to 25 thousand samples. In contrast, our dataset doesn't use multiple in-training snapshots of a single adapter's weights as samples.

Putterman et al. (2024) [19] draw on geometric deep learning, focusing on designing GL-invariant and equivariant architectures for predicting fine-tuned model properties, e.g. accuracy on downstream tasks, training data membership and training data attributes. Their models are trained on a curated less-noisy subset of CelebA [20] dataset to predict physical binary attributes of the person's face diffusion model was personalized to. In contrast, our dataset comprises of diverse hierarchical concepts not constrained to physical attributes of a human face. Their dataset comprises of about 8 thousand adapters' weights in total.

Dravid et al. (2024) [21] focus on representing adapters in the principal component space of rank-one LoRA weights, enabling linear edits for visual concept transfers, sampling new adapters and inverting images into the weight space. Their dataset has roughly 60 thousand samples.

Horwitz et al. (2024) [22] propose a representation learning approach by propagating learnable vectors through a frozen dense matrix and subsequently projecting and aggregating the outputs. Although they demonstrate the efficacy of their method on a classification task analogous to our experimental setup, their empirical validation is limited to 5,000 LoRA adapters. Their methodology, although theoretically extensible, relies on selecting a single LoRA adapter weight matrix to derive model-wide representations, a choice that is determined through hyperparameter optimization on a validation set of 500 samples. Furthermore, their supervised learning framework, which jointly optimizes the representation space and classification head, introduces potential limitations in scenarios where labeled data is scarce, a common constraint in weight-space modeling.

3   Proposed Dataset

We present a comprehensive dataset of fine-tuned LoRA adapters along with their associated training metadata. The code used in the dataset creation process is available at https://github.com/JacekDuszenko/weightspace-lora-for-diffusion

Source Images. The dataset is constructed using a carefully curated subset of ImageNet [23], leveraging its hierarchical structure of visual concepts. We selected 10 distinct hierarchies from ImageNet, encompassing general concepts such as dog, cat, airplane, car, fruit, and vegetable. The leaf nodes within each hierarchy correspond to specific instances of these concepts, such as particular dog or cat breeds.

Sample Size and LoRA Hyperparameters. For each leaf node, we sample between 4 and 15 images without replacement to create training sets for LoRA adaptation, resulting in inputs for 200,000 distinct LoRA adapters. These source images are made available as a separate dataset which we open source at https://hf.co/datasets/jacekduszenko/weightspace-images. Using the well known `Stable-Diffusion-v1-5` as our base model, we perform Dream-Booth fine-tuning without prior preservation loss to create the LoRA adapters. Each adapter is trained using consistent hyperparameters: $rank = \alpha = 1$ and a learning rate of lr $= 1e-4$, with training conducted for 200 iterations.

Fine-Tuning. The training process employs a standardized textual prompt format: `a photo of sks CLASS`, where `CLASS` represents one of the ten general categories, and `sks` serves as a unique identifier for each specific concept, following standard DreamBooth methodology. The adapters are configured to modify the query (W_q) and value (W_v) projections within the attention layers of the Stable Diffusion U-Net. For each adapter, we store both the A and B LoRA matrices for each layer. Given the 64 layers in the architecture, this results in

128 LoRA matrices per adapter, with matrices represented as $320, 640, 768$ or 1024-dimensional vectors due to the rank-1 configuration.

The dataset maintains balanced representation across categories, with adapters trained on each general category comprising 10% of the total dataset. These trained adapters represent 264 unique concepts corresponding to the leaf classes from which the training images were sampled. For each adapter, we preserve comprehensive metadata including epoch-wise loss values, adapter gradient norms, weight norms, and training duration.

Hardware Resources. The training infrastructure utilized 8 NVIDIA A100 GPUs in a distributed setting, requiring a total of 2096 GPU hours to complete.

Dataset Versions. To facilitate various research applications, we release four versions of the dataset at different scales, with uniform distribution across general class labels. These datasets, including LoRA weights and training metadata, are available at the following URLs:

- LoRA-WS-1k (https://hf.co/datasets/jacekduszenko/lora-ws-1k),
- LoRA-WS-10k (https://hf.co/datasets/jacekduszenko/lora-ws-10k),
- LoRA-WS-50k (https://hf.co/datasets/jacekduszenko/lora-ws-50k),
- LoRA-WS-100k (https://hf.co/datasets/jacekduszenko/lora-ws-100k).

NSFW Classification Toy Dataset. Additionally, we develop a small Not Safe For Work detection dataset comprising LoRA adapters fine-tuned on two distinct classes: not safe for work content of one of the class from [24] and neutral concepts from our ImageNet subset. With the exception of fixed training set size of 5 images, adapters are trained using procedures and parameters identical to those of the main dataset, with complete preservation of the training data and the weights of the trained LoRAs. This supplementary dataset consists of 160 samples of adapters distributed evenly across the binary label and is available at https://hf.co/datasets/jacekduszenko/lora-ws-nsfw.

4   Weight-Space Representations

Traditional representation learning methods build representation (embedding) vectors for various data types such as images, audio, text or graphs [25], often utilizing neural networks combined with a carefully crafted objective function, to convert the input objects into vector form. In contrast, weight-space representation learning aims at building representation vectors for neural networks. Their weights are processed either directly, by means of simple statistical operations [26], or by other neural networks [27].

In this paper, we focus on direct vector processing and statistical operations, leaving more advanced neural network approaches for future work. In particular, we examine the following approaches (Fig. 1):

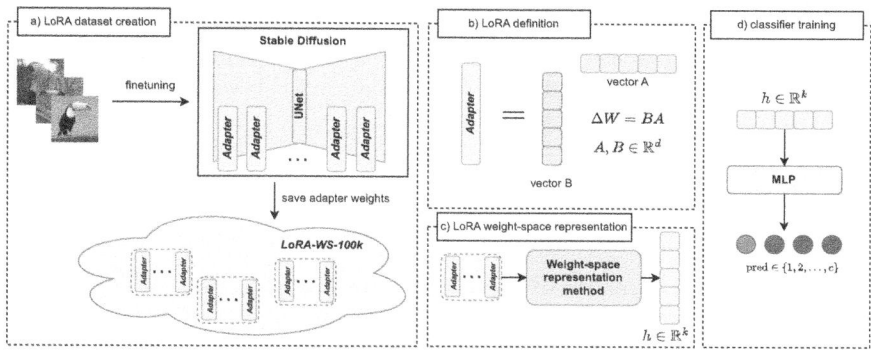

Fig. 1. Overview of our setup. **a)** The LoRA weight-space datasets were created by Stable Diffusion finetuning using low-rank adapters. For each sample of images, we finetune Stable Diffusion and save all the LoRAs from the UNet component. **b)** A single low-rank adapter consists of two vectors A and B (we assume rank = 1, so instead of matrices, we obtain vectors). **c)** The weight-space representation learning methods take as input a collection of LoRAs (from a single Stable Diffusion finetuning instance) and compute a single vector representation h which captures the characteristics of the adapters. **d)** To evaluate our hypothesis that LoRA weights encode sufficient information to retrieve the class of the input finetuning images, we use the weight-space representations to train an MLP classifier and report the classification performance on the test set.

- FlatVec – the most popular baseline approach in weight-space representation learning; for each adapter layer, the weight matrices are flattened into vectors, next all the vectors are concatenated into one long vector; in our case the final vector dimensionality is $dim(\texttt{FlatVec}) = 99648$,
- FlatVecPCA@K – the high dimensionality of the FlatVec approach is unfeasible in most scenarios, therefore, we reduce the dimensionality of FlatVec's output vector by using Principal Component Analysis and keep K components, so that $dim(\texttt{FlatVecPCA@K}) = K$,
- StatsFlatVec – another popular weight-space representation approach is to compute basic statistics of the model weights; previous works [26,27] have shown that such representation provides quite competitive results, even against complex neural network approaches; in our case, we take the FlatVec's output vector and compute the following statistics for it: for the LoRA-WS-1k and LoRa-WS-10k datasets – mean, std, median, min, max, kurtosis, skew, which results in $dim(\texttt{StatsFlatVec}) = 7$, and for LoRA-WS-50k – the same statistics except for kurtosis and skew (due to Out-Of-Memory errors during evaluation), which results in $dim(\texttt{StatsFlatVec}) = 5$,
- StatsLayer – given that each adapter layers consists of two matrices **A** and **B**, we compute the statistics (same as for StatsFlatVec) but for each matrix and adapter layer independently, and finally we concatenate all the vectors, which results in $dim(\texttt{StatsLayer}) = 896$,

- `StatsLayerDense` – we first compute the dense matrix of each adapter layer, i.e., $\Delta\mathbf{W} = \mathbf{BA}$, flatten this matrix and compute the same statistics as before for each adapter layer, finally we concatenate all the vectors; this results in $dim(\texttt{StatsLayerDense}) = 448$.

5   Experiments

We evaluate the performance of the introduced representation methods against our proposed dataset. We perform several experiments to better understand the nature of low-rank adapters and provide insights based on the observed results. Each experiment is described in a separate subsection below.

5.1   Fine-Tuning Dataset Classification

Goal. We investigate whether adapter weights contain sufficient information to predict the class of images on which an adapter was fine-tuned. Specifically, we aim to classify LoRA adapters into one of 10 base classes solely by examining their weight representations.

Setup. For this task, we utilize the five distinct weight representations described in Sect. 4. We partition our dataset using a 70/10/20 split for training, validation, and testing respectively. The classifier architecture consists of a multi-layer perceptron with three hidden layers of size 512 with ReLU activations, trained using cross-entropy loss. We employ the Adam optimizer [28] with a learning rate of $lr = 1\mathrm{e}{-3}$ and train for a maximum of 2000 epochs with early stopping to prevent overfitting. To ensure robust evaluation, we repeat each experiment 10 times with different random seeds and report the mean performance metrics with standard deviations. Furthermore, we conduct separate experiments across multiple dataset scales (1k, 10k, and 50k) to assess how the effectiveness of different representation methods scales with data volume. This systematic approach enables us to identify which weight representation techniques most effectively capture the discriminative information necessary for determining the original fine-tuning class. We report the results in Table 1.

Discussion. For the smallest dataset (`LoRA-WS-1k`), we observe that for all reported metrics, the `FlatVecPCA@200` method clearly outperforms the other approaches by a large margin. It allows achieving up to approx. 86% AUROC, whereas the next best method (`StatsLayerDense`) achieves approx. 81% AUROC. Unsurprisingly, the `FlatVec` method performs poorly – having vectors with almost 100,000 dimensions, while having only 700 training samples makes it hard for any kind of classifier to generalize well. Similarly, the 700 samples do not provide enough information for the `StatsFlatVec` method, which reduces the almost 100,000 numbers into 7 statistics.

Table 1. Classification metrics using different representation learning methods. Results are reported as mean ± standard deviation over 10 runs (different seeds). Higher values are better (↑). Best results are **bolded**, whereas second best results are underlined. (a) Results on LoRA-WS-1k dataset (top), (b) results on LoRA-WS-10k dataset (middle), (c) results on LoRA-WS-50k dataset (bottom).

a) LoRA-WS-1k	Accuracy (↑)	AUROC (↑)	Precision (↑)	Recall (↑)	F1 (↑)
FlatVec	15.62 ± 2.17	55.64 ± 1.94	16.95 ± 3.42	15.50 ± 2.14	14.91 ± 2.21
FlatVecPCA@200	$\mathbf{42.21 \pm 1.20}$	$\mathbf{86.12 \pm 1.01}$	$\mathbf{41.54 \pm 1.55}$	$\mathbf{42.06 \pm 1.19}$	$\mathbf{40.69 \pm 1.05}$
StatsFlatVec	16.22 ± 2.89	55.15 ± 2.26	17.69 ± 3.74	15.94 ± 2.79	15.60 ± 3.04
StatsLayer	33.18 ± 2.10	77.27 ± 0.49	32.83 ± 3.18	33.55 ± 2.20	32.74 ± 2.51
StatsLayerDense	$\underline{37.51 \pm 2.07}$	$\underline{81.62 \pm 0.61}$	$\underline{36.71 \pm 1.85}$	$\underline{37.30 \pm 2.07}$	$\underline{36.41 \pm 1.93}$

b) LoRA-WS-10k	Accuracy (↑)	AUROC (↑)	Precision (↑)	Recall (↑)	F1 (↑)
FlatVec	$\underline{65.83 \pm 1.39}$	$\underline{92.83 \pm 0.51}$	$\underline{66.48 \pm 1.46}$	$\underline{65.81 \pm 1.39}$	$\underline{65.84 \pm 1.42}$
FlatVecPCA@200	$\mathbf{81.36 \pm 1.07}$	$\mathbf{98.14 \pm 0.16}$	$\mathbf{81.63 \pm 1.04}$	$\mathbf{81.33 \pm 1.07}$	$\mathbf{81.41 \pm 1.04}$
StatsFlatVec	64.75 ± 1.66	92.32 ± 0.64	65.44 ± 1.98	64.73 ± 1.68	64.81 ± 1.82
StatsLayer	44.22 ± 1.05	83.50 ± 0.37	44.34 ± 1.14	44.01 ± 1.04	44.07 ± 1.08
StatsLayerDense	52.61 ± 0.63	88.22 ± 0.22	52.43 ± 0.64	52.49 ± 0.63	52.36 ± 0.63

c) LoRA-WS-50k	Accuracy (↑)	AUROC (↑)	Precision (↑)	Recall (↑)	F1 (↑)
FlatVec	$\mathbf{93.26 \pm 2.56}$	$\mathbf{99.63 \pm 0.26}$	$\mathbf{93.28 \pm 2.51}$	$\mathbf{93.26 \pm 2.55}$	$\mathbf{93.26 \pm 2.54}$
FlatVecPCA@1000	78.33 ± 1.37	97.61 ± 0.20	78.47 ± 1.44	78.30 ± 1.37	78.33 ± 1.39
StatsFlatVec	$\underline{93.21 \pm 2.17}$	$\mathbf{99.63 \pm 0.18}$	$\underline{93.20 \pm 2.15}$	$\underline{93.20 \pm 2.17}$	$\underline{93.19 \pm 2.17}$
StatsLayer	63.68 ± 0.38	93.01 ± 0.17	63.57 ± 0.32	63.58 ± 0.38	63.56 ± 0.35
StatsLayerDense	44.49 ± 0.47	84.09 ± 0.26	45.14 ± 0.41	44.44 ± 0.47	44.72 ± 0.44

For the LoRA-WS-10k dataset, we find that the overall metrics have increased substantially compared to the 1k variant, for instance, now the maximum AUROC value is approx. 98% compared to 86% on the 1k dataset; similarly, the Accuracy and F1 metrics have nearly doubled. This confirms our motivation for creating larger datasets for weight-space representation learning on low-rank adapters. Once again, the FlatVecPCA@200 method outperforms other representation methods; however, this time the second best method is FlatVec. Despite having almost 100,000 dimensions, we now have much more samples to fit a classifier and allow it generalize better. Note also that the StatsFlatVec method achieves performs quite similarly to the FlatVec method, while having significantly less dimensional representations.

In case of the LoRA-WS-50k dataset, the FlatVec method obtains the best results for all metrics; however, the StatsFlatVec method performs only slightly worse on all but one metric – for AUROC the mean values are the same with StatsFlatVec having a smaller standard deviation. The data volume (50k samples) allows to achieve even better results than on the 10k dataset variant.

StatsFlatVec being the second best method shows again that despite more samples, the dataset can be efficiently compressed down to only 5 dimensional vectors.

To sum up, we observe that the FlatVec and FlatVecPCA@K methods work the best overall. For the 50k dataset variant, the StatsFlatVec method also works great despite compressing the representations to only 5 dimensions. However, for practical applications, these methods are not suitable. They either use very high dimensional vectors $(dim(\text{FlatVec}) \approx 100,000)$ or they use such vectors as intermediate steps (FlatVecPCA@K and StatsFlatVec), which requires large computational resources. For future work, we would like to focus on representation methods that operate on lower dimensional vectors, allowing for resource efficient data processing.

5.2   Expressiveness of Individual Layers

Goal. To systematically analyze the information content encoded within individual LoRA adapter layers, we isolate and evaluate the predictive power of each adapter layer independently on the LoRA-WS-1k dataset.

Setup. For each LoRA adapter layer l, we extract and flatten its weight matrices adapting $W_q^{(l)}$ and $W_v^{(l)}$ as well as self-attention maps in various attention layers of the U-Net into feature vectors $\mathbf{x}_l \in \mathbb{R}^d$, where d varies based on the layer dimensions. We partition the dataset into training (70%), validation (10%), and test (20%) sets. An MLP classifier is trained on the training set and evaluated on both validation and test sets. The MLP architecture remains consistent with the previous experiment, featuring three hidden 512-dimensional layers with ReLU activations. We employ early stopping based on validation performance and use the Adam optimizer with a learning rate of lr $=$ 1e$-$3. We repeat this process 10 times with different random seeds. We report averaged results across all runs in Fig. 2.

Discussion. We empirically demonstrate that LoRA adapter weights in cross-attention layers encode substantial information about the fine-tuning dataset, with the value projection adapter (W_v) achieving the highest classification accuracy (97.01% on the test set), slightly outperforming the query projection (W_q). This suggests that the learned adaptations in W_v retain stronger dataset-specific signals, potentially due to their direct role in modulating text-conditioned feature integration. In contrast, self-attention layers exhibit significantly lower predictive power, with most performing near chance level (10%) and only a few reaching moderate expressiveness (60%). This disparity highlights the dominant role of cross-attention in shaping model-specific adaptations, while self-attention layers primarily facilitate general feature propagation rather than encoding dataset-specific signatures.

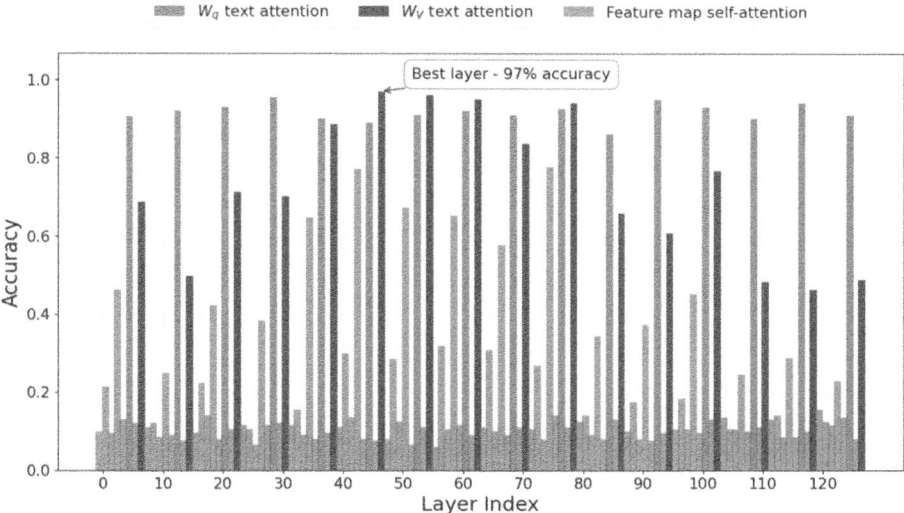

Fig. 2. Classification accuracy across different LoRA adapter layers in the U-Net architecture. The plot demonstrates the predictive power of individual adapter weights for dataset class identification. Notable peaks in accuracy correspond to text conditioning cross-attention layers, particularly in the query (W_q) and value (W_v) projections, achieving up to 97.01% test accuracy. Other attention layers such as self attention of the latent feature maps of the U-net are in gray.

5.3 Dimensionality Reduction

Goal. To elucidate the relationship between representational dimensionality and feature quality, we conduct Principal Component Analysis (PCA) on high-dimensional feature vectors constructed by the `FlatVec` method (note that this setup is essentially the same as the `FlatVecPCA@K` method with different choices of the K value).

Setup. Figure 3 illustrates the relationship between classification accuracy and the number of retained principal components. We extend this experiment to the larger `LoRA-WS-10k` dataset, maintaining the same framework as in our previous analysis, with the key modification being the use of concatenated adapter weights as input features rather than individual layer representations. We try different PCA values (100, 200, 500, 1000, 2000, 3000, 5000 and 7500) and report the mean accuracy achieved in ten runs of the experiment with different random seeds.

Discussion. We observe that the performance on both validation and test splits is quite similar with Validation Accuracy being only a bit worse than the Test Accuracy. Moreover, both metrics highly depend on the choice of the K parameter, ranging from almost 20% for $K = 7500$ up to approx. 85% for $K = 500$

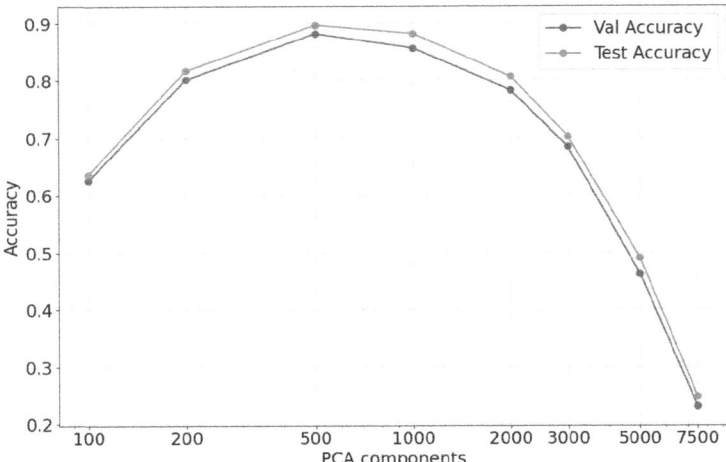

Fig. 3. Classification accuracy as a function of retained PCA components for different LoRA adapter layers. The plot demonstrates how predictive power varies with dimensionality reduction.

(which is the sweet spot for this dataset variant). Using both 200-dim and 1000-dim vectors also provides a decent performance, however, we would recommend going with $K = 200$ as it requires less computational power (due to lower dimensional vectors). Overall, we observe that the data points approximate a convex function, where larger dimensional vectors achieve worse performance that smaller ones. Hence, when using the `FlatVecPCA@K` method, we would recommend using lower dimensional vectors, i.e., smaller values of K and check a few values around $K = 500$.

5.4   Harmful Content Detection

Goal. We want to address a crucial application area of weight-space representation learning for low-rank adapters – detection of potentially harmful LoRAs. Hence, in this experiment we predict whether an adapter was fine-tuned on images depicting harmful content represented as one of the classes from [24].

Setup. We take sexually explicit anime/manga images as concepts to train harmful adapters. As showed empirically in previous experiments, simple classifiers are not doing very well on small datasets. To tackle this problem, we feed a single layer as input to the classifier and choose the best performing layer in evaluation. We use a 80/20% dataset split. We pre-process the weights of adapters to obtain representations described earlier and conduct experiments with each representation. To this end, we train an MLP with three hidden layers of size 1024 and final classification head with two dimensional output. We then backpropagate on the cross-entropy loss. In evaluation, we treat the more probable class as final prediction of the model. We use the learning rate of lr = 1e−4, leverage the Adam optimizer and train with early-stopping on 2000 epochs. We present our results in Table 2.

Table 2. Classification metrics for harmful content detection using different representation learning methods. Results are reported as mean ± standard deviation on the test set over 10 runs (different seeds). Higher values are better (↑). Best results are **bolded**, whereas second best results are underlined.

LoRA-WS-NSFW	Accuracy (↑)	AUROC (↑)	Precision (↑)	Recall (↑)	F1 (↑)
FlatVec	**98.12 ± 1.53**	**99.96 ± 0.12**	**99.41 ± 1.76**	**96.88 ± 3.12**	**98.08 ± 1.57**
FlatVecPCA@20	67.01 ± 4.46	62.46 ± 2.58	63.98 ± 3.42	76.88 ± 7.93	69.74 ± 5.00
StatsFlatVec	66.87 ± 4.00	61.88 ± 1.39	63.84 ± 2.62	<u>77.50 ± 9.35</u>	69.83 ± 4.95
StatsLayerDense	60.34 ± 10.30	60.30 ± 4.41	61.20 ± 4.00	69.01 ± 1.00	65.03 ± 3.38

Discussion. Our results demonstrate that certain adapter weights, primarily the ones concerned with cross-attention to support textual conditioning, contain clear signatures of harmful content training, with flattened vector achieving near-perfect detection accuracy (98.12%). Statistical summaries and dimensionality reduction by PCA perform poorly (66–67%), indicating that feature engineering provides only marginal benefits in the small dataset setting. Our experiments with summary statistics of dense representation yielded poor performance (60.34% accuracy), confirming that joining two layers and aggregating them further degrades classification efficacy. This finding aligns with our previous experimental observations and demonstrates that dense representations have limited expressiveness.

6   Conclusions and Future Work

In this paper, we addressed the problem of learning representations for low-rank adapters based on their weights (so called weight-space representation learning). We contributed several StableDiffusion LoRA datasets varying in size, i.e., LoRA-WS-1k, LoRA-WS-10k, LoRA-WS-50k and LoRA-WS-100k. We performed a variety of experiments to better understand the performance of different weight-space representation learning methods on our dataset. We have shown that low-rank adapters encode sufficient information in their weights to be able to classify the content the adapter was trained on without the need to generate the actual images. Such a setup is crucial when working with potentially harmful adapters, where we do not want to generate the actual images to detect whether it was trained on dangerous content. To that end, we introduced a small dataset LoRA-WS-NSFW and analyzed the performance a several embedding methods. Results show that by just using the weights directly, we can build a well performing classifier.

In future work, we want to extend our experiments to the LoRA-WS-100k dataset. However, to achieve that we need to first focus on developing a dedicated weight-space representation learning method, which does not require operating on almost 100,000-dimensional vectors (as we had to do for the other dataset variants). Working with such large vectors combined with a large dataset size, makes it difficult to train any kind of classification models (due to high resource requirements). Next, we want to also extend our LoRA-WS-NSFW dataset by increasing the dataset size. Finally, we want to introduce noise to the LoRA datasets, which should better reflect real-world conditions and help to test the generalization ability of the weight-space representations. We want to also check scenarios where labels are imperfect and/or classes are imbalanced, which are also likely to be found in real-world cases.

Acknowledgments. We gratefully acknowledge Polish high-performance computing infrastructure PLGrid (HPC Center: ACK Cyfronet AGH) for providing computer facilities and support within computational grant no. PLG/2025/018054.

References

1. Rombach, R., Blattmann, A., Lorenz, D., Esser, P., Ommer, B.: High-resolution image synthesis with latent diffusion models. arXiv preprint arXiv:2112.10752 (2022)
2. Hu, E.J., et al.: LoRA: low-rank adaptation of large language models. In: International Conference on Learning Representations (2022). https://openreview.net/forum?id=nZeVKeeFYf9
3. CivitAI: A community-driven platform for sharing AI-generated models and content. https://civitai.com. Accessed 18 Feb 2025
4. Hugging Face: An open platform for machine learning models and datasets. https://huggingface.co. Accessed 18 Feb 2025
5. Goodfellow, I., et al.: Generative adversarial nets. In: Advances in Neural Information Processing Systems (NeurIPS), vol. 27, pp. 2672–2680 (2014). arXiv preprint arXiv:1406.2661
6. Ho, J., Jain, A., Abbeel, P.: Denoising diffusion probabilistic models. In: Advances in Neural Information Processing Systems (NeurIPS), vol. 33, pp. 6840–6851 (2020). arXiv preprint arXiv:2006.11239
7. Dhariwal, P., Nichol, A.: Diffusion models beat GANs on image synthesis. In: Advances in Neural Information Processing Systems (NeurIPS), vol. 34, pp. 8780–8794 (2021). arXiv preprint arXiv:2105.05233
8. Radford, A., et al.: Learning transferable visual models from natural language supervision. In: Proceedings of the 38th International Conference on Machine Learning (ICML), pp. 8748–8763 (2021). arXiv preprint arXiv:2103.00020
9. Patashnik, O., Wu, Z., Shechtman, E., Cohen-Or, D., Lischinski, D.: StyleCLIP: text-driven manipulation of StyleGAN imagery. arXiv preprint arXiv:2103.17249 (2021)
10. Gal, R., Patashnik, O., Maron, H., Chechik, G., Cohen-Or, D.: StyleGAN-NADA: CLIP-guided domain adaptation of image generators. arXiv preprint arXiv:2108.00946 (2021)

11. Galatolo, F.A., Cimino, M.G.C.A., Vaglini, G.: Generating images from caption and vice versa via CLIP-guided generative latent space search. arXiv preprint arXiv:2102.01645 (2021)
12. Nichol, A., et al.: GLIDE: towards photorealistic image generation and editing with text-guided diffusion models. arXiv preprint arXiv:2112.10741 (2021)
13. Kim, G., Ye, J.C.: DiffusionCLIP: text-guided image manipulation using diffusion models. arXiv preprint arXiv:2110.02711 (2021)
14. Avrahami, O., Lischinski, D., Fried, O.: Blended diffusion for text-driven editing of natural images. arXiv preprint arXiv:2111.14818 (2021)
15. Liu, Z., et al.: AdaLoRA: adaptive budget allocation for parameter-efficient fine-tuning. In: International Conference on Learning Representations (ICLR) (2023). arXiv preprint arXiv:2302.12345
16. Houlsby, N., et al.: Parameter-efficient transfer learning for NLP. In: International Conference on Machine Learning (ICML), vol. 36, pp. 2708–2717 (2019). arXiv preprint arXiv:1902.00751
17. Ruiz, N.J., et al.: DreamBooth: fine-tuning text-to-image diffusion models for subject-driven generation. arXiv preprint arXiv:2303.10745 (2023)
18. Salama, M., Kahana, J., Horwitz, E., Hoshen, Y.: Dataset size recovery from LoRA weights. arXiv preprint arXiv:2406.19395 (2024)
19. Putterman, T., Lim, D., Gelberg, Y., Jegelka, S., Maron, H.: Learning on LoRAs: GL-equivariant processing of low-rank weight spaces for large finetuned models. arXiv preprint arXiv:2410.04207 (2024)
20. Liu, Z., Luo, P., Wang, X., Tang, X.: Deep learning face attributes in the wild. In: Proceedings of International Conference on Computer Vision (ICCV), pp. 3730–3738 (2015)
21. Dravid, A., et al.: Interpreting the weight space of customized diffusion models. arXiv preprint arXiv:2406.09413 (2024)
22. Horwitz, E., Cavia, B., Kahana, J., Hoshen, Y.: Learning on model weights using tree experts. arXiv preprint arXiv:2410.13569 (2024)
23. Deng, J., Dong, W., Socher, R., Li, L.-J., Li, K., Fei-Fei, L.: ImageNet: a large-scale hierarchical image database. In: CVPR (2009). http://www.image-net.org/papers/imagenet_cvpr09.bib
24. Huggingface nsfw_detect dataset. https://huggingface.co/datasets/deepghs/nsfw_detect. Accessed 19 Feb 2025
25. Bengio, Y., Courville, A., Vincent, P.: Representation learning: a review and new perspectives. IEEE Trans. Pattern Anal. Mach. Intell. **35**, 1798–1828 (2013)
26. Schürholt, K., Taskiran, D., Knyazev, B., Giró-i-Nieto, X., Borth, D.: Model zoos: a dataset of diverse populations of neural network models. In: Thirty-Sixth Conference on Neural Information Processing Systems (NeurIPS) Track on Datasets and Benchmarks (2022)
27. Lim, D., Maron, H., Law, M.T., Lorraine, J., Lucas, J.: Graph metanetworks for processing diverse neural architectures. In: ICLR (2024)
28. Kingma, D.P., Ba, J.: Adam: a method for stochastic optimization. arXiv preprint arXiv:1412.6980 (2017). https://arxiv.org/abs/1412.6980

Discovering Governing Equations of Geomagnetic Storm Dynamics with Symbolic Regression

Stefano Markidis$^{(\boxtimes)}$, Jonah Ekelund, Luca Pennati, Andong Hu, and Ivy Peng

KTH Royal Institute of Technology, Stockholm, Sweden
markidis@kth.se

Abstract. Geomagnetic storms are large-scale disturbances of the Earth's magnetosphere driven by solar wind interactions, posing significant risks to space-based and ground-based infrastructure. The Disturbance Storm Time (Dst) index quantifies geomagnetic storm intensity by measuring global magnetic field variations. This study applies symbolic regression to derive data-driven equations describing the temporal evolution of the Dst index. We use historical data from the NASA OMNIweb database, including solar wind density, bulk velocity, convective electric field, dynamic pressure, and magnetic pressure. The PySR framework, an evolutionary algorithm-based symbolic regression library, is used to identify mathematical expressions linking dDst/dt to key solar wind. The resulting models include a hierarchy of complexity levels and enable a comparison with well-established empirical models such as the Burton-McPherron-Russell and O'Brien-McPherron models. The best-performing symbolic regression models demonstrate superior accuracy in most cases, particularly during moderate geomagnetic storms, while maintaining physical interpretability. Performance evaluation on historical storm events includes the 2003 Halloween Storm, the 2015 St. Patrick's Day Storm, and a 2017 moderate storm. The results provide interpretable, closed-form expressions that capture nonlinear dependencies and thresholding effects in Dst evolution.

Keywords: Symbolic Regression · Geomagnetic Storms · Dst Index Prediction · Interpretable Machine Learning

1 Introduction

Space weather investigates the dynamics of the near-Earth space environment driven by solar activity, including solar wind, geomagnetic field disturbances, and energetic particles [1]. One of the most significant events in space weather is the occurrence of geomagnetic storms, which are large-scale disturbances of Earth's

This work is supported by the European Commission, with Automatics in Space Exploration (ASAP), project no. 101082633. The OMNI data were obtained from the GSFC/SPDF OMNIWeb interface at https://omniweb.gsfc.nasa.gov.

magnetosphere caused by increased interactions between the solar wind and the magnetosphere, such as coronal mass ejections. Geomagnetic storms strongly impact human assets in space and on the ground; they can affect satellite operations, navigation systems, power grids, and high-frequency communications [12]. Additionally, increased radiation exposure poses risks to astronauts and those involved in high-altitude flights. Therefore, predicting geomagnetic storms is essential for protecting human life and assets in space and on the ground.

The Disturbance Storm Time (Dst) index measures global geomagnetic activity. Specifically, it reflects Earth's horizontal magnetic field disturbances due to the ring current in the magnetosphere. The Dst index is widely used in space weather to monitor and classify geomagnetic storms according to their value. In this work, we employ symbolic regression to derive data-driven equations that describe the time evolution of the Dst index. To achieve this, we use *i)* data from the NASA OMNIweb database [10], including the Dst index, solar wind parameters (density, bulk velocity, ...), and Interplanetary Magnetic Field (IMF) parameters. This data is obtained from solar wind monitoring spacecrafts, including ACE, Wind, IMP-8, and DSCOVR. *ii)* PySR, a symbolic regression library based on evolutionary algorithms [5]. With PySR, we can search for optimal mathematical expressions that capture the underlying relationships between dDst/dt and key solar wind parameters, such as the convective electric field, dynamic pressure, magnetic pressure, and Dst itself. By varying the equation complexity, we recover a hierarchy of models, ranging from simple empirical-like formulations to more complex expressions that reveal non-linear dependencies. This approach allows us to systematically compare different equations to describe the temporal evolution of the Dst and assess their physical interpretability.

This paper aims to derive data-driven mathematical equations that describe the temporal evolution of the Dst index using symbolic regression. This work uses solar wind, IMF parameters, and Dst historical data from the NASA OMNIweb database. It employs the PySR framework to investigate interpretable relationships between geomagnetic storms and solar wind properties and IMF. The contributions of this work are as follows:

- We employ symbolic regression to derive data-driven equations for the Dst temporal evolution (dDst/dt), using solar wind parameters such as the convective electric field, dynamic pressure, magnetic pressure, and the Dst index, obtained from the OMNIweb database.
- We explore a hierarchy of models, obtained with PySR, by varying equation complexity as input of PySR. We recover equations with increasing physical accuracy.
- We compare the discovered equations with well-established empirical models, such as the Burton-McPherron-Russell [3] and O'Brien-McPherron models [13,14], in accuracy. We show that the best models found with the symbolic regression approach outperform these established models in the cases considered.

– We present the advantages of symbolic regression for geomagnetic storm modeling by providing interpretable, closed-form expressions rather than blackbox predictions, such as those provided by neural networks.

2   Background and Related Work

The Dst index measures variations in Earth's geomagnetic field. It is expressed in nanoteslas as the magnetic fields and primarily reflects the strength of the equatorial ring current, which intensifies during geomagnetic storms. The ring current consists of energetic ions and electrons, which can create a magnetic field in the opposite direction of Earth's intrinsic geomagnetic field. When the ring current increases, such as during geomagnetic storms, it causes a weakening of the Earth's field. A negative Dst value corresponds to a reduction in the Earth's surface magnetic field strength due to the intensification of the ring current. Formally, the Dst index is defined as the horizontal component perturbation on equatorial magnetometers [16], and its first definition dates back to von Humboldt in the early 18th century [11]. The Dst is measured as the longitudinally averaged part of the external field at the geomagnetic dipole equator on the Earth's surface. The World Data Center provides its values for Geomagnetism in Kyoto, and they are also available and tabulated in the NASA OMNIweb database [10], which we use in this work.

One of the most popular definitions of geomagnetic storm is based on a threshold Dst value, e.g., -50 or -100, below which an event is categorized as a geomagnetic storm [2]. The geomagnetic storms can also be classified depending on their minimum Dst value [8] as moderate (min Dst: between -50 nT and -100 nT), intense/great (min. Dst: between -100 nT and -250 nT), and super/extreme (minimum Dst less than -250 nT).

The evolution of the Dst index during a geomagnetic storm follows three phases as depicted in Fig. 1: *(i)* an initial phase with increased Dst, often called Sudden Storm Commencement (SSC), when an interplanetary shock compresses the magnetosphere, frequently associated with a coronal mass ejections *(ii)* a main phase, where the Dst index rapidly drops as solar wind-driven particle injection enhances the ring current, and *(iii)* a recovery phase, where Earth's magnetic field returns to its pre-storm state. The evolution of the Dst index can be compared to a capacitor charging process in an RLC circuit: the magnetosphere can be seen as a capacitor storing energy, and the sudden increase in current is analogous to a transient current increase in an electrical circuit [4].

The time evolution of the Dst index has been studied using a range of models, varying from first-principles physics-based approaches (such as global MHD models coupled with ring current models [7]) to empirical models. Due to their simplicity and physical insights, empirical formulations that approximate dDst/dt based on solar wind parameters are among the most widely used. The Burton-McPherron-Russell (BMR) model [3] is a first and successful example, describing dDst/dt as a balance between solar wind-driven ring current injection and exponential decay due to charge exchange and ionospheric losses. The

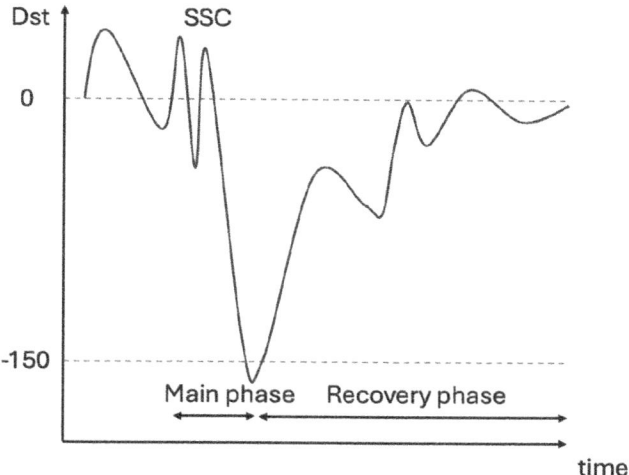

Fig. 1. Typical Dst evolution during a geomagnetic storm.

model uses an input function proportional to the convective electric field, which is linked to the strength of solar wind-magnetosphere coupling and controls the efficiency of energy injection into Earth's magnetosphere. The dynamic pressure of the solar wind also plays an important role by compressing the magnetosphere, modulating the ring current response, and influencing the initial phase of geomagnetic storms. Later improvement, such as the O'Brien-McPherron (OBM) model [13], introduced nonlinear decay terms and improved parameterizations to capture storm-time dynamics better. Neural networks have also been explored for predicting dDst/dt, using their ability to model highly nonlinear relationships [6, 9, 18]. However, while neural networks often achieve high predictive accuracy, they function as black-box models: it is unclear how input variables contribute to the output. AI-driven symbolic regression provides a complementary approach by directly discovering data-driven equations from observational datasets. Unlike neural networks, symbolic regression produces explicit mathematical expressions, allowing for better interpretability and physical insight.

PySR is a symbolic regression framework designed for discovering interpretable mathematical expressions from data using evolutionary search. It is developed primarily in the Julia programming language and provides a Python interface, which is used in this work [5]. PySR uses evolutionary algorithms: it searches for optimal symbolic expressions by iteratively evolving equations, using operations such as addition, multiplication, and exponentiation, as provided by the user. One of the PySR advantages over other symbolic regression frameworks, such as gplearn [15], and AI Feynman [17], is its flexible operator set: this includes conditionals (if-else), min, and max functions. These are essential operators for modeling non-linear and threshold-dependent events in the magnetosphere, such as geomagnetic storms. gplearn is a genetic programming-based tool for symbolic regression. It has a customizable function set, but does not natively include `if-else` logic without user-defined extensions. `AI Feynman` supports a range of nonlinear and algebraic operations, including polynomial and rational functions. However, its support for conditionals and piecewise functions is more limited compared to `PySR`. `AI Feynman` is primarily designed for cases where exact functional relationships exist, such as analytical physics laws. In contrast, `PySR` is more adaptable to complex, noisy datasets where no exact equation is known a priori.

3   Methodology

3.1   OMNIWeb Dataset and Preprocessing

The dataset utilized in this study is obtained from the NASA OMNIweb dataset, a collection of near-Earth solar wind, IMF parameters, and geomagnetic indices. Our dataset ranges from January 1, 1995, to May 31, 2021, and includes measurements of solar wind plasma parameters, IMF components, and geomagnetic activity indicators, such as the Dst, at a 1-hour resolution. Our primary target variable is the rate of change of the Dst index, dDst/dt, expressed in nT/hr. Following the examples of previous empirical, physics-based models of the temporal evolution of the Dst, our input variables include the solar wind speed V_{sw}(km/s), the IMF North-South component in GSM coordinates (B_z(nT), GSM), the solar wind proton density n_p (cm^{-3}), and $|B|$(nT) is the magnitude of the IMF. Data pre-processing involves interpolating missing values using a linear scheme, followed by forward and backward filling to ensure continuity. The time derivative of the Dst index is computed using a central finite difference approximation. We then calculate derived quantities, which have been demonstrated by previous study to impact the temporal evolution of the Dst. These derived quantities are:

– **Convective Electric Field** (E_y), which is defined as $E_y = -V_{SW} B_z \times 10^{-3}$(mV/m). The magnitude of E_y reflects how strongly the solar wind couples with Earth's magnetosphere. In physics-based models of the Dst, E_y is used to estimate the energy injection rate into the ring current.
– **Dynamic Pressure** (P_{dyn}), which is calculated from the solar wind proton density and speed $P_{dyn} = 1.6726 \times 10^{-6} n V_{SW}^2$(nPa), where the pre-factor accounts for the proton mass and necessary unit conversions. The solar wind has two important effects in the context of geomagnetic storms. First, high dynamic pressure leads to a compression of the magnetosphere, e.g., it causes the magnetopause to move inward. Second, strong solar wind shocks with high dynamic pressure can increase ring current formation.
– **Magnetic Pressure** (P_B), which is given by $P_B = \frac{B^2}{2\mu_0}$ with $\mu_0 = 4\pi \times 10^{-7}$ being the permeability of free space. In the context of geomagnetic storms, the magnetic pressure is the contribution of the IMF and Earth's magnetic field to the overall pressure balance in the magnetosphere. The magnetic pressure can affect the magnetopause position by balancing solar wind pressure.

3.2   Exploratory Analysis

We conduct first an exploratory analysis to assess potential relationships among the variables impacting the evolution of the Dst. Figure 2 presents a pairplot comparing $d\text{Dst}/dt$, the previous Dst value (DST_prev), P_{dyn}, E_y, and P_B. In the pairplot, the diagonal entries display the distributions of each variable. The off-diagonal plots show pairwise scatter plots, which show potential correlation or anti-correlation, if the points appear along one line, or non-linear trends. Analyzing Fig. 2, we can first focus on distributions along the diagonal of the pairplot. The distribution shows that $d\text{Dst}/dt$ has peaked near zero, indicating that most variations in the Dst index occur in relatively small increments. The distribution of DST_prev shows that the Dst index peaks on slightly negative values of the Dst with a long tail in the negative Dst direction, which comprises intense and extreme geomagnetic storms. When inspecting the different scatter plots, we can focus on different relationships between the quantities of this study:

- **Relation between dDst/dt and DST_prev.** The panel in the second row and first column shows the relation of dDst/dt and Dst value at the previous measurement, e.g., an hour before. Overall, the relation is linear with a negative slope, corresponding to Pearson's anticorrelation of -0.18. We also note that a number of points in the scatter plots are clearly divided but still aligned along a line with a negative slope. This separation indicates some thresholding phenomenon, such as when Dst reaches an increased value.
- **Relation between dDst/dt and P_{dyn}.** The scatter plot representing the relation of dDst/dt and P_{dyn} does not show any linear behavior (Pearson Correlation: 0.17), hinting at the presence of a non-linear coupling between the Dst evolution and the dynamic pressure. In empirical models, this relation typically follows a square root dependence.
- **Relation between dDst/dt and E_y.** When investigating the relation between Dst/dt and the convective electric field (for instance, the panel in the fourth row and first column), we note an overall linear dependence with a negative slope (Pearson correlation: -0.43). We also identify clear outliers for negative large temporal variations of the Dst index: this might indicate the presence of some thresholding event.
- **Relation between Dst/dt and P_B.** Similarly to the relation between dDst/dt and P_{dyn}, we find a non-linear relationship between the two quantities. However, we observe a larger number of outliers.

3.3   Symbolic Regression with PySR

We run PySR to discover mathematical models that capture the relationship between our input variables and $d\text{Dst}/dt$, which is the target. The overall methodology is shown and summarized in Fig. 3.

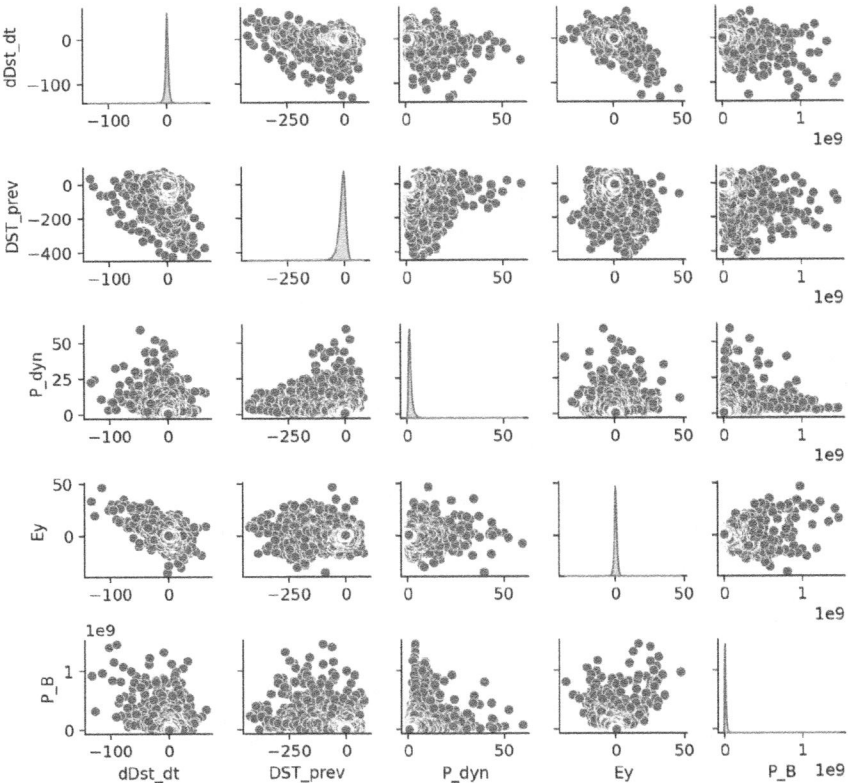

Fig. 2. Pairplot of the primary variables considered in this study: $d\text{Dst}/dt$, DST_prev, P_{dyn}, E_y, and P_B. The diagonal subplots display histograms or density plots for each variable, while the off-diagonal subplots show pairwise relationships.

The PySR regression algorithm searches the space of candidate expressions constructed using operators the user selects. In our study, we select the following operators which could be used in our $d\text{Dst}/dt$ model: $+$, $-$, $\times$, $\div$, max, min, exp, log, $\sqrt{\cdot}$, $(\cdot)^2$, and sign$(\cdot)$. The explored models are evaluated with an L1 loss function, also called Mean Absolute Error (MAE). Since L1 loss is based on absolute differences, it is less sensitive to outliers.

Two crucial concepts in PySR are the complexity and parsimony of an equation/model. The PySR complexity of an equation is based on its expression tree: each mathematical operation, constant, or variable contributes to the total complexity score, potentially with different weight, e.g. trigonometric functions have higher complexity weights than basic operations, such as sum. PySR uses an evolutionary search for symbolic regression and has two main parameters: the number of populations and iterations. The population size determines the number of different candidate solutions that exist in each generation. Increasing the population size improves exploration but also increases the computational cost. The

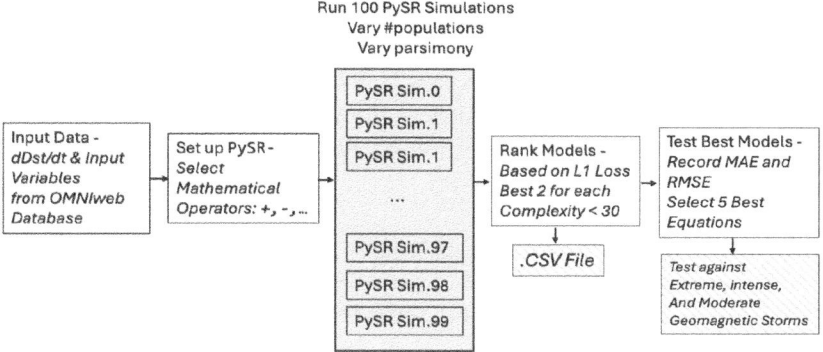

Fig. 3. A diagram showing the methodology for discovering mathematical models that predict dDst/dt. The data includes data input, symbolic regression with varying hyperparameters, and final model selection based on evaluation metrics.

number of iterations defines how many times the evolutionary process updates the population to find better equations. A high number of iterations allows the algorithm to refine solutions further. All the equations discovered have a PySR complexity value.

Parsimony is the preference for simpler mathematical expressions over more complex ones during the model fitting process. PySR implements parsimony through a complexity penalty, which is added to our L1 loss. The level of parsimony can be increased by setting a hyperparameter, which acts as a weight to the complexity penalty in the loss function.

The symbolic regression is performed over 100 independent PySR runs to account for the stochastic nature of evolutionary algorithms. During each optimization run, hyperparameters such as the parsimony coefficient (randomly selected from $[0.0, 0.9]$) and the population size (randomly chosen between 20 and 120) are varied. The test data set used for the regression test spans from January 1, 1995, to March 31, 2021.

After the 100 PySR runs, the candidate equations are extracted and filtered based on their loss and complexity metrics, neglecting all equations with complexity greater than 30. Duplicate equations across optimization runs are removed, and the remaining candidates are ranked according to their L1 loss, where lower values indicate a better fit to the test dataset. The final ensemble of candidate models is collected in a CSV file, providing an overview of the discovered equations for dDst/dt. The performance of each candidate equation is evaluated using the Root Mean Square Error (RMSE) and Mean Absolute Error (MAE), computed from May 1, 2021, to October 1, 2021. The Python code for the symbolic regression, the dataset from the OMNI database, and the discovered equations are available on GitHub[1].

---

[1] GitHub repository: https://github.com/smarkidis/Dst-Symbolic-Regression.

4   Physical Model Hierarchy from PySR

When analyzing the equations generated by PySR, we identify the best equations (with the lowest L1 norm) for different complexities. Table 1 shows how increasing equation complexity increases the mathematical description of dDst/dt and the physical description.

Table 1. Equations for dDst/dt with increasing complexity and their corresponding physical interpretation. The progression from a basic exponential decay to a fully coupled model shows how more complex equations introduce additional physics.

Comp.	Equation	Physical Interpretation
3	$-0.031\,Dst$	Basic exponential decay.
5	$-0.041\,Dst - Ey$	Adding Ey to account for solar wind forcing.
7	$-0.05\,Dst - \max\big(Ey, -0.16\big)$	Introducing thresholding to cap Ey's effect, simulating saturation phenomena.
9	$-0.062\,Dst$ $- \max\big(-0.062,\ Ey/0.638\big)$	Rescaling Ey, and refining the coupling efficiency and threshold limits.
10	$-0.057\,Dst$ $- \max\big(\sqrt{P_{\mathrm{dyn}}}\,Ey, -0.098\big)$	Incorporating $\sqrt{P_{\mathrm{dyn}}}$ to modulate Ey, to reflect the influence of solar wind dynamic pressure.
12	$\min\Big\{\big[-0.05\,Dst - Ey\big]$ $\times\sqrt{P_{\mathrm{dyn}}}, -0.055\,Dst\Big\}$	Balancing a dynamic pressuremodulated driver with pure decay, letting the dominant effect prevail.
19	$\big[-0.036(P_{\mathrm{dyn}} + Dst)$ $- \max(-0.008\,Dst, Ey)\big]$ $\times\sqrt{P_{\mathrm{dyn}} + 1.278} + 0.319$	Combining Dst and P_{dyn} with refined thresholding and a constant offset, representing a fully coupled system with background effects.

With complexity 3, the simplest model captures a basic exponential decay of the Dst index, representing a basic relaxation process. This term corresponds to the decay of the ring current over time due to various loss processes. Note that -1/0.031 corresponds to the decay constant found by the regression (1/0.031 h $\approx$ 32 h). The model begins to consider external forcing effects by introducing the solar wind electric field E_y in the equation with complexity 5. In particular, E_y represents the rate at which energy from the solar wind is injected into the Earth's magnetosphere, increasing the ring current. Further *physical* refinements are included in the equations with complexities 7 and 9: these incorporate thresholding nonlinearities, adjust the impact of E_y, and simulate saturation effects in the magnetospheric response. Including dynamic pressure P_{dyn} in complexity 10 modulates the influence of E_y. This reflects the increased energy transfer under varying solar wind conditions. Finally, the most complex model in Table 1 with complexity 19 combines these elements–coupling DST and P_{dyn}, applying thresholding, and including a baseline offset to include terms for modeling quiet times. Note that magnetic pressure is not included in any of the equations discovered by PySR. This suggests that the symbolic regression process identifies the primary geophysical drivers of dDst/dt, in alignment with physics-based models in space physics, where E_y and P_{dyn} are the dominant contributors to geomagnetic activity.

5   Evaluation and Ranking of Data-Driven DST/DT Models

To assess and rank the equations obtained from PySR, we carry out a comparative evaluation using a subset of the NASA OMNIweb dataset, spanning from May 1, 2021, to October 1, 2021. The tests use 2,000 randomly selected initial conditions. For each valid initial condition, a 48-hour prediction of the Dst index is generated using 11 different equations, ranging in complexity from 25 down to 15.

The 48-hour prediction is achieved by iteratively integrating the dDst/dt equation using an Euler method-like method. We start from an initial actual Dst value, calculate the dDst/dt at each hourly step by including the current values of E_y and P_{dyn}, rather than relying on their initial values. The performance of each equation is then quantified by computing the RMSE and MAE over these 48-hour intervals, providing an evaluation of the different models. Figure 4 shows the RMSE and MAE values for various data-driven models, discovered by PySR, categorized by their complexity (on the x-axis). The error bars represent the standard deviation of the different equations.

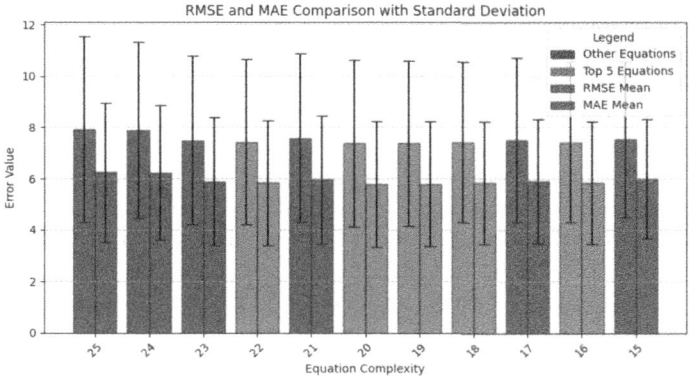

Fig. 4. RMSE and MAE comparison with standard deviation for data-driven Dst/dt models of different complexities. The error bars indicate the standard deviation across the test dataset. The red bars show the five best-performing models (lowest RMSE and MAE). (Color figure online)

Table 2. Equations for Dst/dt: BMR, OBM, and data-driven models (DDM) #1,#2, #3, #4, #5 with different levels of complexity (C).

Model	Equation
DDM#1 (C:19)	$\left[-0.036\,(P_{dyn}+Dst)-\max(-0.008\,Dst,\,Ey)\right]\sqrt{P_{dyn}+1.278}+0.319$
DDM#2 (C:20)	$\left[-0.042\,(P_{dyn}+Dst)-\max(0.168,\,Ey)\right]\times\sqrt{P_{dyn}+\max(0.097,\min(Ey,3.385))}+0.381$
DDM#3 (C:18)	$\min\left(-0.0443\,Dst,\,(0.621+\sqrt{P_{dyn}})\left[-0.0443\,Dst-\max(Ey,-0.728)\right]\right)+0.194$
DDM#4 (C:16)	$\min\left(-0.0443\,Dst,\,(0.621+\sqrt{P_{dyn}})\left[-0.0443\,Dst-Ey\right]\right)+0.194$
DDM#5 (C:22)	$\min\left(\sqrt{P_{dyn}+1.058}\left[-0.0434\,Dst-Ey\right],\,-0.0434\,Dst\right)+0.136\,(2.537-0.735\,P_{dyn})$
BMR	$-0.13\left(Dst-0.2\sqrt{P_{dyn}}+20.0\right)+\begin{cases}-5.4\,(Ey-0.5) & \text{if } Ey\geq 0.5,\\ 0 & \text{if } Ey<0.5\end{cases}$
OBM	$-\dfrac{1}{\tau}\left(Dst-0.2\sqrt{P_{dyn}}+20.0\right)+\begin{cases}-5.4\,(Ey-0.5) & \text{if } Ey\geq 0.5,\\ 0 & \text{if } Ey<0.5\end{cases},\qquad \tau=\begin{cases}7.7 & (Ey<0.5)\\ 3.5 & (Ey\geq 0.5)\end{cases}$

Among the models analyzed, the five `PySR` best-performing models using the test dataset, are indicated in red and explicitly presented in Table 2, which also includes the equations for the established dDst/dt model, BMR and OBM.

We use the five best-performing `PySR` models to predict the Dst evolution of geomagnetic storms and compare them to established models.

5.1 Performance of Data-Driven Models in Real Geomagnetic Storm Scenarios

To assess the accuracy of the five best data-driven models, discovered with `PySR` and presented in Table 2, in predicting geomagnetic storms, we test five data-driven models against the BMR and OBM models using actual storm events. The performance of these models was evaluated by comparing their predicted Dst with actual Dst values over 72-hour periods. The accuracy of each model was quantified using the RMSE and MAE metrics.

The first event we evaluate is the extreme geomagnetic storm that occurred on 29–30 October 2023. This geomagnetic storm has been widely studied in the literature and goes under the name of the Halloween Storm. Figure 5 shows the Dst prediction for the best `PySR` models and compares with BMR and OM models on the top panel, and RMSE and MAE in the bottom panel. When investigating the Dst prediction, we note that the BMR model overestimates the lowest Dst value during the geomagnetic storm, and data-driven models #4 and #5 do not correctly model the recovery from the second Dst minimum. The models with the lowest MAE errors are, in order, data-driven model #1 (complexity: 19) and data-driven model #2 (complexity: 20), followed by the OBM.

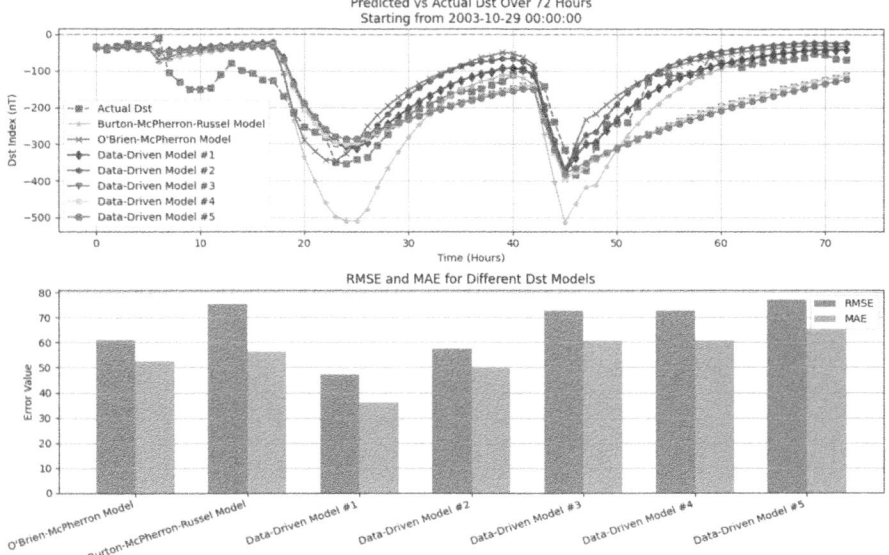

Fig. 5. Comparison of Dst predictions for the extreme Halloween Storm (October 29–30, 2003). The top panel shows the predicted vs. actual Dst, while the bottom panel presents RMSE and MAE errors.

As the second event, we select an intense/great geomagnetic storm that occurred on March 17, 2015, the so-called St. Patrick's Day Storm of 2015. The top panel of Fig. 6) shows the Dst prediction and good performance of all the data-driven models compared to the actual Dst. The BMR model underestimates the Dst minimum value as in the previous event, while the OBM does not correctly capture the recovery phase. In this Dst prediction, all the data-driven models outperform the BMR and OBM models.

Finally we asses the accuracy of different models against a moderate storm occurred on September 27, 2017.

Figure 7) shows the Dst prediction, along with the RMSE and MAE, for the different models. For this moderate storm, as in the case of intense storms, the data-driven models consistently outperform the BMR and OBM models. This is likely because the number of moderate and intense geomagnetic storms is higher than that of extreme geomagnetic storms in the training dataset. However, in extreme cases, such as the 2003 Halloween event, only two equations discovered by PySR provided better accuracy than the BMR and OBM models.

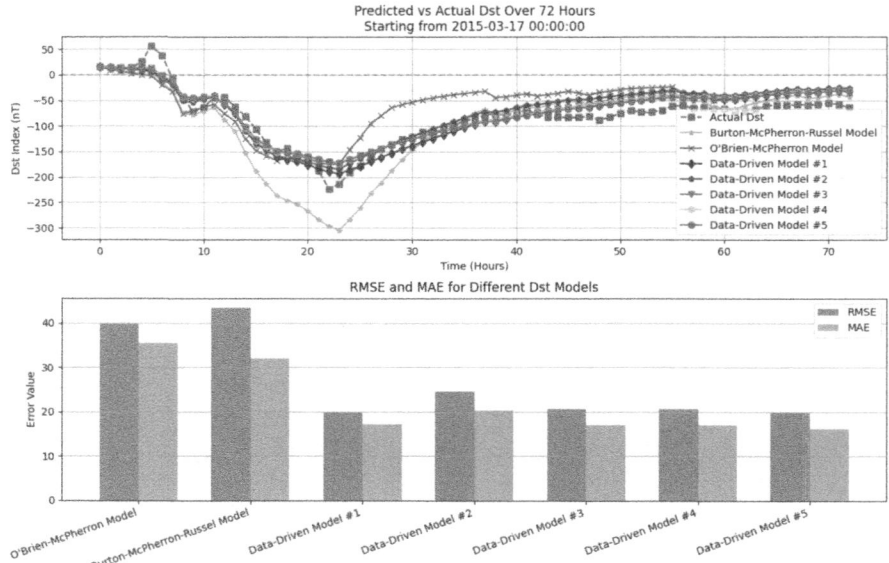

Fig. 6. Comparison of Dst predictions for the intense St. Patrick's Day Storm (March 17, 2015) in the top panel and MAE and RMSE errors for different models in the bottom panel. The data-driven models, found by `PySR`, outperform the BMR and OBM models in terms of MAE and RMSE.

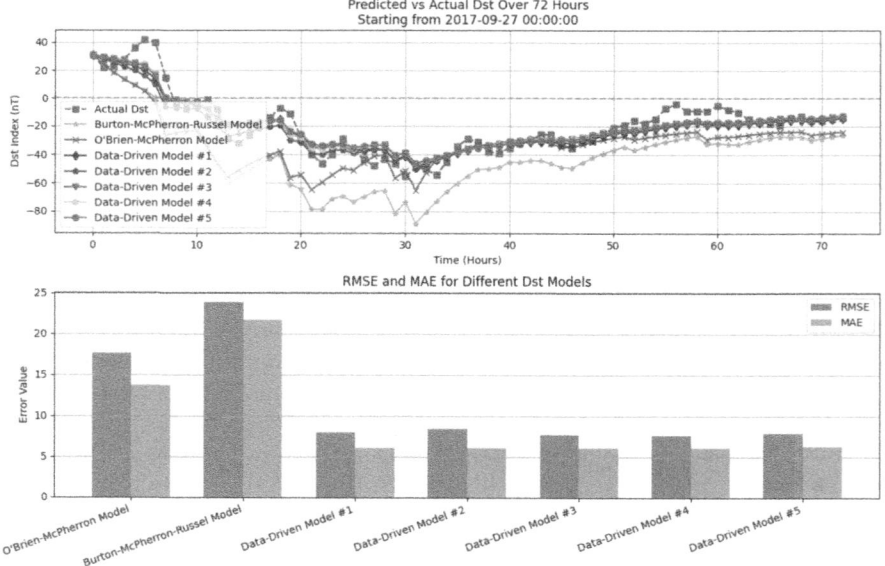

Fig. 7. Comparison of Dst predictions for the moderate geomagnetic storm on September 27, 2017 over 72 h. The data-driven models outperform the semi-empirical BMR and OBM models for moderate geomagnetic storms.

6 Conclusion

This work applied symbolic regression to derive data-driven equations for the temporal evolution of the Dst index, quantifying the evolution of geomagnetic storms. Using NASA OMNIweb data and `PySR`, we explored models and equations of increasing complexity and compared them with established empirical models for dDst/dt prediction. The methodology used `PySR` to identify interpretable expressions linking dDst/dt to solar wind parameters, IMF, and the Dst index itself.

The equations discovered by `PySR` showed increased accuracy compared to traditional empirical models, such as the BMR and OBM models. The best-performing data-driven models, ranked based on RMSE and MAE, captured non-linear dependencies, including thresholding effects and dynamic pressure modulation. A key aspect is that models discovered by `PySR` provide high performance while maintaining physical interpretability, unlike other AI-based methods that rely on neural networks.

Performance evaluation against real geomagnetic storms, including the 2003 Halloween Storm, the 2015 St. Patrick's Day Storm, and a moderate storm in 2017, showed that the data-driven equations consistently outperformed BMR and OBM in most cases. However, only the highest-ranked symbolic regression models demonstrated superior accuracy during extreme geomagnetic storms.

We found that symbolic regression can lead to physically interpretable models for geomagnetic storm prediction. The discovered equations provide physical insight and an alternative to black-box neural network approaches.

References

1. Baker, D.: What is space weather? Adv. Space Res. **22**(1), 7–16 (1998)
2. Borovsky, J.E., Shprits, Y.Y.: Is the DST index sufficient to define all geospace storms? J. Geophys. Res. Space Phys. **122**(11), 11–543 (2017)
3. Burton, R.K., McPherron, R., Russell, C.: An empirical relationship between interplanetary conditions and DST. J. Geophys. Res. **80**(31), 4204–4214 (1975)
4. Cid, C., Saiz, E., Cerrato, Y.: Physical models to forecast the DST index: a comparison of results. In: Solar Wind 11/SOHO 16, Connecting Sun and Heliosphere, vol. 592, p. 601 (2005)
5. Cranmer, M.: Interpretable machine learning for science with PYSR and symbolicregression.jl. arXiv preprint arXiv:2305.01582 (2023)
6. Gleisner, H., Lundstedt, H., Wintoft, P.: Predicting geomagnetic storms from solar-wind data using time-delay neural networks. In: Annales Geophysicae, vol. 14, p. 679 (1996)
7. Glocer, A., et al.: CRCM+ bats-r-us two-way coupling. J. Geophys. Res. Space Phys. **118**(4), 1635–1650 (2013)
8. Gonzalez, W., et al.: What is a geomagnetic storm? J. Geophys. Res. Space Phys. **99**(A4), 5771–5792 (1994)
9. Gruet, M.A., Chandorkar, M., Sicard, A., Camporeale, E.: Multiple-hour-ahead forecast of the DST index using a combination of long short-term memory neural network and gaussian process. Space Weather **16**(11), 1882–1896 (2018)

10. King, J., Papitashvili, N.: Solar wind spatial scales in and comparisons of hourly wind and ace plasma and magnetic field data. J. Geophys. Res.: Space Phys. **110**(A2) (2005)
11. Malin, S., Barraclough, D.R.: Humboldt and the earth's magnetic field. Q. J. R. Astron. Soc. **32**(3/SEP), 279 (1991)
12. Moldwin, M.: An Introduction to Space Weather. Cambridge University Press, Cambridge (2022)
13. O'Brien, T.P., McPherron, R.L.: Forecasting the ring current index DST in real time. J. Atmos. Solar Terr. Phys. **62**(14), 1295–1299 (2000)
14. O'Brien, T., McPherron, R.: Seasonal and diurnal variation of DST dynamics. J. Geophys. Res.: Space Phys. **107**(A11), SMP–3 (2002)
15. Stephens, T.: Genetic programming in Python with a scikit-learn inspired API: gplearn (2019). https://github.com/trevorstephens/gplearn
16. Sugiura, M.: Hourly values of equatorial DST for the IGY. Technical report (1963)
17. Udrescu, S.M., Tegmark, M.: AI Feynman: a physics-inspired method for symbolic regression. Sci. Adv. **6**(16), eaay2631 (2020)
18. Watanabe, S., Sagawa, E., Ohtaka, K., Shimazu, H.: Prediction of the DST index from solar wind parameters by a neural network method. Earth Planets Space **54**, e1263–e1275 (2002)

Combining Shape and Trajectory Features for Human Action Classification Using a Neural Network and Synthetic Data

Katarzyna Gościewska$^{(\boxtimes)}$ and Dariusz Frejlichowski

Faculty of Computer Science and Information Technology, West Pomeranian University of Technology, Żołnierska 52, 71-210 Szczecin, Poland
{kgosciewska,dfrejlichowski}@zut.edu.pl

Abstract. Human activity recognition systems using visual content analysis algorithms use data collected from a variety of sensors, the most popular of which are RGB cameras. Highly accurate motion information can be recorded using motion capture and then used to generate synthetic human body models. The advantage of such data is the absence of other objects in the background, the visualization accuracy and the anonymity of a person. This paper proposes a modified action recognition approach which creates action representations using simple features observed over time, including shape measurements, ratios and centroid trajectory. A feature vector consists of shape descriptors calculated separately for each video frame. These are then normalised, transformed to the frequency domain and supplemented with trajectory information. Action representations are classified using a feed-forward neural network with one hidden layer and varying number of hidden neurons. The high effectiveness values obtained in the experiments show that the appropriate composition of elementary features of moving objects brings considerable benefits.

Keywords: action classification · shape features · centroid trajectory · feed-forward neural network

1 Introduction

Human Activity Recognition (HAR) systems incorporate algorithms related to artificial intelligence, both classical machine learning algorithms and deep learning solutions. They benefit from many approaches, including pattern recognition, signal processing, video content analysis, neural networks and visual computing. HAR systems replace traditional observer-guarded surveillance systems, and simultaneously offer accelerated task execution and new functionalities, thanks to the availability of increased computational capabilities. Human movements vary in complexity—from elementary, single gestures and primitive actions to longer activities and behaviours [7]. A gesture can be, for example, a raise of the hand or a tilt of the head, while an action consists of multiple gestures performed

M. H. Lees et al. (Eds.): ICCS 2025, LNCS 15903, pp. 137–150, 2025.
https://doi.org/10.1007/978-3-031-97626-1_10

over a short period of time. Actions are characterised by simple movement patterns and certain poses of the human silhouette may occur cyclically [6]. If a silhouette shape is used, human movement is seen as a continuous change of human pose. Shapes extracted from video frames can be described by shape descriptors and used to create action representations. Classes of actions include, but are not limited to, walking, bending, waving or jumping.

Numerous attempts to classify action recognition methods have been presented, but despite a broad view of the state of the art, it is difficult to include all existing solutions. Many recent taxonomies distinguished between learning-based methods and traditional machine learning hand-crafted approaches, e.g. [17,20]. On the other hand, taking the type of input data as a criterion, the area of interest includes visual data extracted from video sequences, depicting human movement. Video sequences provide a lot of information, including the poses and location of the person over a period of time. It is possible to use low-cost, off-the-shelf RGB cameras (laptop, smartphone) that allow the video to be recorded without the need to carry any additional equipment such as wearable sensors. This is related to the proposed use case scenario, which matches current trends in human activity recognition, including automatic physical exercise recognition based on video sequences. For experimental purposes, when real data is limited, synthetic data may be used—namely visualizations of the human model based on points recorded using motion capture. A very large number of calculations is required to simulate experimental conditions, including visual data. However, the problem of classifying physical exercises can be solved without any direct human involvement.

Our previous publications take into account the use of the Weizmann dataset and the method of action recognition based on the combination of shape features of individual silhouettes and two-stage classification of actions in subgroups—this method is described in [12] and [11], among others. Here a modification of this approach is proposed. Object trajectories are used in a different way and the manner of constructing action descriptors is changed. The effectiveness of the modified approach is verified on a larger number of data than before, extracted from the AMASS dataset [16]—the classes correspond to selected physical exercises. The main idea of the experiments is to test numerous versions of the modified method in order to find its most effective variant for the action recognition procedure in the assumed application. A physical exercise is characterized more in terms of the type of human movement than the meaning or the purpose of a person's behaviour. As an example, the following physical exercises are recommended for the elderly [24]: aerobic exercise, e.g. walking, jumping jack (activities in which the body's large muscles move in a rhythmic manner); balance training, e.g. sideways, vertical jumping, jumping forward (activities increasing lower body strength); flexibility exercise, e.g., hand waving, bending (activities preserving or extending motion range around joints).

The rest of the paper is organised as follows: Sect. 2 presents selected related works concerning action recognition based on binary shapes and neural networks. Section 3 describes the modified approach for action recognition. Section 4 presents experiments and their results. Section 5 summarizes the paper.

2   Related Works

In recent years, interest in action recognition methods has shifted towards learning-based approaches. Convolutional neural network is one of the more commonly used model based on supervised learning. It is a hierarchical architecture with multiple hidden layers of different types, which process the input data into output categories (classes) [27]. The entire recognition process is carried out without the knowledge of an expert and features are extracted directly from data [25]. On the other hand, there are hand-crafted techniques, including traditional descriptors and classifiers prepared manually [1]. Examples of taxonomies separating hand-crafted methods from learning-based algorithms can be found in [20] or [17], among others. There are also hybrid methods combining hand-crafted features and neural network-based classifiers [15]. These approaches can achieve similar activity recognition rates as deep neural networks used alone, especially in tasks where a small amount of data is available [23].

One of the simpler examples of a unidirectional neural network is the multilayer perceptron, the use of which for activity recognition is presented in [5]. The input data are binary human silhouettes, extracted from a video sequence and described using a set of values based on the bounding rectangle, area and centre of gravity of the object. In [21] an action recognition method using two convolutional neural networks is proposed. The task of the first network is to evaluate each frame from the sequence and create a normalized histogram showing how often a certain class is indicated. In turn, the second network estimates the dense optical flow for consecutive frames, and the results of the analysis are stored in the form of a second histogram. After summing the histograms, the final classification result is obtained. In [9], a binary silhouette sequence is represented by a single image that contains a combination of binary masks extracted from the video frames. The masks are collected based on a weighted function that assigns progressively higher weights to more recent frames. Some areas of the silhouette are made lighter, and this is a way of representing the flow of movement.

In [23], three different shallow neural networks with one hidden layer and one output layer are experimentally tested, each differing in the type of cost function and activation function in the hidden layer. It is concluded that the relevant parameters of the most effective solution can be randomly searched for, one of the most important being the number of neurons in the hidden layer. This demonstrates that the use of a shallow neural network in the activity recognition task is effective when the availability of data is limited, such as when a relatively small dataset is involved. In [4], the results of a study are presented, that aims to create and evaluate the potential for automated, unsupervised monitoring of lower back and shoulder physiotherapy exercises performed at home. Key points corresponding to the positions of the joints are extracted from the video sequences (recorded using a smartphone camera). This positions on consecutive frames form a time series, the segments of which are used to train a convolutional neural network. In [8], a new action recognition scheme based on deep learning is proposed. Each detected human silhouette is described by eleven features extracted from its bounding box. An LSTM-type recurrent neural network is used for classification.

3   Modified Method for Action Recognition

In this paper, human action recognition is performed using a modification of the approach presented in [12] and [11]. It is a combination of image processing operations that use binary shape descriptors to create video sequence descriptors (see Fig. 1). The current version of the approach includes video preprocessing, therefore the input data can be RGB video sequences.

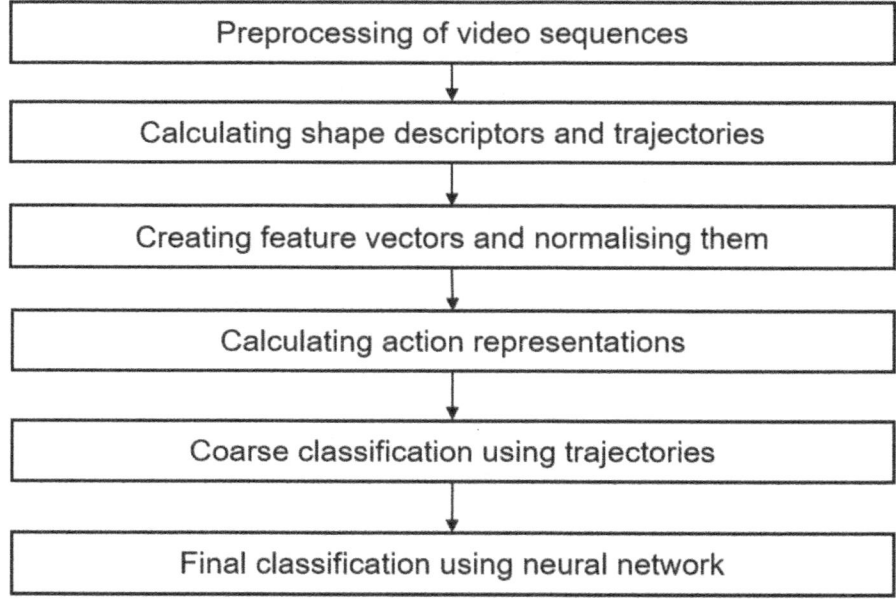

Fig. 1. Main steps of the modified action recognition approach.

3.1   Preprocessing of Video Sequences

Regardless of the type of video frames, the aim of preprocessing is to extract the area corresponding to the moving person in the foreground. Static background subtraction methods are applied if a background is simple. Resulting foreground masks are thresholded and converted to binary images—a foreground shape is white, and the background is black. A video sequence is stored as a set of binary foreground masks and denoted as $BM_i = \{bm_1, bm_2, ..., bm_n\}$, where $i = 1, 2, ..., n$ and n is the number of video frames.

3.2   Calculating Shape Descriptors and Trajectories

Each binary mask bm_i is individually represented using selected shape descriptor, sd_i. Various shape measurements and shape ratios are used for shape description. Since they are scalar values, a reduction of the two-dimensional binary image to a single number is obtained. Shape descriptors can be calculated based on any binary shape—a human silhouette, its convex hull or a bounding box enclosing all shape points. Four subgroups of simple shape descriptors are considered in this study, and these corresponds to the following shapes (shape descriptors are given in brackets):

1. A silhouette (area, perimeter, eccentricity [14], elongation [2], ellipticity ratio and circularity ratio [26], compactness [19], and Feret diameters including X Feret, Y Feret, XY Feret and Max Feret);
2. A convex hull, CH (area, perimeter, convexity [19], solidity [26], and the difference between the area of convex hull and the shape);
3. An axis-aligned minimum bounding rectangle, AABR (area, perimeter, horizontal side and vertical side as number of pixels, elongation [26], rectangularity [26]);
4. An arbitrarily oriented minimum bounding rectangle, OBR (area, perimeter, shorter side and longer side as number of pixels, elongation [26], rectangularity [26]).

At the same time, the coordinates of the centre of gravity of the foreground object are determined and the trajectory of the object is constructed. The trajectory is analysed within the field of view of the camera.

3.3   Creating Feature Vectors and Normalising Them

Shape descriptor values are concatenated into feature vectors. Each set of binary masks BM is now represented by a set of descriptors $SD_i = \{sd_1, sd_2, ..., sd_n\}$. Each feature vector is normalized using min-max normalization [13]. Values are scaled to a range—the smallest feature value is represented by 0 and the highest feature value corresponds to 1. This facilitates vector comparison and reveals the characteristics of feature variation over time. In addition, the influence of shape size, especially for basic shape measurements, is eliminated.

3.4   Deriving Action Representations

At this stage, the dataset consists of vectors of different lengths, and these vectors can be treated as signals. Therefore, the one-dimensional Fourier transform is applied for two main reasons—the number of resulting coefficients can be declared in advance and periodicity in the data is highlighted. The number of predefined coefficients equals 128 or 256. It is the closest power of 2 compared to the number of frames in the longest video sequence. A m-point one-dimensional discrete Fourier transform is computed to obtain action representations $AR = \{ar_1, ar_2, ..., ar_m\}$, where m is the predefined number of resultant Fourier coefficients. If m is greater than n, the feature vectors are zero-padded in the time domain, which corresponds to interpolation in the frequency domain [22]. Otherwise, the feature vectors are first truncated. The result of the transform is a Fourier spectrum with complex numbers—only the magnitude of the spectrum is used, and its values are normalised by dividing all the coefficients by the value of the first coefficient. For real signals, the Fourier spectrum is a two-sided spectrum, therefore the use of half of the spectrum is also tested.

3.5   Coarse Classification Using Trajectories

Each AR vector corresponds to a trajectory of the centre of gravity. The length of the trajectory is used to determine coarse classifications. If the length of the trajectory exceeds 20% of the video frame width, the corresponding AR vector is assigned to the group of actions performed with changing location, and is marked by appending the value 1. Otherwise, AR vectors belong to the group of actions performed in place and are appended with the value 0. Updated action representations are not divided into subgroups as previously [11], instead they are just labelled accordingly.

3.6   Final Action Classification Using Neural Network

Updated action representations are classified using a pattern recognition network, which is a feed-forward network with a single hidden layer and multiple hidden neurons. The dataset is divided into training, validation and testing sets. The layer initialisation function uses the Nguyen-Widrow algorithm, which generates a different weight and bias each time the function is called. By eliminating this randomness it is possible to focus on adjusting other parameters of the proposed approach—the shape descriptor and the number of neurons. Multiple initial experiments were carried out to check the validity of individual variables, which showed that high effectiveness is repeated for random initialisation parameters. Therefore, it is possible to use a stored set of random values in the experiments in order to provide as many fixed parameters as possible, and to ensure the reproducibility of the results. In addition to this, other training functions and different proportions of data partitioning were tested. In the end, the use of the conjugate gradient method and a data partitioning ratio of $70/15/15$ worked best. Due to the large number of possible parameter combinations, the main criterion for evaluating the results is the percentage effectiveness, which reflects the correspondence between the network's indications and the original action classes.

4 Experiments and Results

4.1 Dataset

The test data set is constructed based on examples selected from the AMASS dataset [16]—Archive of Motion Capture as Surface Shapes—which refers to a unified synthetic dataset generated from other datasets containing motion information captured by optical markers (motion capture, mocap). The mocap data are converted into realistic 3D meshes representing a model of the human body. Synthetic visualisations for more than 20 datasets have already been created within the AMASS framework and are available for research purposes [3]. Their main advantage is the availability of renderings in the form of video sequences containing individuals performing various activities.

For the purpose of this research, sequences rendered using the mocap data from the MoVi—Motion and Video dataset [10]—are selected. This dataset is a collection of 21 activity classes (20 predefined and one arbitrary) performed by 90 actors. The activities in the MoVi dataset [18] were recorded in a continuous manner, one after the other, however, in the AMASS dataset [3] they are already divided into video sequences of a few seconds. From the available classes, five were selected, and they include hand waving, walking, sideways, vertical jumping and jumping jacks—there are 75, 77, 71, 70 and 77 video sequences, respectively, and 370 in total. Each sequence depicts one person performing an action—the silhouette of this person is rendered based on the previously recorded mocap points. Video sequences used in the experiments are characterized by: a resolution of 2048×1600 pixels (scaled to 256×200), 24 frames per second, a duration of less than 10 s, a black background and a camera following a moving person. Examples of video frames for five selected action classes from the AMASS dataset are provided in Fig. 2.

Coordinates of the centre of gravity of the moving object are stored separately. Figure 3 shows example trajectories for five different classes—in this case, the values range from almost zero to around 3.5. The length of the trajectory is calculated as a distance between the most distant points (horizontal axis) using Euclidean distance. The differences are evident—actions with short trajectory (less than 1) are referred to as actions performed in place, while the long trajectories (more than 1) correspond to actions with changing location of a silhouette.

Video sequences from the dataset are converted into binary masks, as specified in the preprocessing step. These masks (see examples in Fig. 4) are used to calculate shape features. A single video sequence is represented as a cell array with dimensions $256 \times 200 \times n$, where n corresponds to the number of frames.

4.2 Description of the Experiments

The purpose of the experiments is to test the effectiveness of the modified action recognition method and to select its parameters in the proposed use case scenario. The research is carried out in the Matlab R2022b environment and using a laptop

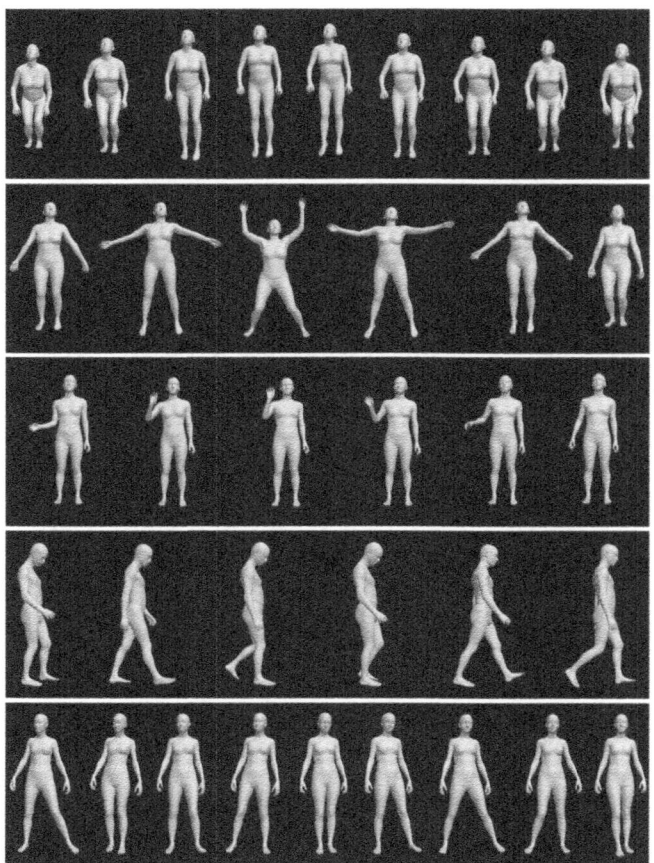

Fig. 2. Example frames corresponding to: vertical jumping, jumping jacks, hand waving, walking and sideways [16].

computer equipped with 32 GB of RAM. A set of binary images is used to prepare representations of actions, which are then classified using a neural network-based classifier. A number of experiments were carried out and each was repeated for all descriptors and varying number of neurons in a hidden layer.

The main experimental guidelines for a single experiment are summarised in the list below:

1. Binary masks are represented using shape descriptors and trajectories are calculated;
2. Feature vectors are created and values in each feature vector are normalised to a range [0–1];

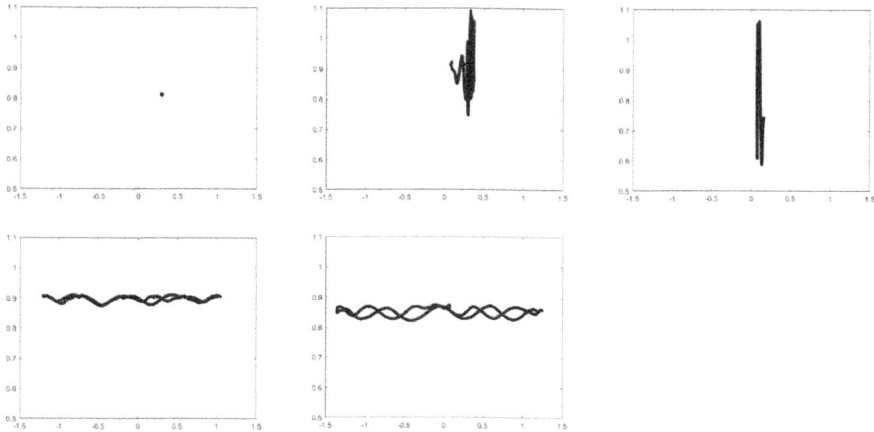

Fig. 3. Trajectories calculated for five example actions from the AMASS dataset, given in Fig. 2. The first row corresponds to hand waving, jumping jacks and vertical jumping, while the second row illustrates the trajectory of sideways and walking actions.

3. Feature vectors are transformed using the one-dimensional Fourier transform—the predefined number of the absolute spectrum coefficients equals 128 or 256, and all values in the vector are divided by the value of the first coefficient;
4. Action representations are created using resultant coefficients (whole or half of the spectrum), and a number is appended which determines the type of trajectory;
5. A simple feed-forward neural network is used for classification:
   (a) neural network input consists of 370 vectors of action representations;
   (b) for a single neural network input there is a single output representing the degree of similarity to classes under recognition;
   (c) there is one hidden layer with a variable number of neurons (1 to 50);
   (d) the activation function is based on the conjugate gradient method;
   (e) the input data is divided into training, validation and testing sets in a 70/15/15 ratio;
   (f) target classes are encoded into one-hot vectors and there are five unique classes;
   (g) the initialization parameters of a network are a set of random values that is usually generated again each time the network is trained. In order to be able to repeat the results of the experiments, a separate set of parameters is retained and used for all tests.

4.3   Results and Analysis

Through preliminary experiments, some parameters of the neural network-based classifier have been established. However, the number of possible versions of the modified action recognition approach is very large. With 28 descriptors

Fig. 4. Examples of binary masks corresponding to the following actions: walking, hand waving, vertical jumping, sideways and walking.

and 50 hidden layer sizes, this gives 1,400 combinations. Any change in action representation vectors will multiply this value. Hence, the effectiveness is taken as the main measure for evaluating classification results—it is a percentage of predicted classes that coincide with real class labels. Table 1 presents the experimental results for which the highest effectiveness is obtained. In all experiments the same action representation type is applied. For each shape descriptor, a series of tests is performed for a different number of hidden neurons, and then the value that corresponds to the highest effectiveness is selected (see Table 1).

According to the results provided in Table 1, the highest classification effectiveness is 99.72% in the experiment that used the perimeter of the convex hull. Such a result indicates that only one action is misclassified—in this case, an action from the class hand waving is incorrectly labelled as vertical jumping. In some other tests, the effectiveness exceeded 97%—this includes the area of the convex hull and the difference of the areas of the convex hull and the shape, as well as the area and perimeter of the axis aligned minimum bounding rectangle. This allows us to conclude that the calculation of shape descriptors on the basis of their bounding shapes is sufficient and improves the results. An important aspect that determines the quality of the results is the size of the action descriptor and the way it is prepared—the results discussed here were obtained thanks to action representations with the following properties:

1. The normalized feature vectors are transformed using the Fourier transform with a declared number of resulting coefficients equal to 256;
2. The real part of the spectrum is selected and normalized with respect to its first coefficient;
3. Action representation uses half of the spectrum and an indication regarding coarse classification is appended at the end of the vector, therefore each final action representation consists of 129 elements.

Table 1. The results of the best experiments with an indication of the effectiveness values and the number of hidden neurons.

Shape descriptors		Classification effectiveness	Number of neurons
Silhouette	Area	95,68 %	25
	Perimeter	92,97 %	48
	Eccentricity	91,62 %	46
	Elongation	94,32 %	32
	Ellipticity ratio	92,43 %	17
	Circularity ratio	94,86 %	47
	Compactness	94,59 %	35
	X Feret	94,32 %	47
	Y Feret	91,89 %	24
	XY Feret	91,35 %	38
	Max Feret	92,70 %	39
Convex hull (CH)	CH area	97,57 %	46
	CH perimeter	**99,73 %**	34
	CH convexity	95,95 %	19
	CH solidity	96,76 %	26
	CH area difference	97,84 %	21
Axis-aligned minimum bounding rectangle (AABR)	AABR area	97,57 %	31
	AABR perimeter	97,84 %	31
	AABR horizontal side	94,32 %	47
	AABR vertical side	91,89 %	24
	AABR elongation	92,70 %	42
	AABR rectangularity	94,59 %	44
Arbitrarily-oriented minimum bounding rectangle (OBR)	OBR area	95,68 %	21
	OBR perimeter	96,76 %	47
	OBR shorter side	94,05 %	41
	OBR longer side	88,11 %	43
	OBR elongation	94,59 %	31
	OBR rectangularity	95,14 %	44

5   Summary

This paper presents a modified approach to action recognition that combines simple shape features and a classifier based on a unidirectional neural network. Video sequences are converted into binary images and then into vectors of shape descriptors, whose values are scaled to the interval [0-1]. These vectors are then transformed using the one-dimensional Fourier transform. The magnitude of the spectrum is taken and then all spectrum coefficients are normalized with respect to the first coefficient. The object trajectory is used to encode a coarse classification labels, and final classification is performed using simple feed-forward neural network. Experimental studies prove that it is possible to obtain high classification effectiveness using small action representations (129 elements) that are simple to compute. In addition, the approach is adaptable and can be used in other applications, using both real and synthetic video data. Future work includes comparative studies using other neural network architectures, such as Recurrent Neural Networks, while using more activity classes and video sequence examples.

References

1. Aggarwal, C.C.: Machine learning with shallow neural networks. In: Aggarwal, C.C. (ed.) Neural Networks and Deep Learning, pp. 53–104. Springer, Cham (2018). https://doi.org/10.1007/978-3-319-94463-0_2
2. Aktaş, M.A., Žunić, J.: Measuring shape ellipticity. In: Real, P., Diaz-Pernil, D., Molina-Abril, H., Berciano, A., Kropatsch, W. (eds.) Computer Analysis of Images and Patterns, pp. 170–177. Springer, Heidelberg (2011)
3. AMASS: AMASS: Archive of Motion Capture as Surface Shapes (2019). https://amass.is.tue.mpg.de/download.php. Accessed 20 May 2024
4. Arrowsmith, C., Burns, D., Mak, T., Hardisty, M., Whyne, C.: Physiotherapy exercise classification with single-camera pose detection and machine learning. Sensors **23**(1) (2023). https://doi.org/10.3390/s23010363
5. Babiker, M., Khalifa, O.O., Htike, K.K., Hassan, A., Zaharadeen, M.: Automated daily human activity recognition for video surveillance using neural network. In: 2017 IEEE 4th International Conference on Smart Instrumentation, Measurement and Application (ICSIMA), pp. 1–5 (2017). https://doi.org/10.1109/ICSIMA.2017.8312024
6. Borges, P., Conci, N., Cavallaro, A.: Video-based human behavior understanding: a survey. IEEE Trans. Circuits Syst. Video Technol. **23**(11), 1993–2008 (2013)
7. Chaaraoui, A.A., Climent-Pérez, P., Flórez-Revuelta, F.: A review on vision techniques applied to human behaviour analysis for ambient-assisted living. Expert Syst. Appl. **39**(12), 10873–10888 (2012). https://doi.org/10.1016/j.eswa.2012.03.005
8. Cob-Parro, A.C., et al.: A new framework for deep learning video based human action recognition on the edge. Expert Syst. Appl. **238**, 122220 (2024). https://doi.org/10.1016/j.eswa.2023.122220
9. Dobhal, T., Shitole, V., Thomas, G., Navada, G.: Human activity recognition using binary motion image and deep learning. Procedia Comput. Sci. **58**, 178–185 (2015). https://doi.org/10.1016/j.procs.2015.08.050. Second International Symposium on Computer Vision and the Internet (VisionNet 2015)

10. Ghorbani, S., et al.: MoVi: a large multi-purpose human motion and video dataset. PLoS ONE **16**(6), e0253157 (2021)

11. Gościewska, K., Frejlichowski, D.: A combination of moment descriptors, Fourier transform and matching measures for action recognition based on shape. In: Krzhizhanovskaya, V.V., et al. (eds.) ICCS 2020. LNCS, vol. 12138, pp. 372–386. Springer, Cham (2020). https://doi.org/10.1007/978-3-030-50417-5_28

12. Gościewska, K., Frejlichowski, D.: The analysis of shape features for the purpose of exercise types classification using silhouette sequences. Appl. Sci. **10**(19) (2020). https://doi.org/10.3390/app10196728

13. Han, J., Kamber, M., Pei, J.: 3: data preprocessing. In: Han, J., Kamber, M., Pei, J. (eds.) Data Mining, 3rd edn., pp. 83–124. The Morgan Kaufmann Series in Data Management Systems, Morgan Kaufmann, Boston (2012). https://doi.org/10.1016/B978-0-12-381479-1.00003-4

14. Haralick, R.M., Shapiro, L.G.: Computer and Robot Vision. Addison-Wesley Longman Publishing Co., Inc., USA (1992)

15. Khan, M.A., Sharif, M., Akram, T., Raza, M., Saba, T., Rehman, A.: Hand-crafted and deep convolutional neural network features fusion and selection strategy: an application to intelligent human action recognition. Appl. Soft Comput. **87**, 105986 (2020). https://doi.org/10.1016/j.asoc.2019.105986

16. Mahmood, N., Ghorbani, N., Troje, N.F., Pons-Moll, G., Black, M.J.: AMASS: archive of motion capture as surface shapes. In: International Conference on Computer Vision, pp. 5442–5451 (2019)

17. Morshed, M.G., Sultana, T., Alam, A., Lee, Y.K.: Human action recognition: a taxonomy-based survey, updates, and opportunities. Sensors **23**(4) (2023). https://doi.org/10.3390/s23042182

18. MoVi: MoVi: A Large Multipurpose Motion and Video Dataset (2020). https://doi.org/10.5683/SP2/JRHDRN. https://borealisdata.ca/dataset.xhtml?persistentId=. Accessed 20 May 2024

19. Peura, M., Iivarinen, J.: Efficiency of simple shape descriptors. In: Arcelli, C., Cordella, L.P., di Baja, G.S. (eds.) Advances in Visual Form Analysis, pp. 443–451. World Scientific (1997)

20. Saif, S., Tehseen, S., Kausar, S.: A survey of the techniques for the identification and classification of human actions from visual data. Sensors **18**(11) (2018). https://doi.org/10.3390/s18113979

21. Silva, V., Vidal, F., Romariz, A.: Human action recognition based on a two-stream convolutional network classifier. In: 2017 16th IEEE International Conference on Machine Learning and Applications (ICMLA), pp. 774–778 (2017). https://doi.org/10.1109/ICMLA.2017.00-64

22. Smith, J.: Mathematics of the Fourier Transform (DFT): With Audio Applications. W3K Publishing, BookSurge Publishing (2007)

23. Suto, J., Oniga, S.: Efficiency investigation from shallow to deep neural network techniques in human activity recognition. Cogn. Syst. Res. **54**, 37–49 (2019). https://doi.org/10.1016/j.cogsys.2018.11.009

24. Thaxter-Nesbeth, K., Facey, A.: Exercise for healthy, active ageing: a physiological perspective and review of international recommendations. West Indian Med. J. **67**(5), 351–356 (2018). https://doi.org/10.7727/wimj.2018.177

25. Ullah, H.A., Letchmunan, S., Zia, M.S., Butt, U.M., Hassan, F.H.: Analysis of deep neural networks for human activity recognition in videos–a systematic literature review. IEEE Access **9**, 126366–126387 (2021). https://doi.org/10.1109/ACCESS.2021.3110610

26. Yang, M., Kpalma, K., Ronsin, J.: A survey of shape feature extraction techniques. Pattern Recogn. 43–90 (2008)
27. Yao, G., Lei, T., Zhong, J.: A review of convolutional-neural-network-based action recognition. Pattern Recogn. Lett. **118**, 14–22 (2019). https://doi.org/10.1016/j. patrec.2018.05.018

Improving Object Detection Quality in Football Through Super-Resolution Techniques

Karolina Seweryn[1]([✉])(iD), Gabriel Chęć[2], Szymon Łukasik[1,2](iD), and Anna Wróblewska[3](iD)

[1] NASK - National Research Institute, Warsaw, Poland
karolina.seweryn@nask.pl
[2] AGH University of Krakow, Krakow, Poland
[3] Faculty of Mathematics and Information Science,
Warsaw University of Technology, Warsaw, Poland

Abstract. This research examines the effectiveness of super-resolution techniques in improving object detection accuracy in football. Given the sport's fast pace and the need for precise tracking of players and the ball, super-resolution can offer significant improvements. The study applies super-resolution techniques to SoccerNet football videos and evaluates their impact on Faster R-CNN detection accuracy. Findings reveal a significant boost in object detection accuracy following the application of super-resolution preprocessing. Enhancing object detection by integrating super-resolution techniques provides substantial advantages, particularly in low-resolution settings, with a 12% rise in mean Average Precision (mAP) at an IoU (Intersection over Union) range of 0.50:0.95 for 320×240 pixel images when the resolution is quadrupled using RLFN. As the image dimensions grow, the extent of improvement becomes less pronounced; however, a consistent enhancement in detection quality remains clear. Moreover, the implications of these results for real-time sports analytics, player tracking, and the overall viewing experience are discussed.

Keywords: object detection · super-resolution · low-quality videos

1 Introduction

The tracking of players and the ball in football matches is a critical aspect of sports analytics, significantly impacting the tactical elements of the game. Player tracking data provides valuable insights into individual and team performance, enabling coaches to analyze movements, formations, and strategies more effectively [5,16]. For instance, tracking data can be used to assess player fitness, monitor fatigue, and reduce the risk of injury by understanding players' physical demands during a match. Furthermore, ball-tracking technology plays a key role in enhancing the spectator experience, contributing to more accurate decisions through technologies like goal-line technology and video assistant referee (VAR) [31].

M. H. Lees et al. (Eds.): ICCS 2025, LNCS 15903, pp. 151–163, 2025.
https://doi.org/10.1007/978-3-031-97626-1_11

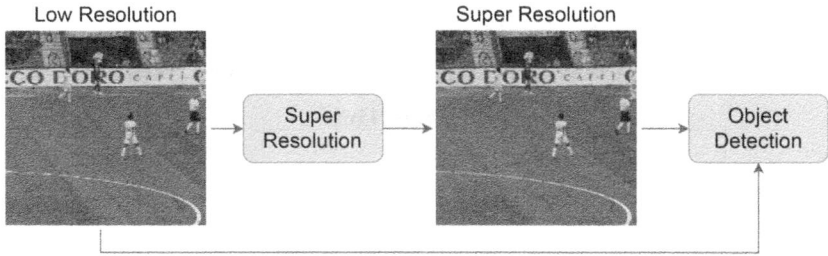

Fig. 1. An image is enhanced by a super-resolution model and fed to an object detector.

Despite the wealth of research and annual publications on tracking football players and the ball [16–18,20,21], the precise tracking of small, fast-moving objects like a football ball remains a challenge. These advanced tracking solutions frequently depend on the availability of high-definition video input to ensure appropriate results. However, a significant portion of football clubs, ranging from grassroots clubs to professional organizations, struggle with the provision of such high-quality data. This situation is not solely a consequence of the prohibitive costs associated with state-of-the-art recording equipment. Still, it is also exacerbated by the prevalent use of drones and portable devices that, while offering versatility and convenience, often compromise on video quality. Additionally, the widespread practice of compressing video content for distribution on platforms like YouTube or other streaming services further degrades the fidelity of footage available for analytical purposes, presenting substantial hurdles to the deployment and effectiveness of sophisticated tracking models in real-world settings. This disparity in technological access not only affects the competitiveness of smaller clubs but also limits the overall development of sports analytics by constraining data diversity and research opportunities.

While high-quality video costs may decrease in the future, alternative solutions remain valuable. This research explores the potential of super-resolution techniques to enhance object detection in football, reducing dependence on high-resolution input data. By integrating super-resolution with modern detection models, teams can lower costs, minimize data storage requirements, and eliminate the need for expensive equipment. These methods not only enhance detection performance on low-resolution footage but also improve accuracy even when high-resolution images are available, making super-resolution a valuable preprocessing step.

2   Related Work

Initial approaches to extracting tracking data relied on views from multiple cameras [6,7,15,17] for precise player localization. While effective, these approaches came with significant drawbacks, including high installation and maintenance costs, as well as complex technical challenges such as camera calibration and object re-identification [7]. These limitations made widespread adoption difficult, particularly for teams with limited resources.

More and more latest models find the coordinates of players and the ball based on just one camera [8,9]. Still, locating small objects like the ball remains a challenge due to occlusion, false detection, lighting, and frame rate variations [10]. Also, interactions between players, and players and the ball can cause complex problems. Additionally, solutions for detecting players and a ball often rely on high-resolution input videos or zoom-in views and their trajectory imputations, i.e. inferring their locations based on adjacent frames in which their detections were feasible [11]. Tracking in a football match can be classified as a multi-object tracking task, and it remains challenging due to factors like abrupt object appearance differences and even severe object shadings and obscurations [22]. The latest solutions enhance neural architectures to capture small objects, e.g. adding focal loss to YOLOv7 [23], or using neural radiance fields [24].

A similar issue of tracking football players in low-quality videos has been addressed in the works [21] and [19]. In both cases, the researchers focus on adapting systems to challenging visual conditions typical of football broadcasts. The authors of [21] adapt advanced multi-object tracking systems for use with low-quality videos. At the same time, the other research concentrates on detecting and tracking small, less visible players without the need for manual data annotation. All those approaches highlight the increasing capability of sports analytics to handle visual challenges in football.

In recent years, super-resolution approaches have been significantly improved [12,13]. Currently, a bunch of research copes with applying and adapting super-resolution techniques to many detection tasks, such as satellite imagery [25], underwater object detection [27], and occluded small commodities [26]. In [33], the authors explored the effects of using super-resolution as a preprocessing step in object classification. They demonstrated that this approach significantly improves the detection of small objects. However, the analysis of performance in this work is somewhat limited, as it focuses solely on the number of true positives and the mean probability returned by the model. In sports analytics, only one study has explored super-resolution for ball detection, focusing on tennis [28]. Instead of integrating super-resolution into the model, it was used as a preprocessing step to enhance frame resolution before labeling, improving annotation accuracy. This approach markedly diverges from the use of super-resolution techniques within model architectures, as it was applied to refine the input data before annotation.

3 Experiments

Data. Our study uses the SoccerNet dataset [3], consisting of 12 full football matches recorded from the main camera in 1080p (1920×1080). It includes 100 clips, each 30 s long, spanning various games and seasons to ensure a diverse evaluation of object detection methods. The dataset is split into 42,750 training frames and 36,750 test frames, with no match overlap. We followed this division, using the training set for model training and the test set for evaluation. The dataset classifies on-field elements into 8 categories (player team left, player team right, goalkeeper team left, goalkeeper team right, main referee, side referee, staff, and ball) for object detection. For simplicity, our experiments grouped them into two: ball and person.

For super-resolution model training, we used the SoccerNet train set and UHDSR8K [14], which includes 2,100 images in 4K resolution. This dataset's high resolution offers a different range of challenges and opportunities for super-resolution model training. The UHDSR8K dataset is part of a study that aimed to benchmark single image super-resolution (SISR) methods on Ultra-High-Definition (UHD) images, including both 4K and 8K resolutions. It was used to evaluate the performance of SISR methods under various settings, contributing to the development of baseline models for super-resolution. In our experiments, we specifically utilized the 4K images from this dataset.

Super-Resolution. Choosing the super-resolution network architecture for image reconstruction affects its quality, speed, and efficiency. The architecture selection in this study was based on the NTIRE 2022 Challenge on Efficient super-resolution results [1]. The challenge emphasized designing networks capable of efficiently converting single images to higher resolutions, with a focus on metrics like runtime, parameter count, floating-point operations, and memory usage while maintaining or exceeding a Peak Signal-to-Noise Ratio (PSNR) [29] of 29 on the DIV2K dataset, where PSNR measures the quality of image reconstruction. The winning architecture, Residual Local Feature Network (RLFN) [4] was chosen for this study due to its compact size, fast learning, and effective high-resolution reconstruction capabilities. RLFN combines convolutional layers with attention mechanisms to extract local features and maintain global pixel relationships, crucial for detailed and accurate high-resolution image reconstruction.

Object Detection. We used Faster-RCNN [2] with ResNet50 backbone for object detection, chosen for its pre-trained availability in PyTorch, widespread adoption, and robust performance across diverse datasets. Faster R-CNN works by integrating two key components: a Region Proposal Network (RPN) and a Fast R-CNN detector. The RPN first scans the input image with a sliding window approach to generate object proposals, identifying regions that most likely contain an object. These proposals are then passed to the Fast R-CNN detector, which extracts features using a convolutional neural network, classifies the objects within these regions, and refines their bounding boxes for precise localization.

Experimental Setup. In our experiments using the SoccerNet dataset, images from successive frames were degraded to lower resolutions by reducing their original dimensions of 1920×1080 by factors of 2, 3, 4, and 6. For UHDSR8K images with an original size of 3840×1920, similar reductions were applied.

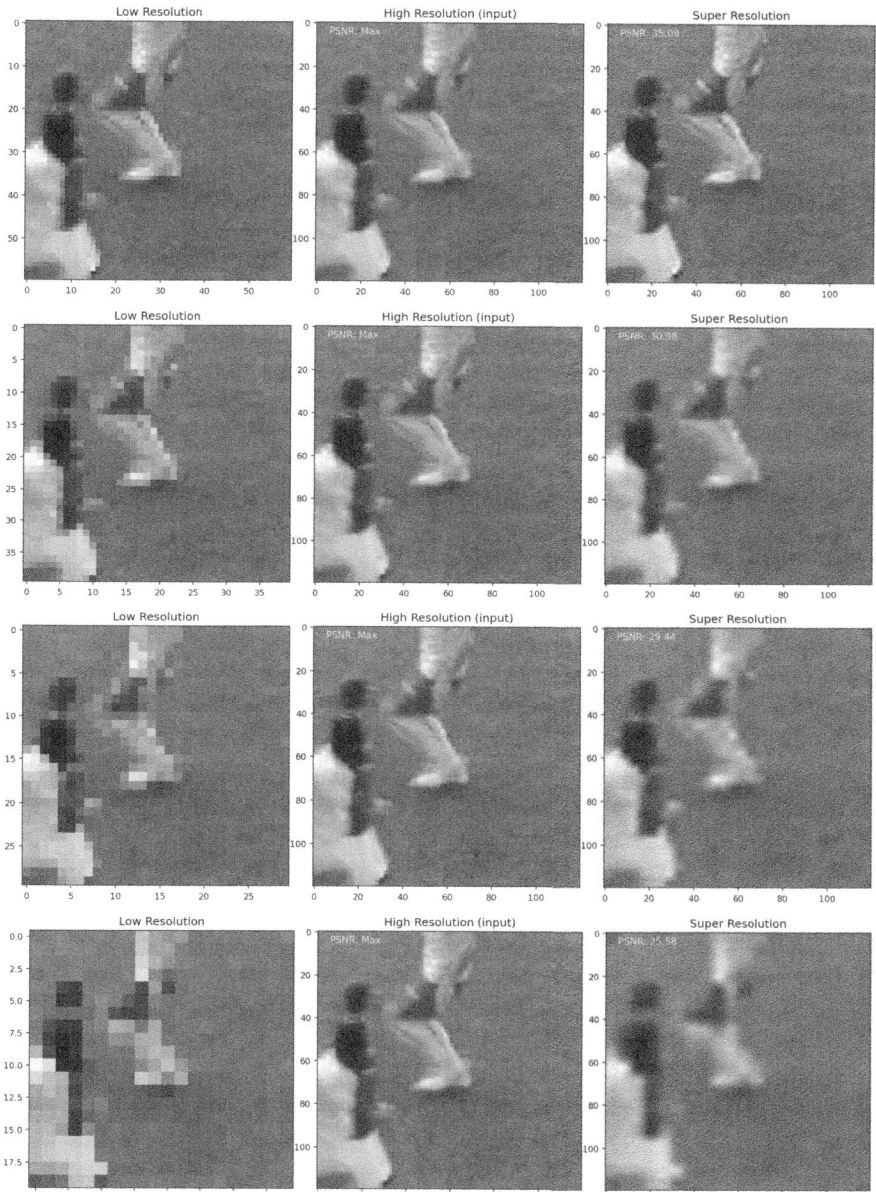

Fig. 2. Sequential visualizations of super-resolution enhancements showcasing progressive upscaling factors. Rows of images illustrate the enhancements achieved through super-resolution RLFN techniques for scales of x2, x3, x4, and x6, respectively. The middle column displays the original image (ground truth image), the left column shows the low-resolution version, and the right column presents the enhanced result after applying super-resolution to the low-resolution image.

We tested the object detection performance of Faster R-CNN on these low-resolution images and images enhanced by the super-resolution network RLFN, trained on both SoccerNet and UHDSR8K datasets (see Fig. 1). Experiments were conducted using NVIDIA A100-40GB GPUs. Training of super-resolution models and object detection was performed with a batch size of 32. The super-resolution model was trained with a learning rate ranging from 1e-5 to 0.01 using the Adam optimizer [32].

4   Results

Figure 2 presents the results of the RLFN model applied to a SoccerNet test sample. The middle column illustrates the original input image, which was resized by factors of 2, 3, 4, and 6 (left), followed by processing through the super-resolution network (right). With a small - double reduction, the network performs well in reconstructing the original image. A PSNR value over 35 indicates a high quality of reconstruction. However, the smaller the dimensions of the low-resolution image, the more challenging it becomes to replicate the input. In the last case - a sixfold reduction - the PSNR drops to 25.58. This example demonstrates that the network is capable of fairly accurately reproducing objects despite having highly blurred pixels as input.

Table 1. Impact of the input image shape on detection performance of Faster RCNN.

Image Shape	mAP		mean IoU	
	IoU $= 0.5{:}0.95$	IoU $= 0.5$	$\tau_{IoU} = 0.5$	$\tau_{IoU} = 0.9$
(320×240)	24.3	46.7	78.7	92.7
(640×480)	27.6	52.3	79.6	92.8
(720×576)	27.5	51.8	79.6	92.8
(1280×720)	29.5	56.3	79.9	92.9
(1920×1080)	29.5	56.0	79.9	92.9

Table 1 presents the results of Faster R-CNN model tested on SoccerNet with various image shapes. Increasing the input size improves mAP, with a sixfold increase (from 320×240 to 1920×1080) boosting mAP by over 21% at IoU 0.50:0.95 and around 20% at mAP@IoU=0.50. The average IoU remained consistent across various thresholds. These results confirm that higher image resolution enhances detection quality.

After training super-resolution models, we evaluated their performance on SoccerNet using PSNR and MSE metrics (Table 2). The analysis reveals that larger reductions in image size (denoted as xN) before input into the super-resolution model correlate with increased challenges in image reconstruction. Notably, the RLFN model trained on the SoccerNet dataset achieved higher

Table 2. Quality of super-resolution models on SoccerNet and UHDSR8K. 'x[N]' indicates that the image's dimensions were downscaled by a factor of N before processing by the super-resolution network. ↑ - the higher the better, ↓ - the lower the better.

	Trained on UHDSR8K		Trained on SoccerNet	
SR	PSNR ↑	MSE ↓	PSNR ↑	MSE ↓
RLFN x2	33.15	36.47	34.63	26.44
RLFN x3	29.73	79.43	30.54	66.76
RLFN x4	27.83	119.82	29.02	92.82
RLFN x6	25.48	204.47	26.45	166.41

Table 3. Detection performance (mAP) of Faster R-CNN model for various super-resolution scales and training datasets.

Input Shape	Train Dataset	Output Shape		mAP		
				@IoU = 0.50:0.95	@IoU = 0.50	@IoU = 0.75
320×240	-	-	-	24.3	**46.7**	22.7
	SoccerNet	640×480	x2	26.0	39.8	31.6
		960×720	x3	26.9	40.8	31.9
		1280×960	x4	**27.3**	41.4	**32.2**
		1920×1440	x6	26.6	40.8	31.2
	UHDSR8K	640×480	x2	24.7	40.2	27.3
		960×720	x3	24.9	39.7	28.8
		1280×960	x4	25.3	40.3	29.1
		1920×1440	x6	27.2	41.3	31.9
640×480	-	-	-	27.6	**52.3**	26.4
	SoccerNet	1280×960	x2	27.8	40.1	30.6
		1920×1440	x3	28.1	40.3	31.1
		2560×1920	x4	**28.5**	40.5	31.6
		3840×2880	x6	28.3	40.4	**32.1**
	UHDSR8K	1280×960	x2	28.3	40.9	31.7
		1920×1440	x3	28.0	40.1	31.6
		2560×1920	x4	**28.5**	40.3	**32.2**
		3840×2880	x6	28.4	40.0	**32.2**
1280×720	-	-	-	29.5	**56.3**	27.5
	SoccerNet	2560×1440	x2	30.1	42.0	34.3
		3840×2160	x3	30.1	41.8	33.5
		5120×2880	x4	30.3	42.4	34.0
		7680×4320	x6	29.8	41.8	32.8
	UHDSR8K	2560×1440	x2	30.4	42.8	34.3
		3840×2160	x3	30.3	42.8	34.3
		5120×2880	x4	**30.6**	42.6	**34.5**
		7680×4320	x6	29.6	41.8	32.2

PSNR values on its test set, underscoring the advantage of training on images akin to the target dataset for optimal results. A PSNR over 30 is regarded as good quality. Thus, the RLFN models with x2 and x3 reductions on SoccerNet, and the x2 reduction on UHDSR8K, met the criterion of good quality.

Table 4. Mean IoU results of Faster R-CNN model for various super-resolution scales and training datasets, evaluated at $\tau_{IoU} = 0.5$ and $\tau_{IoU} = 0.9$. τ_{IoU} denotes the IoU threshold that the model uses to determine if an object is predicted correctly.

Input Shape	Train Dataset	Output Shape	Upscale Factor	mean IoU	
				$\tau_{IoU} = 0.5$	$\tau_{IoU} = 0.9$
320×240	-	-	-	78.7	92.7
	SoccerNet	640×480	x2	81.1	91.9
		960×720	x3	81.6	92.6
		1280×960	x4	81.8	92.7
		1920×1440	x6	81.1	**93.1**
	UHDSR8K	640×480	x2	80.2	92.2
		960×720	x3	80.4	91.7
		1280×960	x4	80.8	91.8
		1920×1440	x6	**82.1**	92.5
640×480	-	-	-	79.6	92.8
	SoccerNet	1280×960	x2	83.7	92.9
		1920×1440	x3	83.9	92.7
		2560×1920	x4	84.0	**93.0**
		3840×2880	x6	83.8	92.8
	UHDSR8K	1280×960	x2	83.7	92.9
		1920×1440	x3	84.1	92.8
		2560×1920	x4	84.2	**93.0**
		3840×2880	x6	**84.3**	92.8
1280×720	-	-	-	79.9	92.9
	SoccerNet	2560×1440	x2	84.5	92.9
		3840×2160	x3	84.2	92.9
		5120×2880	x4	84.2	92.9
		7680×4320	x6	84.2	93.0
	UHDSR8K	2560×1440	x2	84.4	92.8
		3840×2160	x3	84.3	93.0
		5120×2880	x4	**84.8**	92.9
		7680×4320	x6	84.3	**93.1**

Table 3 and Table 4 present our findings on the impact of super-resolution techniques on object detection. We evaluated an object detection model in its original form, without any additional training on the dataset. Applying super-resolution to 320×240 images leads to an approximate 12% improvement in the mAP @IoU=0.50:0.95 metric, while the mAP@IoU=0.50 metric shows a decline. A similar trend is observed with 640×480 images, where the mAP @IoU=0.50:0.95 metric increased from 27.6 to 28.5 and 1280×720, where we observed an increase from 29.5 to 30.6. Super resolution can reduce performance at lower IoU thresholds (e.g., 0.50), indicating a trade-off between detection sensitivity and localization accuracy. The mAP@IoU=0.75 results further confirm that super-resolution primarily benefits higher-precision localization, especially for very low-resolution inputs (e.g., 320×240), but also brings noticeable improvements at higher resolutions—for instance, from 26.4 to 32.1 and from 27.5 to 34.3. Super-resolution also enhanced average IoU values across evaluated thresholds, improving bounding box precision. However, increasing image size does not always yield better detection performance. A sixfold upscaling did not show a clear advantage over a fourfold increase, possibly due to distortions introduced by the super-resolution network. Nonetheless, super-resolution consistently outperformed the baseline without image enhancement.

Although Table 2 suggests that super-resolution models trained on the SoccerNet dataset have superior quality, the relation to their detection efficacy remains ambiguous. The detector's performance was similar regardless of the training dataset. When upscaling 320×240 images by six times, the UHDSR8K-trained model achieved a higher mAP (27.2 vs. 26.6 for SoccerNet). Interestingly, even for 1280×720 images quality improved, but the most significant detection enhancement occurred at 320×240, with a 12.3% mAP@IoU=0.50:0.95 increase, compared to 3.3% for medium and 3.7% for large images.

In Table 5, the impact of classification of each class is considered for the original image with dimensions of 320×240. Notably, the table reveals that the application of super-resolution enhances the detection of players (person class). However, the detection of smaller objects, such as a ball, continues to pose challenges; in the original image, the mean Average Precision (mAP) for the ball was already low at 11. After applying super-resolution, this metric declined further to 0. Detecting small objects remains a significant hurdle, as discussed in [34], and unfortunately, the proposed method did not yield improvements in the detection quality of small objects.

Table 5. Detection performance of Faster R-CNN model for various super-resolution scales trained on SoccerNet for input of shape 320×240.

Output Shape	Scale	mAP @IoU = 0.50		mAP @IoU = 0.50:0.95	
		Person	Ball	Person	Ball
-	-	45.1	**3.5**	82.3	**11.2**
(640, 480)	x2	52.0	0.0	79.7	0.0
(960, 720)	x3	53.9	0.0	81.7	0.0
(1280, 960)	x4	**54.6**	0.0	**82.8**	0.0
(1920, 1440)	x6	53.1	0.0	81.6	0.0

5 Discussions and Conclusions

Our study highlights the crucial role of advanced super-resolution (SR) techniques in improving object detection in football. Integrating SR methods with object detection models resulted in 12% increase in mAP for low-quality images. We also highlights the importance of selecting a suitable dataset for training SR models. Our analysis revealed that training the SR algorithm with football-specific data slightly enhanced the quality of super-resolution on SoccerNet, as compared to SR models that were trained on generic data.

The practical implications of these enhancements are far-reaching. For instance, in automated video analysis for coaching and tactical evaluation, higher detection accuracy enables more detailed and nuanced analysis of players' movements and actions. Similarly, in the context of automated officiating and player tracking, improved object detection can contribute to more reliable and fair decision-making processes. Furthermore, these improvements can lead to cost reductions as there is less need to invest in high-quality cameras; the super-resolution technology compensates by enhancing the quality of the images captured by more standard equipment.

Although this study focuses solely on object detection with super-resolution in the context of football, its findings can be extended to other sports and even entirely different domains. For example, they can enhance facial recognition, license plate reading, and suspicious activity detection in low-resolution CCTV footage, thereby improving public safety and investigative capabilities. Similarly, in manufacturing, particularly in quality control, super-resolved images can help identify defects in products or machinery even when using standard cameras.

The limitation of our approach is the lack of real-time processing capability, as the SR preprocessing introduces computational overhead that may be unsuitable for latency-sensitive applications such as live broadcasts; however, it remains valuable in offline scenarios, which are common in professional sports environments where post-game analysis is typically conducted.

Looking ahead, our research opens several avenues for future exploration. One promising direction is the investigation of real-time SR and object detection algorithms that can operate efficiently in a live broadcast environment. Another area of interest is the exploration of domain-specific SR techniques that are optimized for varying weather conditions and lighting environments typical of football matches. In our future work, we also plan to explore various super-resolution techniques and object detection models. We aim to evaluate the effectiveness of cutting-edge methods in enhancing image quality and accurately identifying objects. Our goal is to find optimal solutions for diverse applications. What is more, we would like to analyse how fine-tuning object detectors and using sliding windows [30] affects detection quality.

We also acknowledge potential ethical concerns. Enhanced tracking may raise privacy issues, particularly outside sports contexts. Additionally, detection models can exhibit biases, such as varying accuracy across skin tones or lighting. Future work should include fairness testing and privacy impact assessments to ensure responsible deployment.

In conclusion, the integration of advanced super-resolution methods into object detection frameworks presents a significant advancement in the analysis of football imagery. By addressing the challenges posed by low-resolution images, this approach enhances the accuracy of object detection, thereby offering valuable insights for both tactical analysis and the development of automated officiating systems.

Acknowledgments. A.W. was supported by OMINO grant (no 101086321) funded by the European Union under the Horizon Europe and by the Polish Ministry of Education and Science within the International Projects Co-Financed program. (However, the views and opinions expressed are those of the authors only and do not necessarily reflect those of the European Union or the European Research Executive Agency. Neither the European Union nor the European Research Executive Agency can be held responsible for them.)

Disclosure of Interests. The authors have no competing interests to declare that are relevant to the content of this article.

References

1. Li, Y., et al.: NTIRE 2023 challenge on efficient super-resolution: methods and results. In: Proceedings of the IEEE/CVF Conference on Computer Vision and Pattern Recognition, pp. 1922–1960 (2023)
2. Ren, S., He, K., Girshick, R., Sun, J.: Faster R-CNN: towards real-time object detection with region proposal networks. IEEE Trans. Pattern Anal. Mach. Intell. **39**(6), 1137–1149 (2016)
3. Deliege, A., et al.: Soccernet-v2: a dataset and benchmarks for holistic understanding of broadcast soccer videos. In: Proceedings of the IEEE/CVF Conference on Computer Vision and Pattern Recognition, pp. 4508–4519 (2021)
4. Kong, F., et al.: Residual local feature network for efficient super-resolution. In: Proceedings of the IEEE/CVF Conference on Computer Vision and Pattern Recognition, pp. 766–776 (2022)
5. Torres-Ronda, L., Beanland, E., Whitehead, S., Sweeting, A., Clubb, J.: Tracking systems in team sports: a narrative review of applications of the data and sport specific analysis. Sports Med.-Open **8**(1), 15 (2022)
6. Olagoke, A.S., Ibrahim, H., Teoh, S.S.: Literature survey on multi-camera system and its application. IEEE Access **8**, 172892–172922 (2020)
7. Wang, X.: Intelligent multi-camera video surveillance: a review. Pattern Recogn. Lett. **34**(1), 3–19 (2013)
8. Breitenstein, M.D., Reichlin, F., Leibe, B., Koller-Meier, E., Van Gool, L.: Online multiperson tracking-by-detection from a single, uncalibrated camera. IEEE Trans. Pattern Anal. Mach. Intell. **33**(9), 1820–1833 (2010)
9. Ma, C., Yang, F., Li, Y., Jia, H., Xie, X., Gao, W.: Deep trajectory post-processing and position projection for single & multiple camera multiple object tracking. Int. J. Comput. Vis. **129**, 3255–3278 (2021)
10. Akan, S., Varlı, S.: Use of deep learning in soccer videos analysis: survey. Multimed. Syst. **29**(3), 897–915 (2023)

11. Kim, H., Choi, H.-J., Kim, C.J., Yoon, J., Ko, S.-K.: Ball trajectory inference from multi-agent sports contexts using set transformer and hierarchical bi-LSTM. In: Proceedings of the 29th ACM SIGKDD Conference on Knowledge Discovery and Data Mining, pp. 4296–4307 (2023)
12. Chauhan, K., et al.: Deep learning-based single-image super-resolution: a comprehensive review. IEEE Access **11**, 21811–21830 (2023)
13. Lepcha, D.C., Goyal, B., Dogra, A., Goyal, V.: Image super-resolution: a comprehensive review, recent trends, challenges and applications. Inf. Fusion **91**, 230–260 (2023)
14. Zhang, K., et al.: Benchmarking ultra-high-definition image super-resolution. In: Proceedings of the IEEE/CVF International Conference on Computer Vision, pp. 14769–14778 (2021)
15. Zhang, R., Wu, L., Yang, Y., Wu, W., Chen, Y., Xu, M.: Multi-camera multi-player tracking with deep player identification in sports video. Pattern Recogn. **102**, 107260 (2020)
16. Manafifard, M., Ebadi, H., Abrishami Moghaddam, H.: A survey on player tracking in soccer videos. Comput. Vis. Image Underst. **159**, 19–46 (2017)
17. Komorowski, J., Kurzejamski, G.: Graph-based multi-camera soccer player tracker. In: 2022 International Joint Conference on Neural Networks (IJCNN), pp. 1–8 (2022)
18. Naik, B.T., Hashmi, M.F., Geem, Z.W., Bokde, N.D.: DeepPlayer-track: player and referee tracking with jersey color recognition in soccer. IEEE Access **10**, 32494–32509 (2022)
19. Martins, E., Brito, J.H.: Soccer player tracking in low quality video. arXiv preprint arXiv:2105.10700 (2021)
20. Cioppa, A., et al.: Soccernet-tracking: multiple object tracking dataset and benchmark in soccer videos. In: Proceedings of the IEEE/CVF Conference on Computer Vision and Pattern Recognition, pp. 3491–3502 (2022)
21. Hurault, S., Ballester, C., Haro, G.: Self-supervised small soccer player detection and tracking. In: Proceedings of the 3rd International Workshop on Multimedia Content Analysis in Sports, pp. 9–18 (2020)
22. Luo, W., Xing, J., Milan, A., Zhang, X., Liu, W., Kim, T.-K.: Multiple object tracking: a literature review. Artif. Intell. **293**, 103448 (2021)
23. Vicente-Martínez, J.A., Márquez-Olivera, M., García-Aliaga, A., Hernández-Herrera, V.: Adaptation of YOLOv7 and YOLOv7_tiny for soccer-ball multi-detection with DeepSORT for tracking by semi-supervised system. Sensors **23**(21), 8693 (2023)
24. Lewin, S., Vandegar, M., Hoyoux, T., Barnich, O., Louppe, G.: Dynamic NeRFs for soccer scenes. In: Proceedings of the 6th International Workshop on Multimedia Content Analysis in Sports, pp. 113–121 (2023)
25. Wang, Y., et al.: Remote sensing image super-resolution and object detection: benchmark and state of the art. Expert Syst. Appl. **197**, 116793 (2022)
26. Dong, H., et al.: Detection of occluded small commodities based on feature enhancement under super-resolution. Sensors **23**(5), 2439 (2023)
27. Ji, X., Liu, G.-P., Cai, C.-T.: Collaborative framework for underwater object detection via joint image enhancement and super-resolution. J. Mar. Sci. Eng. **11**(9), 1733 (2023)
28. Hashmi, M.F., Naik, B.T., Keskar, A.G.: BDTA: events classification in table tennis sport using scaled-YOLOv4 framework. J. Intell. Fuzzy Syst. **44**(6), 9671–9684 (2023)

29. Korhonen, J., You, J.: Peak signal-to-noise ratio revisited: is simple beautiful? In: 2012 Fourth International Workshop on Quality of Multimedia Experience, pp. 37–38 (2012)
30. Akyon, F.C., Altinuc, S.O., Temizel, A.: Slicing aided hyper inference and fine-tuning for small object detection. In: 2022 IEEE International Conference on Image Processing (ICIP), pp. 966–970 (2022)
31. Held, J., Cioppa, A., Giancola, S., Hamdi, A., Ghanem, B., Van Droogenbroeck, M.: VARS: video assistant referee system for automated soccer decision making from multiple views. In: Proceedings of the IEEE/CVF Conference on Computer Vision and Pattern Recognition, pp. 5086–5097 (2023)
32. Kingma, D.P., Ba, J.: Adam: a method for stochastic optimization. arXiv preprint arXiv:1412.6980 (2014)
33. Na, B., Fox, G.C.: Object classifications by image super-resolution preprocessing for convolutional neural networks. In: Advances in Science, Technology and Engineering Systems Journal (ASTESJ), vol. 5, no. 2, pp. 476–483 (2020)
34. Cheng, G., et al.: Towards large-scale small object detection: survey and benchmarks. IEEE Trans. Pattern Anal. Mach. Intell. **45**(11), 13467–13488 (2023)

Flexible User-Defined Domain Decomposition in Kilometer-Scale E3SM Land Model Simulation

Zhuowei Gu[1]([✉]), Dali Wang[2], Dawei Gao[3], Yunhe Feng[3], and Qinglei Cao[1]

[1] Saint Louis University, St. Louis, USA
zhuowei.gu@slu.edu
[2] Oak Ridge National Laboratory, Oak Ridge, USA
[3] University of North Texas, Denton, USA

Abstract. The Energy Exascale Earth System Model (E3SM) Land Model (ELM) has been extended to kilometer-scale (km-ELM) resolutions, enabling high-fidelity simulations of terrestrial processes at $1\,km \times 1\,km$ grid spacing. In ELM, domain decomposition partitions the computational domain across processors, ensuring efficient parallel execution. Currently, round-robin decomposition is applied, providing a straightforward way to distribute computational workload. As ELM continues evolving at the kilometer-scale (km-scale), particularly with integrating lateral flow modeling, decomposition strategies must also account for the increased workload and data movement. This paper introduces a flexible user-defined domain decomposition framework, allowing users to customize domain partitioning based on application requirements. The impact of different decomposition strategies is evaluated across various applications concerning computation, communication, and I/O. Results demonstrate that while 1D partitioning yields superior I/O performance, k-nearest neighbors (KNN) clustering effectively reduces inter-process communication overhead. This study lays the groundwork for scalable partitioning in large-scale land surface simulations, enhancing next-generation Earth system modeling.

Keywords: Earth System Modeling · E3SM Land Model (ELM) · Kilometer-Scale ELM · Domain Decomposition · Load Balancing · I/O

1 Introduction

The Energy Exascale Earth System Model (E3SM) [1] is a state-of-the-art, high-resolution Earth system model developed to simulate and project climate change with a focus on water cycle, biogeochemical, and cryospheric processes. Designed for exascale computing, E3SM integrates multiple components, including atmosphere, ocean, land, and ice, to provide high-fidelity climate predictions. The land component of E3SM, known as the E3SM Land Model (ELM), is responsible for simulating terrestrial processes such as energy balance, water fluxes,

M. H. Lees et al. (Eds.): ICCS 2025, LNCS 15903, pp. 164–178, 2025.
https://doi.org/10.1007/978-3-031-97626-1_12

vegetation dynamics, and biogeochemical cycles. ELM has evolved as a sophisticated land surface model, supporting simulations at various spatial resolutions and facilitating studies on climate-land interactions.

With the advancement of high-performance computing (HPC) and the increasing demand for more granular climate simulations, ELM has been extended to kilometer-scale ELM (km-ELM) [2,3]. Unlike traditional land models that operate at coarser resolutions (e.g., 10–100 km), km-ELM enables fine-scale simulations down to 1 km × 1 km grid spacing, capturing small-scale heterogeneities in land surface processes. This capability is crucial for improving regional climate assessments, hydrological modeling, and ecosystem studies, particularly in complex terrains and highly heterogeneous landscapes.

However, achieving efficient and scalable kilometer-scale (km-scale) simulations poses significant challenges due to the sheer volume of data and computational workload. First, the computational cost increases drastically as the number of grid cells grows, requiring efficient load balancing to prevent idle processors and ensure optimal utilization of computing resources. Second, km-ELM requires lateral flow processes. Unlike traditional column-based land models where each grid cell operates independently, lateral flow requires frequent data exchanges between neighboring subdomains to simulate surface and subsurface water transport. Third, the high-resolution nature leads to substantial I/O demands, particularly in restart files and history files [3].

These challenges can cause potential bottlenecks in km-ELM in load balancing, data movement, and I/O, especially when considering the current simple approach of round-robin decomposition strategy [4,5]. Efficient domain decomposition strategies, a technique used to partition the vast computational domain into smaller subdomains for parallel processing, are critical to ensuring that computational workloads are evenly distributed across processors while minimizing communication overhead, allowing km-ELM to leverage modern supercomputing architectures effectively.

In this paper, we introduce a flexible user-defined domain decomposition framework (FUDD) in km-scale ELM, allowing users to customize domain partitioning based on specific application requirements. Unlike the current round-robin strategy, FUDD provides adaptability, enabling optimized load balancing, efficient communication, and improved I/O performance across different km-ELM use cases. The contributions of this paper are as follows:

- Introducing a flexible user-defined domain decomposition framework integrated into km-scale ELM.
- Demonstrating domain decomposition using static predefined strategies, adaptive data-driven methods, and watershed-based partitioning.
- Evaluating decomposition strategies' impact in multiple applications on different clusters in terms of computation, communication, and I/O.

To the best of our knowledge, this is the first work to systematically study the effect of domain decomposition in kilometer-scale ELM simulations.

The structure of this paper is as follows. Section 2 discusses prior research relevant to our work. Section 3 provides the necessary background on km-scale

ELM along with domain partitioning and gridcell grouping. The user-defined decomposition framework is described in Sect. 4. A performance evaluation and analysis is then conducted in Sect. 5, and finally, Sect. 6 outlines our conclusions and future directions.

2   Related Work

2.1   Km-Scale E3SM Land Model Development

E3SM [1] is a state-of-the-art climate modeling system built to capture intricate interactions among atmospheric, oceanic, terrestrial, and cryospheric processes with fine-grained spatial and temporal fidelity. Within this framework, ELM plays a pivotal role by offering a sophisticated representation of surface processes, including energy fluxes, hydrological cycles, vegetation dynamics, and biogeochemical transformations [6].

The ELM codebase is extensive, comprising nearly half a million lines of code. It employs a diverse range of specialized data structures, including roughly 2,000 globally referenced multidimensional arrays and more than 1,000 subroutines [7,8]. In response to the increasing demand for precision in land surface simulations, researchers have recently introduced ultrahigh-resolution ELM (uELM) models. These leverage Exascale computing capabilities to conduct highly detailed simulations at continental and global scales. Cutting-edge computational frameworks have been devised to optimize uELM execution on hybrid Exascale platforms, further enhancing the efficiency and accuracy of these large-scale simulations [2,9]. In one notable study from 2022, a high-resolution ELM simulation at a 1 km × 1 km scale was applied to such a limited domain [10]. This investigation utilized a streamlined ELM model, incorporating a reduced number of subgrid components alongside a satellite phenology submodel, aiming to examine subgrid topographic influences on land-atmosphere interactions in mountainous regions. The increasing availability of high-resolution datasets, including detailed climate forcing and soil property information [11–13], has significantly improved the feasibility of conducting km-scale ELM simulations. These datasets provide the necessary spatial resolution to better capture complex geographical features and extreme meteorological phenomena. The ongoing development of a kilometer-scale (km-scale) ELM is being pursued alongside other E3SM components, including high-resolution atmosphere and ocean models, to ensure optimal performance on Exascale systems [14,15].

Despite these advancements, transitioning to high-resolution simulations can present substantial computational challenges. The increased grid density significantly amplifies resource demands, necessitating robust computing power and storage solutions to accommodate the expansive data volumes.

2.2   Domain Decomposition in E3SM

Domain decomposition is a critical technique for improving the scalability and efficiency of land surface models, particularly in high-resolution Earth system simulations. The scalability of land models in Earth system simulations has been a long-standing challenge. Hoffman et al. [4] analyzed parallel efficiency and domain decomposition strategies for climate models, emphasizing the importance of load balancing in large-scale simulations. More recent work has further optimized land model performance at high resolutions. Tang et al. [5] introduced a km-scale ELM framework for large-domain simulations, demonstrating the scalability of ELM at continental scales. They employed a basic round-robin domain decomposition, which, while effective in load distribution, presents inefficiencies when lateral interactions, such as hydrological processes, are introduced. However, existing decomposition strategies, such as round-robin, often struggle with heterogeneous land cover and variable workloads [3]. While these approaches ensure equal partitioning, they do not account for communication costs associated with lateral flow or optimize I/O performance for large-scale km-ELM simulations. Our work introduces a flexible user-defined domain decomposition approach, allowing users to tailor partitioning strategies based on specific application needs, addressing both load balancing and communication overhead.

3   Background

3.1   Km-Scale ELM

This work utilizes ELM in *land-only mode*, isolating it from atmospheric feedback mechanisms. Instead of interactive coupling, it employs observational datasets to impose atmospheric forcing, enabling the simulation of ecosystem dynamics under past climatic conditions [16–19]. By adopting this configuration, the study ensures that ecosystem responses to climate fluctuations are examined independently, free from interactions with other components of the Earth system.

ELM functions as a terrestrial ecosystem model driven by data, where each gridcell is processed independently. Its execution requires numerous computational cycles spanning gridcells and their respective subgrid elements. Though the model consists of more than 1000 subroutines, none impose a significant computational burden. The model advances in half-hourly or hourly increments. At every step, computations are performed over each *clump*, a structured grouping of gridcells and their active subgrid components, to update ecosystem state variables and flux exchanges among terrestrial processes. This process relies on more than 3,000 global arrays to manage state variables and nutrient cycles. While most calculations related to subgrid and gridcell dynamics—such as the nutrient cycles of carbon (C), nitrogen (N), and phosphorus (P)—are conducted independently, certain operations introduce dependencies.

ELM's computational demands stem from handling extensive global datasets while rigorously upholding the conservation principles of mass and energy. To maintain these conservation laws, ELM systematically compiles all pertinent state variables and rigorously verifies their consistency at every timestep throughout all gridcells. At the end of a simulation, ELM produces a substantial volume of output data. By default, each monthly history file records more than 550 variables, encapsulating diverse terrestrial ecosystem dynamics such as hydrological fluxes, energy exchanges, and the biogeochemical interactions of carbon, nitrogen, and phosphorus across all land grid cells.

3.2 Domain Partitioning and Gridcell Grouping

At the beginning of a simulation, the ELM model systematically traverses the computational domain, assigning unique identifiers to individual land grid cells. These cells are then distributed among MPI (Message Passing Interface) processes following a round-robin assignment approach, which helps maintain an even computational load across processes [4]. Within each MPI process, a collection of specialized datatypes is utilized to delineate the computational domain (see Table 1). After distribution, the assigned grid cells are further organized into structures referred to as *clumps* within each MPI process. Every clump maintains metadata, including the MPI rank of the corresponding processor and information about the total count of subgrid components, along with their respective index boundaries. To optimize memory usage, each MPI process preallocates contiguous memory blocks (arrays) to store ELM variables corresponding to its assigned grid cells and their subgrid components. ELM variables are globally allocated and initialized as arrays to facilitate uniform access throughout the execution. Each grid cell is provisioned with a predefined upper limit on the number of subgrid components, and during each timestep, a filtering mechanism identifies and tracks the active subgrid components within individual grid cells.

Table 1. Customized Data Structures for Defining the Computational Domain

Datatype	Description
npes	Represents the total number of MPI processes used in the simulation.
nclumps	Represents the total number of clumps in the simulation.
begg & endg	Determine offset for each clump assigned to each process
clumps	Contains the owner process ID, size of grid cells (including subgrid components), and the starting and ending indices of these subgrid elements within each clump. Each process can have multiple clumps.
lcid	A mapping structure used to store the processor ID
gdc2glo	1D mapping array that stores the global 2D position of each grid cell, each processor manages the range from its begg to endg
procinfo	Stores information on the number of clumps, clump identifiers, size of grid cells, and the start and end indices of subgrid elements within the corresponding MPI process.
bounds_type	Manages data related to the total number of subgrid components in grid cells, including their starting and ending indices within either clumps or MPI processes.

4 User-Defined Decomposition Framework

4.1 Framework Description

As mentioned hereinbefore, in the default ELM decomposition framework, domain partitioning relies on a static round-robin distribution of land grid cells across processors to maintain load balance. However, this approach lacks flexibility in defining custom partitioning strategies, potentially leading to inefficiencies in computation, communication, and I/O at the kilometer scale.

To address this limitation, we introduce a user-defined partition framework that provides a flexible mechanism, allowing users to modify the amask values (user-provided decomposition) in the domain file to control domain decomposition. The amask is a 1D array where each grid cell is assigned a value, indicating whether it is a land unit (≥ 1 denotes land, and 0 indicates ocean) and defining the affinity or clump assignment of each grid cell (ranging from 1 to nclumps). This method directly maps amask to lcid, which records the clump assignment for each grid cell (see Table 1). This mapping enables fine-grained control over decomposition, allowing users to apply various partitioning strategies without modifying the core algorithm in ELM. Algorithm 1 illustrates the process of iterating through each grid cell, reading amask values, and assigning them to lcid to ensure proper workload distribution. Here, lns represents the total number of grid cells, while numg, lcid, procinfo, and clumps are utilized later when updating gcd2glo.

The implementation of this framework requires modifications to ELM's domain decomposition logic. A key modification involves adapting the gdc2glo mapping, which converts between 1D and 2D grid structures. Rather than relying on predefined round-robin assignments, the framework employs lcid to determine the number of land units assigned to each clump of each processor. This

---

Algorithm 1: User-Defined Domain Decomposition in Fortran.

```
1  for ln = 1 to lns do
       // Loop over each grid cell; lns represents the 1D domain size
2      if amask(ln) > 0 then
           // Check if the current grid cell is a land unit
3          numg = numg + 1 // Count active land grid cells
4          lcid(ln) = amask(ln) // Assign processor ID from mask (amask is
               the mask value read from the domain file)
5          if iam == clumps(lcid(ln))%owner then
6              procinfo%ncells = procinfo%ncells + 1 // Update processor
                   cell count
7          end
8          clumps(lcid(ln))%ncells = clumps(lcid(ln))%ncells + 1 // Update
               clump cell count
9      end
10 end
```

method ensures precise updates of `begg` and `endg`, which define the computational range (refer to Table 1). In detail, `lcid` provides the 1D position of each assigned land unit, facilitating efficient indexing. These 1D indices are mapped back to their 2D global positions (i.e., latitude `lni` and longitude `lnj`), ensuring that `gdc2glo` is correctly updated while preserving spatial relationships.

Another complexity arises from ELM's hierarchical grid structure, where each primary land grid cell contains six subgrid levels. A different partition necessitates corresponding updates to the `bounds_type` structures (see Table 1) at each subgrid level, which governs memory allocation and communication. To ensure consistency across different configurations, rigorous validation is also performed.

4.2   Partition Strategies for Demonstration

Different partitioning strategies impact computation, communication, and I/O performance differently. Therefore, we select six representative partitioning strategies across three categories: (i) static, predefined partitioning, (ii) adaptive, data-driven methods, and (iii) watershed-based strategy. These strategies represent different workload distributions and data access patterns, as visualized in Fig. 1, where a 10×10 grid is distributed across 10 clumps.

Static, Predefined Partitioning. 1D Partition divides the computational domain into contiguous segments, ensuring that each processor is assigned a continuous block of grid cells. Regular round robin distributes grid cells sequentially across all processors in a cyclic manner. Block round robin provides an intermediate approach between these two, offering a comparative perspective. It

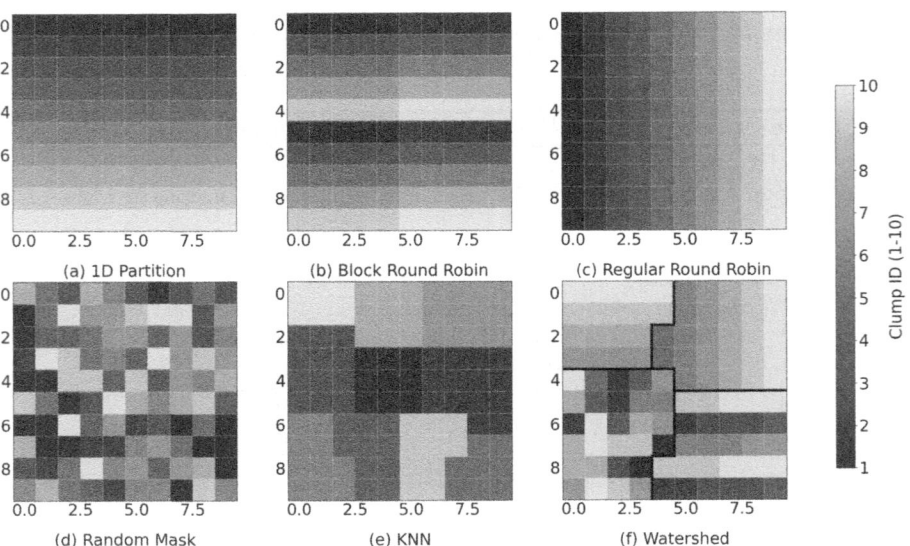

Fig. 1. Visualization of various partitioning strategies using ten clumps.

assigns gridcells in fixed-size blocks, cyclically distributing them among processors. The block size (e.g., 5 in Fig. 1(b)) determines how many consecutive grid cells are assigned to a single processor before moving to the next.

Adaptive, Data-Driven Partitioning. Random Mask allows users to directly assign processors to specific grid cells by modifying the land mask randomly, presenting an extreme case in partitioning strategies. K-Nearest Neighbors (KNN) employs a clustering algorithm based on spatial proximity or feature similarity to group neighboring gridcells together.

Watershed-Based Partitioning. Watershed-based partitioning divides the computational domain based on the demographic partitioning of the Tennessee Valley Authority (TVA) region (details in Sect. 5.2). Within these subdomains, the independent partitioning strategies mentioned above are applied to assess effectiveness in grouping grid cells based on hydrological characteristics.

5   Performance Results and Analysis

5.1   Experimental Settings

The experiments are conducted on two high-performance computing systems from the Oak Ridge Leadership Computing Facility (OLCF):

- `Baseline`: A CPU-based system composed of 180 compute nodes. Each node is equipped with two AMD EPYC 7713 processors, providing a total of 128 cores running at 2.0 GHz. The nodes are configured with either 256 GB or 512 GB of memory. It offers 2.3 PB of shared storage through the Wolf2 GPFS filesystem, making it suitable for large-scale simulations with high I/O throughput requirements.
- `Frontier`: A GPU-accelerated supercomputer utilizing AMD hardware, comprising 9,408 compute nodes. Each node integrates a 64-core AMD Optimized 3rd Gen EPYC processor alongside four AMD MI250X GPUs, supported by 512 GB of DDR4 memory. The system is connected to Orion, a parallel filesystem based on Lustre and HPE ClusterStor, featuring a 679 PB usable namespace.

In this study, the MPI binding policy is set to by-core, meaning that each core hosts exactly one process. Without loss of generality, each process includes one clump for simplicity. The block size remains fixed for block round-robin in each application throughout the experiments. Additionally, under the KNN strategy, the number of assigned blocks can be adjusted based on the number of processes. This work utilizes ELM in land-only mode, as detailed in Sect. 3.

5.2   Application Settings

To evaluate different partition strategies, we select four distinct regions: AKSP, AKSPx10, TES_TVA, and TN. These domains allow us to assess the impact of different partition methods.

- **AKSP**: The AKSP domain covers the Seward Peninsula in western Alaska, spanning approximately 330 km in length and 145âĂŞ225 km in width, with a total area of 72,083 km². The simulation employs a high-resolution 1 km × 1 km grid, resulting in 72,083 grid cells.
- **AKSPx10**: This experiment scales up the AKSP domain to evaluate partitioning methods in ELM. The AKSPx10 simulation maintains the same spatial characteristics but expands to 720,830 land grid cells with corresponding subgrid components. Its presentation is identical to that of the AKSP domain.
- **TES_TVA**: Designed to test the Watershed-Based Partition method, this domain is first subdivided into four distinct subdomains within the Tennessee Valley Authority (TVA)—Mississippi (MS), Ohio (OH), Tennessee (TN), and Gulf (GULF)—each representing a hydrologically significant region. The domain consists of 10,357 grid cells, reflecting natural watershed boundaries.
- **TN**: This focuses on the TN region within the TVA watershed, encompassing 6,317 grid cells and capturing its hydrological and ecological characteristics.

5.3   Decomposition Map

Figure 2 presents the partition map of different decomposition strategies (see Fig. 1) in real applications (AKSP and TN) generated on `Baseline` with 128 processes. Grid cells in E3SM are stored in a structured 1D format but are mapped to a 2D computational grid using row and column indices. The column index corresponds to the x-axis direction (longitude), while the row index represents the y-axis direction (latitude). These indices are derived by normalizing

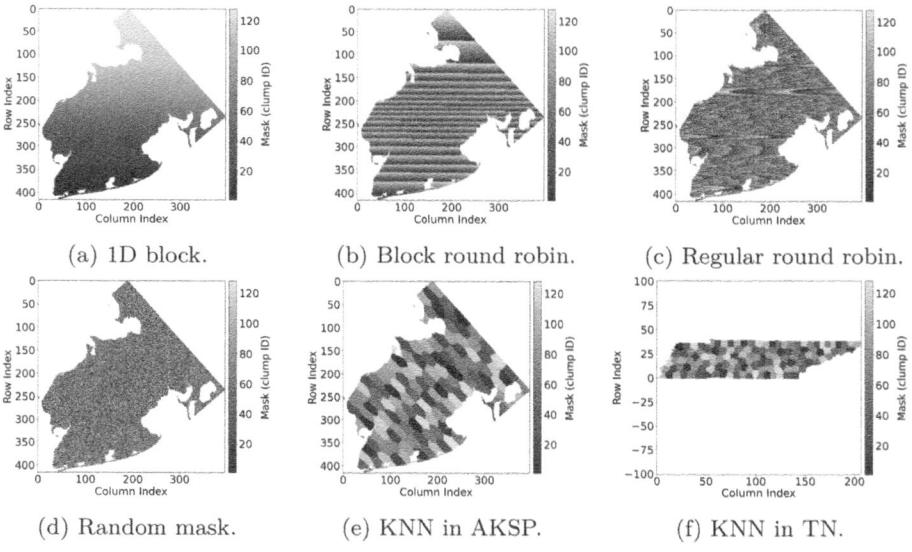

(a) 1D block.   (b) Block round robin.   (c) Regular round robin.

(d) Random mask.   (e) KNN in AKSP.   (f) KNN in TN.

Fig. 2. Decomposition map of AKSP ((a)-(e)) and TN ((f))

spatial coordinates relative to their minimum values and dividing by the grid resolution. In Fig. 2, grid cells are ordered (1) starting from the left corner for AKSK in Fig. 2(a)–(e) (i.e., (column index, row index): (0, largest_value_in_y-axis)) and (2) starting from the left center for TN in Fig. 2(f) (i.e., (column index, row index): (0, 0)), which are then progressed row by row. AKSP is based on an older mapping system, so its actual starting point does not align with modern geographic conventions, resulting in a rotated and flipped representation. From these figures, we observe slight differences in the decomposition results compared to Fig. 1, primarily due to irregularities in the domain file and the skipping of ocean grid cells (see Algorithm 1). These results illustrate the decomposition of real-world applications under different strategies, emphasizing the complexities introduced by domain-specific constraints.

5.4   Impact of Partitioning Methods on Computation

Table 2 presents a comparison of computational time across different partitioning strategies for various test cases on `Baseline`. Each case type is evaluated under different node configurations. The computational time is determined by subtracting lnd component I/O time from the lnd component total running time. While the similarity in computational times across different partitioning methods within each case suggests effective load balancing, slight variations are observed. Static, predefined partitioning methods (1D partition, block round robin, and regular round robin) distribute grid cells evenly among processors, resulting in better workload balance and reduced computational time. In contrast, adaptive, data-driven partitioning methods (random mask and KNN) exhibit slightly higher computational times due to variations in workload distribution. For instance, Random Mask may lead to certain processors receiving more grid cells than others, causing computational load imbalance.

The results also demonstrate both weak and strong scalability. Consider the 1D Partition as an example. Weak scalability is assessed by comparing AKSP on a single node (337.02 s) with AKSPx10 on 10 nodes (432.12 s), showing that as problem size increases proportionally with computational resources, performance remains stable. Strong scalability is evident in the AKSP cases, where compu-

Table 2. Comparison of Computational Time (Seconds). Red: worst computational time; Blue: best computational time.

Case Type	1D Partition	Block Round Robin	Round Robin	Random Mask	KNN
AKSP 1 node	**337.02**	338.97	338.44	343.16	346.59
AKSP 5 nodes	**63.14**	64.24	64.14	62.02	62.99
AKSP 10 nodes	**31.74**	33.73	33.78	35.11	34.67
AKSPx10 10 nodes	**432.12**	521.32	491.77	511.44	516.67
TVA Watershed 1 node	68.17	63.44	**63.09**	70.82	64.90
TN 1 node	34.61	33.34	32.44	36.55	**32.29**

tational time decreases from 337.02 s (1 node) to 63.14 s (5 nodes) to 31.74 s (10 nodes), indicating improved efficiency with increasing computational resources.

5.5 Impact of Partitioning Methods on Communication

Figure 3(a) visualizes the four distinct subdomains within TVA, and Fig. 3(b)–(f) presents the boundary grid cell heatmap for different partitioning strategies applied to the TVA watershed sub-domain on `Baseline` with 128 processes. The partitioning method for each sub-domain remains consistent in each setting, while each sub-domain can adopt a different decomposition. Each partitioning method results in a unique distribution of boundary grid cells, which are computed using an adjacency-based method, indicating the extent of inter-process communication required under each strategy. For each grid cell, its process/-

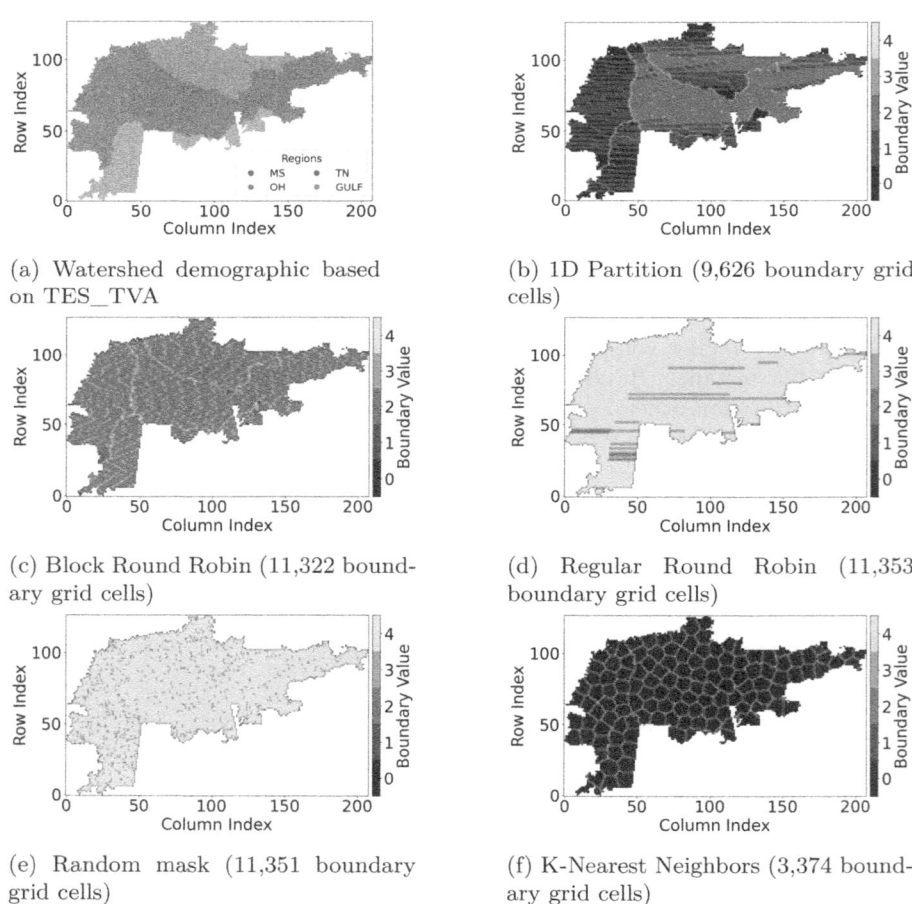

(a) Watershed demographic based on TES_TVA

(b) 1D Partition (9,626 boundary grid cells)

(c) Block Round Robin (11,322 boundary grid cells)

(d) Regular Round Robin (11,353 boundary grid cells)

(e) Random mask (11,351 boundary grid cells)

(f) K-Nearest Neighbors (3,374 boundary grid cells)

Fig. 3. Different partition strategies applied to watershed sub-domain.

clump ID is compared with the four neighboring grid cells (top, bottom, left, and right). If the ID of any neighboring grid cell differs, the boundary count for that grid cell is incremented. Consequently, the boundary value of each grid cell ranges from 0 to 4, indicating the number of communications if lateral flow is enabled. The total number of boundary grid cells is obtained by counting all grid cells where this computed value is greater than zero. From the results shown in the figures and the boundary grid cell counts, KNN exhibits the lowest number of boundary grid cells, thereby introducing the lowest communication overhead when lateral connectivity is activated. The 1D partition performs worse than KNN due to irregularities in the domain file and the omission of ocean grid cells, as discussed in Sect. 5.3.

5.6   Impact of Partitioning Methods on I/O Performance

Table 3. Impacts of Partitioning Methods on I/O (Seconds). Red: worst running time; Blue: best running time.

Time (s)	1D Partition	Block Round Robin	Regular Round Robin	Random Mask	KNN
AKSP 1 node (Baseline)					
TOT Time	387.76	413.83	636.52	1277.80	435.00
LND Time	375.61	393.17	613.96	1263.57	415.85
AKSP 2 node (Frontier)					
TOT Time	364.25	374.16	382.01	390.27	387.72
LND Time	359.15	370.08	363.02	368.02	370.42
AKSP 10 nodes (Baseline)					
TOT Time	93.73	99.59	96.58	113.24	102.32
LND Time	86.71	85.88	83.41	97.61	87.34
AKSPx10 10 nodes (Baseline)					
TOT Time	901.98	915.99	1191.92	1168.17	1171.88
LND Time	866.31	840.01	1060.50	1106.21	1081.51
AKSPx10 20 nodes (Frontier)					
TOT Time	401.23	420.74	432.02	432.03	432.02
LND Time	367.64	379.09	385.38	389.24	385.38
TVA 1 node (Baseline)					
TOT Time	74.54	77.74	82.02	79.74	83.72
LND Time	70.54	72.89	72.62	72.58	78.09
TVA Watershed 1 node (Baseline)					
TOT Time	82.28	75.98	77.61	87.57	81.33
LND Time	76.30	70.85	70.93	79.32	74.10
TN 1 node (Baseline)					
TOT Time	42.02	41.48	41.50	53.45	46.10
LND Time	39.90	38.30	37.29	45.11	39.73

Table 3 compares the effect of different partitioning strategies on I/O, including five settings on `Baseline` and two on `Frontier`. The table reports the total running time (TOT Time) and land component running time (LND Time), where TOT Time includes the time of LND, CPL (coupler), and ATM (atmosphere) [3]. Km-ELM is deployed in land-only mode (details in Sect. 3), so LND contributes to most of the time-to-solution. The computational time across different partitioning strategies remains nearly identical (see Sect. 5.4), and lateral flow is under development in km-ELM, so we use the TOT/LND time as the I/O effects. Additionally, each `Baseline` node consists of 128 cores, while each node on `Frontier` has 64 cores. To maintain consistency in computational resources, all experiments on `Frontier` are conducted using twice the number of nodes as in the `Baseline` setup. Km-ELM optimizes I/O efficiency by employing the *SCORPIO* parallel I/O library [20,21], which organizes distributed data before transferring it to underlying storage libraries such as *PnetCDF* and *NetCDF*. These libraries improve write performance by grouping MPI processes into aggregators, which manage collective data output to parallel file systems like *Lustre* [22] and *GPFS* [23]. The aggregation process reduces file fragmentation and enhances overall throughput.

From this table, partitioning methods significantly impact I/O performance due to differences in data arrangement, where 1D partition usually is the best and random mask the worst. When using a 1D partition strategy, the grid cells are already assigned to processors in a contiguous and sequential manner. As a result, *SCORPIO* can efficiently process the data with minimal rearrangement, reducing I/O overhead. In contrast, methods like random mask introduce non-contiguous data distribution, increasing the time required for *SCORPIO* to restructure the output before writing it to storage. Given the absence of lateral flow in the current experiments, the reduced data reordering time makes 1D partition the most efficient method in most scenarios. Furthermore, comparing the results of AKSPx10 between the `Baseline` and `Frontier` (901.98 s on `Baseline` and 401.23 s on `Frontier` for 1D partition), `Frontier` demonstrates shorter I/O time (515.41 s on `Baseline` and 29.66 s on `Frontier` for 1D partition), indicating the improved filesystem on `Frontier`.

6   Conclusion and Future Work

This study introduces FUDD, a flexible framework for user-defined domain decomposition, and explores its application within km-scale ELM. The analysis examines six distinct partitioning strategies, categorized into three main types: static predefined approaches, adaptive data-driven techniques, and watershed-based methods, each representing different workload distributions and data access behaviors. By applying FUDD in real-world scenarios across two clusters, we evaluate FUDD's impact on computation, communication, and I/O. The results provide a foundation for advancing scalable domain decomposition strategies, contributing to the development of efficient next-generation Earth system modelings. Looking ahead, we plan to extend the capabilities of FUDD by integrating it with the GPU-enabled version of ELM, known as uELM [5], to further explore its effectiveness in heterogeneous environments. Additionally, we intend to evaluate FUDD when the lateral flow in km-ELM becomes available. Furthermore, we aim to develop an automated mechanism for intelligently selecting the most suitable partitioning strategy based on the characteristics of a given application.

Acknowledgments. This research is supported as part of the Energy Exascale Earth System Model (E3SM) project funded by DOE's Office of Science, Office of Biological and Environmental Research. This project leverages the Daymet data product funded by NASA and uses resources of the Oak Ridge Leadership Computing Facility, supported by DOE's Office of Science under Contract No. DE-AC05-00OR22725.

References

1. DOE (2024). https://e3sm.org/
2. Yuan, F., et al.: An ultrahigh-resolution E3SM land model simulation framework and its first application to the Seward peninsula in Alaska. J. Comput. Sci. **73**, 102145 (2023)
3. Wang, D., et al.: Kilometer-scale E3SM land model simulation over North America. arXiv preprint arXiv:2501.11141 (2025)
4. Hoffman, F.M., Vertenstein, M., Kitabata, H., White, J.B., III.: Vectorizing the community land model. Int. J. High Perform. Comput. Appl. 19(3), 247–260 (2005)
5. Schwartz, P.D., Wang, D., Yuan, F., Thornton, P.E.: Developing ultrahigh-resolution E3SM land model for GPU systems. In: Computational Science and Its Applications – ICCSA 2023, vol. 13956, pp. 277–290. Springer, Cham (2023)
6. Burrows, S.M., et al.: The DOE E3SM v1. 1 biogeochemistry configuration: description and simulated ecosystem-climate responses to historical changes in forcing. J. Adv. Model. Earth Syst. **12**(9), e2019MS001766 (2020)
7. Yang, X., Wang, D., Janjusic, T., Wei, W., Pei, Yu., Yao, Z.: A web-based visual analytic framework for understanding large-scale environmental models: a use case for the community land model. Procedia Comput. Sci. **108**, 1731–1740 (2017)
8. Zheng, W., Wang, D., Song. F.: Xscan: an integrated tool for understanding open source community-based scientific code. In: International Conference on Computational Science, pp. 226–237. Springer, Cham (2019)
9. Wang, D., Schwartz, P., Yuan, F., Thornton, P., Zheng, W.: Towards ultra-high-resolution e3sm land modeling on exascale computers. Comput. Sci. Eng. **01**, 1–14 (2022)
10. Hao, D., et al.: Impacts of sub-grid topographic representations on surface energy balance and boundary conditions in the E3SM land model: a case study in Sierra Nevada. J. Adv. Model. Earth Syst. **14**(4), e2021MS002862 (2022)
11. Thornton, P.E., et al.: Daymet: daily surface weather data on a 1-km grid for North America, version 2. Technical report, Oak Ridge National Lab.(ORNL), Oak Ridge, TN (United States) (2014)
12. Kao, S.-C., Thornton, M., Thornton, P.E., Shrestha, R.: Gridded sub-daily climate forcings for North America based on daymet and gswp3 (daymet-gswp3). Technical report, Oak Ridge National Laboratory (ORNL), Oak Ridge, TN (United States). Oak (2024)

13. Han, Q., et al.: Global long term daily 1 km surface soil moisture dataset with physics informed machine learning. Sci. Data **10**(1), 101 (2023)
14. Donahue, A.S., et al.: To exascale and beyond—the simple cloud-resolving e3sm atmosphere model (scream), a performance portable global atmosphere model for cloud-resolving scales. J. Adv. Model. Earth Syst. **16**(7), e2024MS004314 (2024)
15. Omega-future E3SM ocean model. https://climatemodeling.science.energy.gov/news/omega-future-e3sm-ocean-model. Accessed 30 Sept 2010
16. Viovy, N.: Cruncep version 7-atmospheric forcing data for the community land model. Research Data Archive at the National Center for Atmospheric Research, Computational and Information Systems Laboratory (2018)
17. Qian, T., Dai, A., Trenberth, K.E., Oleson. K.W.: Simulation of global land surface conditions from 1948 to 2004. Part I: forcing data and evaluations. J. Hydrometeorol. **7**(5), 953–975 (2006)
18. Yoshimura, K., Kanamitsu, M.: Incremental correction for the dynamical downscaling of ensemble mean atmospheric fields. Mon. Weather Rev. **141**(9), 3087–3101 (2013)
19. Thornton, P.E., et al.: Gridded daily weather data for North America with comprehensive uncertainty quantification. Sci. Data **8**(1), 1–17 (2021)
20. Krishna, J., Wu, D.: Software for caching output and reads for parallel I/O (2020). https://github.com/E3SM-Project/scorpio
21. Hartnett, E., Edwards, J.: The Parallelio (PIO) C/Fortran libraries for scalable HPC performance (2021)
22. SUN. High-performance storage architecture and scalable cluster file system. Technical report, Sun Microsystems, Inc. (2007)
23. Schmuck, F., Haskin, R.: GPFs: a shared-disk file system for large computing clusters. In: Proceedings of the 1st USENIX Conference on File and Storage Technologies, FAST 2002, pp. 19–es. USENIX Association, USA (2002)

A Deeper Look into the Limitations of Early-Exit Architectures for Single and Multi-label Classification

Klaudia Bałazy[1,3(✉)], Julian McAuley[2], and Jacek Tabor[1]

[1] Jagiellonian University, Kraków, Poland
[2] University of California San Diego, San Diego, USA
[3] Doctoral School of Exact and Natural Sciences at Jagiellonian University,
Kraków, Poland
klaudia.balazy@doctoral.uj.edu.pl

Abstract. In this study, we explore the limitations of early-exit architectures, which are designed to enhance computational efficiency in neural networks, focusing particularly on both single and multi-label classification tasks within the computer vision domain. We introduce a systematic evaluation framework that not only advances research in this area but also bridges an important gap in understanding how these architectures perform within the complexities of multi-label settings. Our findings reveal a significant challenge: while early-exits improve efficiency without compromising accuracy in single-label tasks, they struggle to offer similar benefits in multi-label classification, necessitating uniquely tailored strategies. Further insights from our ablation suggest that the difficulty in achieving benefits from early-exits in multi-label classification may stem from the varying complexities of processing distinct classes within a single instance. This work lays a solid foundation for future research focused on developing early-exit strategies that effectively handle the complexities of diverse classification contexts.

Keywords: Computational Efficiency · Dynamic Neural Networks · Early-Exit Architecture · Multi-Label Classification

1 Introduction

Deep neural network models have shown remarkable performance in various fields, including computer vision and natural language processing. The significant improvements in accuracy and predictive capability often come at the cost of increased computational complexity and resource demands [1]. Early-exit strategies [2–5], which dynamically adjust the computation time based on the complexity of the input, have emerged as a promising approach to mitigate these challenges. Early-exit architectures introduce internal classifiers at different depths within a neural network. These classifiers, also known as "heads", can independently make predictions and stop the inference process early if certain conditions, such as a predefined confidence threshold, are met. By enabling

M. H. Lees et al. (Eds.): ICCS 2025, LNCS 15903, pp. 179–193, 2025.
https://doi.org/10.1007/978-3-031-97626-1_13

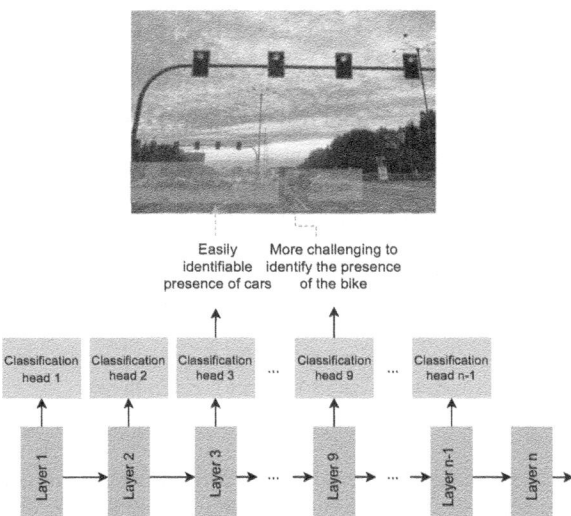

Fig. 1. The underlying complexity of leveraging early-exit models for multi-label classification can be illustrated by this example image with easily identifiable cars and a harder-to-spot bike. Our experiments (see Sect. 4.3) suggest that while the model may quickly detect easier to notice classes (like cars in this example), identifying the bike could require further processing. This highlights the challenge of finding the optimal exit point when multiple objects with varying recognition difficulty are present (this problem is absent in single-label classification).

models to exit inference early under certain conditions, these strategies can significantly improve computational efficiency.

While early-exit models have been the subject of growing interest, several key aspects remain underexplored. Primarily, the evaluation of these strategies is still a challenging task due to the lack of a universally agreed-upon framework for comparison. Although the Efficient Language Understanding Evaluation (ELUE) [6] has been proposed for natural language processing tasks allowing for evaluation of early-exit methods with different exit thresholds, a broader framework applicable across diverse domains is still lacking. Secondly, the understanding of the limitations and benefits of early-exit strategies needs deeper investigation. Finally, despite considerable research on early-exits for single-label tasks, studies focusing on multi-label tasks are noticeably missing.

In this study, we address these gaps by introducing a systematic framework for the evaluation of early-exit strategies. Our proposed framework offers a structured guideline for future assessments and comparisons within this field. It incorporates a robust evaluation metric that quantifies the performance differences between training each model's head independently and using a collective training approach with a selected early-exit strategy. This approach allows for a deeper understanding of early-exit strategies and can serve as a cornerstone for subsequent research in this field.

For single-label classification, we demonstrate that early-exits are advantageous regardless of the number of output classes. The advantages are particularly pronounced in the initial layers of the network and gradually diminish in later layers. Moreover, our analysis adds nuance to the "big label" problem, a phenomenon identified by Liu et al. [7] in the context of the WOS-46985 dataset, where early-exits struggle with single-label classification tasks involving a large number of output classes. Our investigation reveals that this issue is not universal but dataset-dependent. Specifically, our single-label experiments with multiple output classes did not reproduce the "big label" problem, emphasizing the need for context-aware analyses of early-exit models.

Furthermore, we delve into the analysis for multi-label classification tasks, an area that we have identified as presenting considerable challenges in terms of deriving benefits from early-exit methods. We provide a solid baseline and contribute with insights into this underexplored area. We examine diverse architecture choices and various exiting criteria, including entropy [8–10], confidence [11–13], patience [14,15], learning-to-exit [16], and hybrid approaches [17]. Our investigation reveals that using a confidence or entropy criterion is an effective approach for multi-label classification. To further improve the performance, we explore a learning-based approach for early-exiting that integrates a modified loss function, resulting in enhanced performance at a specific speedup.

We also design an experiment to gain a deeper understanding of the challenges associated with employing early-exit strategies for multi-label classification. Our hypothesis, supported by our findings, indicates that the varying recognition difficulty associated with each class within an instance could be a key challenge when employing early-exits in a multi-label context (see Fig. 1). This investigation not only enriches our understanding of the complexities inherent in multi-label classification with early-exit strategies but also points towards intriguing directions for future research.

By offering insights into the limitations of early-exit architectures and shedding light on the complexities of applying these strategies to multi-label classification tasks, our study opens up important directions for future research in this domain. We hope that our work will not only contribute to a better understanding of early-exit models but also inspire further advancements in their design and application.

Our contributions can be summarized as follows:

- We propose a systematic framework for evaluating the benefits of early-exit architectures in deep neural network models.
- We conduct an in-depth examination of various exiting criteria and architecture choices for single-label and multi-label tasks within the computer vision domain, establishing a strong baseline for future research.
- We assess the "big label" problem [7] within the context of early-exit architectures. Our findings reveal that the challenges associated with handling a large number of output classes in single-label classification are influenced by

specific dataset characteristics, rather than being inherent limitations of the early-exits.

- We show that applying early-exit models to multi-label classification presents substantial complexities compared to single-label scenarios (see Fig. 3). We provide detailed insights into this phenomenon and offer an intuitive understanding of the underlying factors (see Fig. 1 and Fig. 6).

2 Related Work

Early-exit models are types of dynamic neural networks that can accelerate inference by terminating it at an earlier layer. This section provides an overview of the key developments in early-exit methods, and then discusses the approaches for evaluating and analyzing the performance of these methods.

Early-Exit Methods. Adaptive Computation Time (ACT) [2,3] introduced a trainable halting mechanism for input-adaptive inference. However, training the halting model required extra effort and added more parameters and increased inference costs. To address this issue, BranchyNet [4] utilized the entropy of the prediction probability distribution as a measure of branch classifier confidence to enable early exiting. Shallow-Deep Networks (SDN) [5] employed the softmax scores of branch classifier predictions to counteract the overthinking problem associated with deep neural networks.

Following research in computer vision, there have also been approaches for natural language processing models, as these models are usually large and deep. Some methods rely on predefined confidence thresholds to determine early-exit points. Examples include DeeBERT [8], RightTool [11], FastBERT [9], Rome-BERT [10], and SkipBERT [13]. These approaches typically involve training BERT with internal classifiers and using entropy or other metrics to determine when a model's prediction is confident enough to exit early.

Another line of studies recycle the predictions of internal classifiers to improve overall performance and reduce wasted computation. PABEE [14] uses early stopping from model training to jointly train internal classifiers and exits when k consecutive classifiers make the same prediction. LeeBERT [15] encourages consistency among internal classifiers, while Sun et al. [18] introduce diversity loss and voting mechanisms for ensembling. Ensemble-based methods have been shown to improve both efficiency and robustness. Zero Time Waste (ZTW) [12] adds direct connections between internal classifiers and combines previous outputs in an ensemble-like manner, to improve performance.

In contrast to predefined confidence thresholds, some methods learn the early-exit criterion. BERxiT [16] uses a *learning-to-exit* module to predict the correctness of current internal classifiers, while CAT [19] employs a "meta consistency classifier" to determine conformity with the final classifier.

Evaluation and Analysis of Early-Exits. Evaluating and analyzing early-exit models poses unique complexities, and currently, there is no universally agreed-upon method for comparison and evaluation across various contexts. In the context of natural language processing, the Efficient Language Understanding Evaluation (ELUE) [6] has been proposed as a potential standard, which considers multiple speed-accuracy pairs when different thresholds are selected for early-exit methods. However, it is still crucial to agree on the evaluation framework that can facilitate robust evaluation across various applications and domains.

Analyzing the limitations and benefits of early-exit strategies is a complex task, yet it is critical for understanding the conditions under which these models excel or fall short. Liu et al. [7] provide valuable insights into the limitations for single-label classification tasks, uncovering the "big label" issue for the WOS-46985 dataset. The term "big label" problem represents the case in which early-exit models have difficulties accurately processing tasks with a large number of output classes. As a potential solution, they propose a strategy of label reduction to mitigate this issue. Our investigation further deepens the understanding of this issue, demonstrating its dataset-dependent nature and emphasizing the importance of context-aware analyses. While considerable research exists on early-exits for single-label classification tasks, there is a notable gap in the literature regarding multi-label classification tasks. In contrast to existing works, our study comprehensively explores both single- and multi-label classification using early-exit strategies.

3   Evaluation Framework of the Early-Exits Effectiveness

In this section, we outline our evaluation framework for assessing the advantages and limitations of using early-exits to speed up inference in neural networks.

Evaluation Setup. To evaluate the performance of early-exit strategies, we start with a pre-trained classification model with an unfrozen backbone. We enhance this model by adding additional classification heads, also referred to as internal classifiers, after each layer, aligning with methods used in prior studies [5, 12]. Each of these H heads is trained independently using an appropriate loss function, resulting in H distinct models. These models demonstrate the potential of a static exit strategy, where each head is configured to terminate processing after specific layers. Concurrently, we develop an early-exit model in which all heads are trained collectively. This dynamic model requires a carefully defined exit strategy for each example.

Training Objective. In single-label classification tasks, we employ the cross-entropy (CE) loss function coupled with a softmax activation. This setup is chosen for its effectiveness in handling mutually exclusive class predictions typical in single-label scenarios. For multi-label classification, where multiple independent labels may be correct, we use binary cross-entropy (BCE) loss with sigmoid activation at the classification heads. Sigmoid activation allows each label to be treated independently, predicting a label as positive (with a score above a threshold of 0.5) or negative (below this threshold). Alternatively, we also utilize CE loss with softmax activation, requiring the selection of a specific threshold above which a label is considered positive.

Early-Exiting Strategies. We evaluate various commonly used early-exiting strategies: confidence thresholds, entropy thresholds, rank-patience-based strategies, hybrid approaches blending confidence (or entropy) with rank-patience measures, and learning-based (classifier-based) methodologies. For strategies that employ thresholds, we explore a spectrum of potential thresholds, generating a performance-speedup curve in the process. The effectiveness of a strategy is indicated by the area under this curve; the larger the area, the more effective the strategy. In the case of the learning-based approach, we derive a single performance-speedup point.

Confidence-based strategies [5, 11–13] suggest that the model should exit if a certain confidence threshold is surpassed. This threshold applies to the highest probability label in single-label classification and to all positively classified labels in multi-label case.

Entropy-based approaches [8–10] operate on the principle that a model should exit if the entropy of its prediction drops below a particular level (low entropy implies a higher degree of certainty in the prediction).

In *patience-based strategies* [14, 15], the decision to exit depends on the consistency of predictions across successive internal classifiers. We adapt this approach for multi-label classification by introducing a rank-based measure, where the exit decision also takes into account the order of predicted classes based on their probability magnitudes. We name this strategy *rank-patience-based approach*, as the model is permitted to exit if the rank of class probabilities remains adequately consistent across consecutive layers. This modification enhances the applicability of patience-based strategies in the context of multi-label classification.

Hybrid approaches [17] combine confidence (or entropy) measures with patience-based measures (rank-patience-based in our multi-label scenario). In this strategy, the model's exit decision is influenced by the confidence or entropy measure, but it also incorporates a rank-patience factor. This factor assesses the stability of the rank (order) of the predictions, allowing the model to process additional layers before deciding to exit, even if the confidence or entropy threshold has been met.

Learning-based approaches Xin et al. [16] introduce an auxiliary classifier with sigmoid activation at each potential exit point. These binary classifiers, trained to predict whether the model should exit at that head, use the corresponding hidden state or the logits from the respective head classifier as input. We employ the Mean Squared Error (MSE) loss function (following previous works [16]), which calculates the difference between the predicted exit probability p_j and the actual binary label y_j (indicating whether an early exit should occur) across all m training instances.

To impose a heavier penalty on incorrect predictions when the model should continue processing, we introduce an additional regularization loss component.

This component is computed as the MSE of the predicted exit probability p_j in cases where the model should not exit. We also incorporate a weighting factor into the final loss calculation to impose a greater penalty on the earlier heads, as their predictions have the most significant influence on further processing. The weights are determined by $\frac{1}{h_i+1}$, where h_i denotes the head index. Furthermore, we balance the MSE loss and the regularization loss with an additional hyperparameter α. The final loss $\mathcal{L}$ is formulated as the sum of individual losses $\mathcal{L}_{h_i}$ across all H heads:

$$\mathcal{L} = \sum_{i=1}^{H} \mathcal{L}_{h_i}, \tag{1}$$

where the loss $\mathcal{L}_{h_i}$ for each head h_i is defined as:

$$\mathcal{L}_{h_i} = \frac{1}{m(h_i+1)}\left(\alpha \cdot \sum_{j=1}^{m}(p_j - y_j)^2 + (1-\alpha) \cdot \sum_{j=1}^{m}(p_j \cdot (1-y_j))^2\right). \tag{2}$$

Early-Exit Benefits Evaluation Framework. To assess the effectiveness of early-exit architectures, we compare the performance of static classifiers with early-exit models, similarly to the ELUE metric used in NLP [6]. We propose the following systematic approach:

– **Performance evaluation:** We first measure the performance P_{h_j} of each static classifier at head h_j for all n instances in the test set. Simultaneously, we evaluate the performance of the early-exit model $P_{ee}(E)$ at specific average exit layer E. The average exit layer E is calculated as: $E = \frac{\sum_i (e_i)}{n}$, where e_i denotes the exit layer for each instance. The performance $P_{ee}(E)$ represents the aggregated performance across all test instances at their respective exit points.

– **Locating the Nearest Performance Point:** To ensure fair comparisons, we calculate $E_{closest}(h_j)$, which identifies the average exit layer in the early-exit model that most closely matches the computational depth of each static classifier head h_j. We determine the average exit layer that best aligns, in computational terms, with the layer where head h_j is located. This average is computed across all test instances and may vary depending on the thresholds set within our chosen strategy. This alignment ensures that performance comparisons between the early-exit and static models are conducted under equivalent computational conditions:

$$E_{closest}(h_j) = \arg\min_{E} |E - h_j|$$

– **Early-Exits Benefits Quantification:** We then calculate the benefits of early-exit strategies B_{EE_j} for each static head index j by comparing the performance of the early-exit model at the $E_{closest}(h_j)$ average exit layer to the performance of the static classifier at head h_j:

$$B_{EE_j} = P_{ee}(E_{closest}(h_j)) - P_{h_j}$$

If the computed value B_{EE_j} is positive, it indicates that the early-exit strategy enhances performance relative to the static model for an equivalent computational effort. This positive benefit suggests that the early-exit model offers a more efficient processing strategy without sacrificing accuracy. A negative B_{EE_j} indicates no benefit from using early-exit models compared to static networks, while a zero value indicates equivalent performance. Our methodology enables us to systematically evaluate and demonstrate the practical advantages of early-exit strategies across various applications.

4   Experiments

In our study, we employ a pre-trained ResNet50 model [20] as the backbone. We extend this model with internal classifiers, which are appended after core bottleneck layers to enable early-exiting. To establish a baseline for comparison, each internal classifier head is fine-tuned independently on a designated dataset, resulting in H distinct static models. These models serve as benchmarks for evaluating the performance at various depths without early-exiting strategies. In parallel, we also develop an integrated early-exit model where all internal classifier heads are collectively optimized. This model is designed to assess the performance and efficiency of dynamic exiting during inference, contrasting directly with the static models that do not incorporate early-exit functionality.

The performance of the models is assessed on specific test sets through F1 score (averaged over all instances) and accuracy score. For static models with independently trained heads, we compute the scores across the entire test set for each head. In contrast, for the early-exit model, we employ diverse strategies, the specifics of which are covered in the subsequent experiment descriptions.

4.1   Early-Exits for Single-Label Classification

We start by exploring early-exits for single-label classification. We use the ImageNet [21] dataset with a varying number of output classes. We train models using different learning rates: $\lambda \in \{10^{-3}, 10^{-4}, 10^{-5}\}$. For the early-exit models, we employ a widely-adopted confidence strategy [5,12] to determine the exit head for specific examples.

Our findings, illustrated in Fig. 2, highlight the considerable benefits of using early-exit for single-label. However, these advantages seem to decrease in the final layers for this specific dataset. Moreover, the models with 2-output classes exhibit increased instability, likely attributed to the limited volume of training and test data. Based on these experiments, we draw the conclusion that the incorporation of early-exit architectures for single-label offers significant advantages. The "big label" problem identified for WOS-46985 [7], which refers to poor early-exits performance with many output classes, appears to be dataset-dependent and did not occur in our experiments.

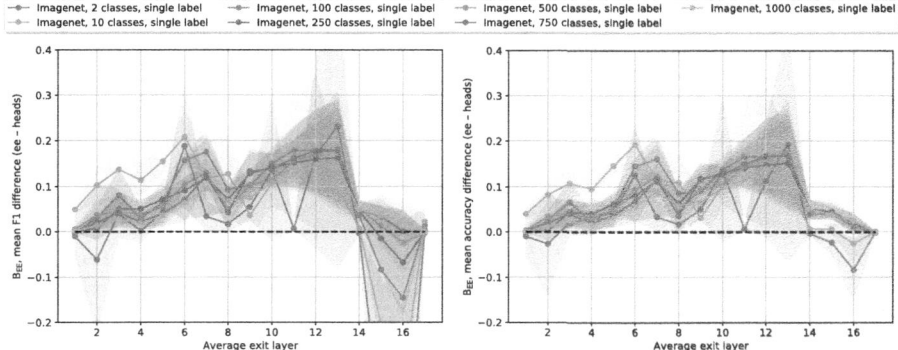

Fig. 2. Benefits of early-exits for single-label ImageNet classification across varying number of output classes. B_{EE} shows the mean F1 and accuracy difference between static exits and a confidence-based early-exit model. The graph shows mean benefits and standard deviations for each exit layer speedup across models trained with different learning rates. Regions above the dashed line indicate favorable performance-speedup trade-offs, highlighting the effectiveness of early-exits for single-label classification.

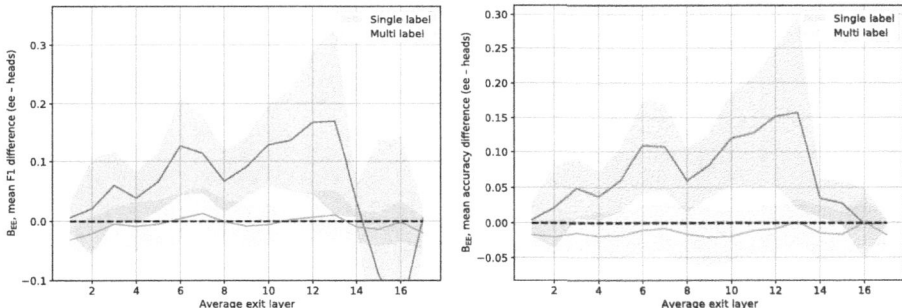

Fig. 3. The disparity in benefits of early-exit architectures for single- and multi-label classification. The graph shows mean benefits B_{EE} and standard deviations from experiments with different hyperparameters. Curves above the dashed line indicate a favorable performance-speedup trade-off. For single-label classification, benefits are predominantly positive, while for multi-label tasks, they are considerably smaller and symmetrically distributed around zero.

4.2   Early-Exits for Multi-label Classification

We next evaluate the effectiveness of early-exit strategies for multi-label classification tasks using three datasets: VOC [22], COCO [23], and a modified version of ImageNet [21]. To adapt the ImageNet dataset for multi-label classification, we randomly select $n = 10$ classes and combine images from m randomly chosen categories ($m \in \{2, 4\}$), creating composite images that belong to multiple classes. We train early-exit models using either sigmoid or softmax activation functions at the outputs of the internal classifiers. For loss functions, we employ binary cross-entropy (BCE) for sigmoid activations and cross-entropy (CE) for softmax activations. In the softmax models, we use varying thresholds {0.01, 0.05, 0.1, 0.2, 0.3} during the evaluation phase to determine positive classifications. For our initial experiments, we evaluate early-exit models using two strategies: an entropy-based strategy for softmax-activation models and both confidence and entropy-based strategies for sigmoid-activation models. All models are trained with learning rates $\lambda \in 10^{-3}, 10^{-4}, 10^{-5}$.

Limited Benefits for Multi-label Classification. Figure 3 illustrates the mean and the standard deviation of performance outcomes achieved by early-exits with basic exiting strategies, such as confidence and entropy thresholds, for VOC, COCO, and ImageNet test sets. The benefits of implementing early-exit architectures for multi-label classification are significantly less pronounced than those observed for single-label scenarios, with statistical significance levels $p \leq 0.05$. These findings indicate that applying early-exit strategies in a multi-label context poses greater challenges compared to their application in single-label tasks.

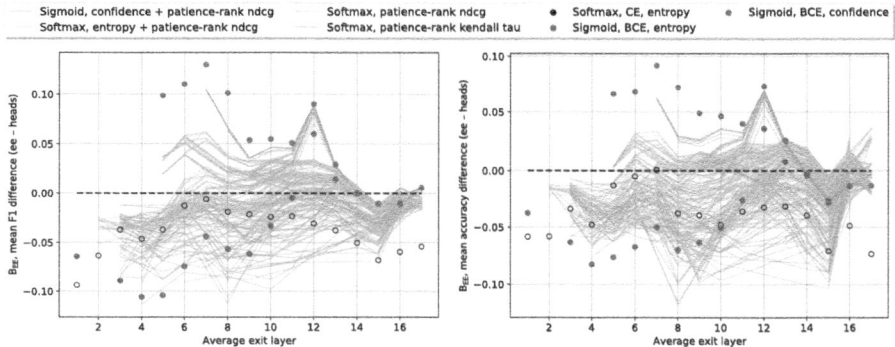

Fig. 4. The benefits B_{EE} of using rank-patience-based early-exiting strategies for multi-label VOC test dataset. Rank-patience-based approaches generally yield less promising results compared to simple confidence-threshold strategies. However, there are notable benefits in employing rank-patience-based approaches to enhance accuracy at higher speedups.

Rank-Patience-Based Early-Exiting Strategy for Multi-label Classification. Recognizing the performance gap outlined in previous experiments, we decided to further explore the alternate early-exit strategies for multi-label classification. Figure 4 presents the performance outcomes when implementing rank-patience-based strategies and hybrid strategies that combine rank-patience information with confidence or entropy thresholds. We employ two metrics to assess rank agreement: Normalized Discounted Cumulative Gain (nDCG) and Kendall's tau. Models are trained using different learning rates ($\lambda \in \{10^{-3}, 10^{-4}, 10^{-5}\}$), patience thresholds ($t \in \{2, 3, 4, 5, 6\}$), and rank-agreement tolerance thresholds (for nDCG scores values below $\{0.001, 0.01, 0.05, 0.1, 0.15\}$ and for Kendall's tau correlation tolerances values above $\{0.7, 0.8, 0.9, 0.95\}$).

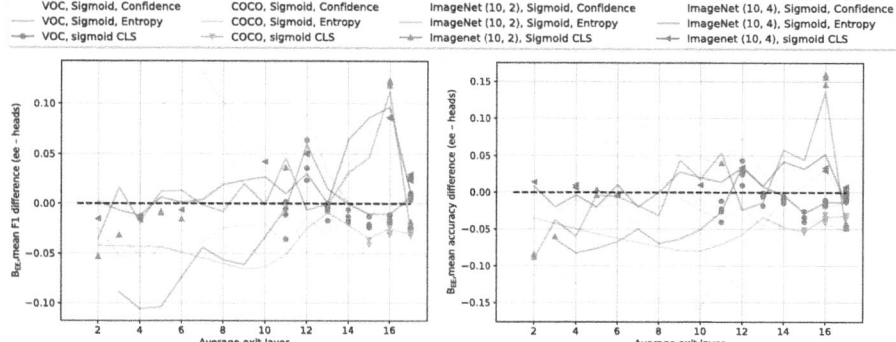

Fig. 5. Benefits B_{EE} of the learning-based early-exit strategy compared to baseline confidence and entropy methods. For each dataset, we show the best results from baseline methods alongside various hyperparameter configurations of the learning-based approach. Each point of the learning-based strategy represents a distinct model setup, illustrating that while there are modest performance improvements at certain speedups, these gains vary depending on the dataset.

Figure 4 showcases the results from the VOC dataset, where we tested various learning rates, consecutive head agreements, and prediction order tolerance thresholds. Our analysis reveals that despite rigorous testing, strategies based on patience and predicted probability order did not surpass basic methods such as the confidence-based strategy, with all results being statistically significant ($p \leq 0.05$). However, certain experiments demonstrated modest accuracy improvements at higher average exit layers, likely due to enhanced decision consensus in deeper layers, resulting in more accurate and reliable predictions.

Learning-Based Early-Exiting. We further explore the learning-based (classifier-based) early-exiting strategy. We introduce an auxiliary classifier added to each classification head, deciding whether an example should exit at that head. During evaluation, the output value (following sigmoid activation) ranges from 0 to 1, with a value exceeding 0.5 indicating a decision to exit. This single-point decision criterion directly determines the model's performance and specific speedup. While this strategy eliminates extra threshold selection, it also restricts the flexibility of controlling the performance-speedup ratio, offered by other strategies.

For the input to the early-exit decision classifiers, we use the logits from the corresponding classification head outputs (experiments with hidden states did not notably improve performance). We evaluated several common binary classification loss functions: Binary Cross-Entropy (BCE), Mean Squared Error

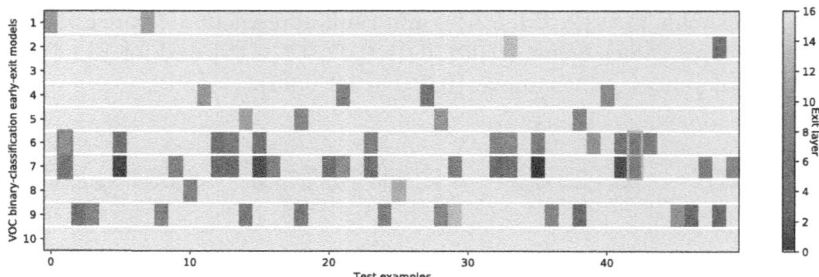

Fig. 6. Illustration of the challenge in early-exit models for multi-label classification. The x-axis shows test examples; the y-axis, ten class-specific binary models. Colors indicate the chosen exit head, highlighting varied recognition difficulty across classes. Rare instances where two different models exit at the same layer are shown in orange, while an example with objects of significantly different recognition difficulty, is highlighted in red. (Color figure online)

(MSE), L1 loss, and Hinge loss, across different hyperparameter configurations. Mean Squared Error loss with regularization, as shown in Eq. (1), yielded the best performance. To maintain conciseness, we present only the results from this loss function. Notably, the parameters of the main model remain frozen while training the early-exit decision classifiers, allowing these classifiers to be trained independently.

Figure 5 illustrates the performance of the learning-based early-exiting strategy in comparison to the baseline confidence and entropy strategies across different multi-label datasets: VOC, COCO, and ImageNet. While the learning-based approach achieves performance improvements at certain speedups, the extent of these improvements varies significantly across datasets. This variability highlights the importance of customizing early-exit strategies to fit specific dataset characteristics. Further exploration of learning-based strategies with different architectures and learning schemes may lead to more robust and universally effective solutions for multi-label classification tasks.

4.3   Ablation Study

Our exploration of the early-exit limitations in multi-label classification reveals that they provide fewer benefits compared to single-label scenarios. We suppose this complexity arises from the distinct recognition difficulty associated with each class within a single example. In other words, when we have a lot of classes present in one image, the network recognizes the presence of objects from different classes at different processing stages. To validate our hypothesis, we conduct a simple experiment, depicted in Fig. 6.

We train ten early-exit models, each dedicated to a unique class from the VOC dataset. Each model consists of binary classifiers heads that determine the presence or absence of its designated class. If the classifier recognizes its class as present, it exits at the current head. We selected test examples that included at least two of the ten classes we focused on. In Fig. 6, each point on the x-axis represents a test example, with its color indicating the exit head chosen by the respective model on y-axis. We highlight with orange color when two distinct models, each tasked with a different class, exit at the same layer. The diversity of exit heads across the models illustrates that different classes within an example indeed present varying degrees of recognition difficulty, thereby reinforcing our initial hypothesis.

Our findings suggest that in a multi-label settings, the idea of early-exit must extend beyond the traditional notion of determining an optimal exit point based on the overall prediction confidence. Instead, it should consider the varied recognition difficulty of each class within an example, hence posing a more nuanced problem. Our intuition is that the key characteristic impacting the effectiveness of early-exits could be the similarity of the classes within a dataset. When classes are very similar, the problem resembles multi-label case, making it more challenging for the model to distinguish between them, thus requiring more processing. Conversely, when classes are very distinct, they are easier to recognize, potentially benefiting more from early-exit strategies.

5   Conclusions

We introduced a systematic framework for evaluating early-exit architectures in single- and multi-label classification tasks, demonstrating its effectiveness in computer vision. We found that early-exits reduce computational time in single-label tasks with minimal accuracy loss, while their benefits in multi-label tasks are limited due to the complexity of recognizing multiple classes (see Sect. 4.3). We revisited the "big label" problem, suggesting that challenges in single-label tasks with many output classes arise more from dataset characteristics than from early-exit limitations. For single-label tasks, widely used confidence-based strategies proved highly effective. In contrast, multi-label classification presents greater challenges, underscoring the need for adaptive exit strategies tailored to class-specific difficulties. This presents a promising direction for future research aimed at improving both performance and efficiency.

Acknowledgements. The work of Klaudia Bałazy was carried out within the research project "Bio-inspired artificial neural network" (grant no. POIR.04.04.00-00-14DE/18-00) within the Team-Net program of the Foundation for Polish Science co-financed by the European Union under the European Regional Development Fund. This research was partially funded by the National Science Centre, Poland, grants no. 2023/49/B/ST6/01137 (work by Jacek Tabor). Some experiments were performed on servers purchased with funds from the flagship project entitled "Artificial Intelligence Computing Center Core Facility" from the DigiWorld Priority Research Area within the Excellence Initiative – Research University program at Jagiellonian University in Krakow.

References

1. Xu, C., McAuley, J.J.: A survey on dynamic neural networks for natural language processing. In: EACL (Findings), pp. 2325–2336. Association for Computational Linguistics (2023)
2. Graves, A.: Adaptive computation time for recurrent neural networks. arXiv preprint arXiv:1603.08983 (2016)
3. Dehghani, M., Gouws, S., Vinyals, O., Uszkoreit, J., Kaiser, L.: Universal transformers. In: ICLR (2019)
4. Teerapittayanon, S., McDanel, B., Kung, H.T.: Branchynet: fast inference via early exiting from deep neural networks. In: 2016 23rd International Conference on Pattern Recognition (ICPR), pp. 2464–2469. IEEE (2016)
5. Kaya, Y., Hong, S., Dumitras, T.: Shallow-deep networks: Understanding and mitigating network overthinking. In: International Conference on Machine Learning, pp. 3301–3310. PMLR (2019)
6. Liu, X., et al.: Towards efficient nlp: a standard evaluation and a strong baseline. arXiv preprint arXiv:2110.07038 (2021)
7. Liu, W., Zhao, X., Zhao, Z., Ju, Q., Yang, X., Lu, W.: An empirical study on adaptive inference for pretrained language model. IEEE Trans. Neural Netw. Learn. Syst. **34**(8), 4321–4331 (2021)
8. Xin, J., Tang, R., Lee, J., Yu, Y., Lin, J.: Deebert: dynamic early exiting for accelerating BERT inference. In: ACL, pp. 2246–2251. Association for Computational Linguistics (2020)
9. Liu, W., Zhou, P., Wang, Z., Zhao, Z., Deng, H., Ju, Q.: Fastbert: a self-distilling BERT with adaptive inference time. In: ACL, pp. 6035–6044. Association for Computational Linguistics (2020)
10. Geng, S., Gao, P., Fu, Z., Zhang, Y.: Romebert: robust training of multi-exit bert. arXiv preprint arXiv:2101.09755 (2021)
11. Schwartz, R., Stanovsky, G., Swayamdipta, S., Dodge, J., Smith, N.A.: The right tool for the job: Matching model and instance complexities. In: ACL, pp. 6640–6651. Association for Computational Linguistics (2020)
12. Wołczyk, M., et al.: Zero time waste: recycling predictions in early exit neural networks. Adv. Neural. Inf. Process. Syst. **34**, 2516–2528 (2021)
13. Wang, J., Chen, K., Chen, G., et al.: Skipbert: efficient inference with shallow layer skipping. In: ACL (2022)
14. Zhou, W., Canwen, X., Ge, T., McAuley, J., Ke, X., Wei, F.: Bert loses patience: fast and robust inference with early exit. Adv. Neural. Inf. Process. Syst. **33**, 18330–18341 (2020)
15. Zhu, W.: Leebert: learned early exit for BERT with cross-level optimization. In: ACL-IJCNLP, pp. 2968–2980. Association for Computational Linguistics (2021)
16. Xin, J., Tang, R., Yu, Y., Lin, J.: Berxit: early exiting for BERT with better fine-tuning and extension to regression. In: EACL, pp. 91–104. Association for Computational Linguistics (2021)
17. Zhang, Z., Zhu, W., Zhang, J., Wang, P., Jin, R., Chung, T.S.: PCEE-BERT: accelerating BERT inference via patient and confident early exiting. In: Findings of the Association for Computational Linguistics: NAACL 2022, Seattle, United States, July 2022, pp. 327–338. Association for Computational Linguistics (2022). https://doi.org/10.18653/v1/2022.findings-naacl.25. https://aclanthology.org/2022.findings-naacl.25

18. Sun, T., et al.: Early exiting with ensemble internal classifiers. arXiv preprint arXiv:2105.13792 (2021)

19. Schuster, T., Fisch, A., Jaakkola, T., Barzilay, R.: Consistent accelerated inference via confident adaptive transformers. arXiv preprint arXiv:2104.08803 (2021)

20. He, K., Zhang, X., Ren, S., Sun, J.: Deep residual learning for image recognition. In: Proceedings of the IEEE Conference on Computer Vision and Pattern Recognition, pp. 770–778 (2016)

21. Deng, J., Dong, W., Socher, R., Li, L.J., Li, K., Fei-Fei, L.: Imagenet: a large-scale hierarchical image database. In: 2009 IEEE Conference on Computer Vision and Pattern Recognition, pp. 248–255 (2009). https://doi.org/10.1109/CVPR.2009.5206848

22. Everingham, M., Van Gool, L., Williams, C.K.I., Winn, J., Zisserman, A.: The PASCAL Visual Object Classes Challenge 2007 (VOC 2007) Results (2007). http://www.pascal-network.org/challenges/VOC/voc2007/workshop/index.html

23. Lin, T.-Y., et al.: Microsoft COCO: common objects in context. In: Fleet, D., Pajdla, T., Schiele, B., Tuytelaars, T. (eds.) ECCV 2014. LNCS, vol. 8693, pp. 740–755. Springer, Cham (2014). https://doi.org/10.1007/978-3-319-10602-1_48

Comparative Analysis of Black-Box Optimization Methods for Weather Intervention Design

Yuta Higuchi[1]([envelope])[iD], Rikuto Nagai[2][iD], Atsushi Okazaki[3][iD],
Masaki Ogura[1]([envelope])[iD], and Naoki Wakamiya[2][iD]

[1] Hiroshima University, 1-3-2 Kagamiyama, Higashi-hiroshima,
Hiroshima 739-8511, Japan
{higuchi141226,oguram}@hiroshima-u.ac.jp
[2] Osaka University, 1-5 Yamadaoka, Suita, Osaka 565-0871, Japan
{rk-nagai,wakamiya}@ist.osaka-u.ac.jp
[3] Chiba University, 1-33 Yayoi, Inage, Chiba, Chiba 263-8522, Japan
atsushi.okazaki@chiba-u.jp

Abstract. As climate change increases the threat of weather-related disasters, the importance of weather control research is growing. The goal of weather control is to mitigate disaster risks by applying interventions with optimal timing, location, and intensity. This study considers a simulation-based control framework in which interventions in the wind velocity field serve as inputs and accumulated precipitation as the output. The objective is to identify optimal interventions that minimize total precipitation. However, the optimization process is highly challenging due to the vast scale and complexity of weather phenomena, which introduces two major challenges. First, obtaining accurate gradient information for optimization is difficult. In addition, numerical weather prediction models demand enormous computational resources, necessitating parameter optimization with minimal function evaluations. To address these challenges, this study proposes a method for designing weather interventions based on black-box optimization, which enables efficient exploration without requiring gradient information. The proposed method is evaluated in two distinct control scenarios: one-shot initial value intervention and sequential intervention based on model predictive control. Furthermore, a comparative analysis is conducted among four representative black-box optimization methods in terms of total rainfall reduction. Experimental results show that Bayesian optimization achieves higher control effectiveness than the others, particularly in high-dimensional search spaces. These findings suggest that Bayesian optimization is a highly effective approach for weather intervention computation.

Keywords: Black-Box optimization · Model Predictive Control · Warm Bubble Experiment · Real Atmosphere Experiment

M. H. Lees et al. (Eds.): ICCS 2025, LNCS 15903, pp. 194–209, 2025.
https://doi.org/10.1007/978-3-031-97626-1_14

1   Introduction

As global warming progresses, weather-related disasters such as hurricanes, floods, and torrential rain have become increasingly frequent and severe across many regions of the world. Over the 35 years leading up to 2014, the number of weather-related loss events approximately tripled, with total economic losses reaching up to US\$125 billion in 2005 [1]. Various studies on weather control have been conducted under this background [2,3]. However, because weather control involves complex and large-scale meteorological phenomena, it faces the following three major challenges: (1) identifying effective interventions, (2) selecting feasible interventions, and (3) reducing the computational time required for intervention calculations. To overcome these challenges, the application of control theory is required; however, traditional control theory has limitations in handling nonlinear, high-dimensional, and complex models such as weather systems [4].

The problem of identifying an effective intervention reduces to the optimization of its parameters such as timing, location, and intensity. However, the optimization process is extremely challenging due to the vast and complex nature of meteorological phenomena [5]. First, obtaining accurate gradient information of the objective function is difficult. Moreover, state-of-the-art numerical weather prediction (NWP) models demand enormous computational resources for each simulation run, thereby enforcing the necessity of minimizing function evaluations during parameter optimization. These constraints collectively suggest that black-box optimization methods [6], which operate exclusively on input-output data, offer a promising approach for designing interventions. Specifically, such methods iteratively explore the parameter space to seek optimal input values that maximize or minimize the objective function without recourse to derivative information. Their capacity to leverage evaluation results for efficient exploration renders them particularly well-suited to the demands of weather intervention optimization. However, to the best of our knowledge, no previous studies have applied black-box optimization techniques to the design of weather interventions, leaving the question of the most effective algorithm unresolved.

Therefore, in this paper, we design a weather intervention computation method using black-box optimization and evaluate its effectiveness through simulations using NWP models. We formulate two control scenarios: one that permits intervention only at a single time point (initial value intervention) and another that enables sequential interventions via model predictive control. Furthermore, by combining these control problems with two experimental settings that differ in scale and complexity, we conduct a comprehensive evaluation of several representative black-box optimization methods. Specifically, we compare the performance of Bayesian optimization, random search, particle swarm optimization, and genetic algorithms.

The contributions of this work are summarized as follows. This study proposes a black-box optimization framework for weather intervention optimization and demonstrates that effective interventions can be identified with minimal function evaluations. Furthermore, a rigorous comparative analysis of four representative black-box optimization algorithms is performed in terms of total rain-

Table 1. Overview of warm bubble experiment and real atmosphere experiment.

	Warm bubble experiment	Real atmosphere experiment
Objective	Ideal settings to reproduce cumulus convection and localized phenomena	Reproduce large-scale meteorological phenomena using real atmospheric conditions
Computational domain	Small (e.g., $10\,\mathrm{km}^2$)	Large (e.g., $3\,240\,000\,\mathrm{km}^2$)

fall reduction, with a detailed examination of their operational characteristics. Notably, the experimental results provide evidence that Bayesian optimization achieves superior performance, particularly in high-dimensional search spaces. Overall, the findings support Bayesian optimization as a highly effective and reliable approach for computing weather interventions, outperforming alternative algorithms.

2   Control Problem Formulation

This section first describes the meteorological model used in this study, Scalable Computing for Advanced Library and Environment Regional Model (SCALE-RM). Next, two weather control problems are formulated using this model and the challenges associated with solving the corresponding optimization problems are discussed.

2.1   SCALE-RM

SCALE-RM is a NWP model specifically developed for climate research. This model is part of the SCALE software library, which supports weather forecasting across various computational platforms [7,8]. Due to its versatility and reliability, SCALE-RM has been widely utilized in studies on weather forecasting and atmospheric science [9]. In weather control research, SCALE-RM plays a vital role in modeling and assessing control methods. This model enables detailed simulations of atmospheric interactions and external interventions, facilitating precise evaluations of control method effectiveness.

In this study, numerical simulations were conducted using two experimental setups provided by SCALE-RM. The first one, the warm bubble experiment [10], employs a two-dimensional model to idealize and simulate convective clouds. The second one, the real atmosphere experiment, uses a three-dimensional model to replicate more realistic atmospheric behavior. An overview of these two experimental settings is provided in Table 1, and the details of each experiment are described in Sects. 4 and 5, respectively.

Let w_t denote the atmospheric state variables at a given time t, such as potential temperature or humidity, and f_0 denote the model representing atmospheric state changes. The weather system can be modeled as

$$w_{t+1} = f_0(w_t), \tag{1}$$

where, w_t is a high-dimensional vector that encapsulates all meteorological state variables distributed in space, and f_0 represents an idealized model assuming no noise. In both experimental setups, the dimensionality of w_t exceeds tens of thousands. The model f_0 exhibits nonlinearity due to the complexity of processes governing meteorological phenomena, such as turbulence and cloud microphysics.

2.2 Weather Control Problem

In this study, interventions to the initial atmospheric state in a numerical weather prediction model are considered as control inputs for weather modification, without assuming specific intervention methods. This approach is based on the framework proposed by Ohtsuka et al. [11]. Let u_t denote the intervention at a given time t, and f denote the model representing atmospheric state change with the intervention. The control system in this study can be modeled as

$$w_{t+1} = f(w_t, u_t). \tag{2}$$

SCALE-RM includes variables MOMX and MOMY, which represent atmospheric momentum [kg · m/s]. In this study, these variables are modified as an intervention in the atmospheric state, corresponding to the manipulation of the wind velocity field. Real-world intervention methods are expected to involve deploying offshore wind turbines [12] or installing obstacles. Furthermore, the objective of the intervention is set to minimization of the total precipitation intensity, expressed in [kg/m^2/s], over a specified surface region from $t = 1$ [s] to $t = T_e$ [s]. NWP models simulate physical phenomena within a spatial framework called the computational domain, which is divided into basic units called grid cells. Let $\mathrm{PREC}(t, x, y)$ denote the precipitation intensity at grid cell $(x, y, 0)$ at time t. Then, define G as the set of grid cells where total precipitation is to be minimized.

In this study, we formulate two control problems: one based on initial value intervention, which applies intervention at a single time step, and another using model predictive control, which enables sequential interventions. The following sections present an overview of each control problem and describe the corresponding optimization formulation.

Initial Value Intervention. In this control problem, intervention is applied only at the initial time. Let the intervention to MOMX and MOMY be denoted as d_X and d_Y, respectively. The intervention is applied to a single grid cell, with the intervention location given by (x, y, z). Then, this intervention is represented as

$$u_0 = (d_X, d_Y, x, y, z). \tag{3}$$

Let U denote the set of grid cells (x, y, z) where interventions can be applied. The optimization problem to be solved can then be formulated as

$$\underset{u_0}{\text{minimize}} \quad \sum_{t=1}^{T_e} \sum_{(x,y)\in G} \text{PREC}(t,x,y)$$

$$\text{subject to} \quad w_{t+1} = f_0(w_t), \quad t = 1, 2, \ldots, T_e - 1 \tag{4}$$

$$w_1 = f(w_0, u_0)$$

$$d_X \in [\underline{d}, \bar{d}], \quad d_Y \in [\underline{d}, \bar{d}], \quad (x, y, z) \in U$$

where $\underline{d}, \bar{d}$ are the boundary values for the amount of change applied to the atmospheric momentum.

Model Predictive Control. MPC [13] is a control method that optimizes control inputs while predicting the future state of the control target at each time step. MPC measures the system output in real time via feedback control and sequentially computes the appropriate control inputs. Due to this characteristic, MPC is considered highly effective, especially for control systems exhibiting chaos and uncertainty, such as meteorological systems.

In this control problem, intervention is applied at multiple time steps. Specifically, we consider a setting in which interventions are applied every T_{step} [s] starting from time $t = 0$ [s]. Here, we define the interval between interventions as one time step. Let $d_{\tau,X}$ and $d_{\tau,Y}$ denote the intervention to MOMX and MOMY at τ steps ahead of the current time, respectively. The intervention location is represented as (x_τ, y_τ, z_τ), and the intervention at step τ is then represented as

$$u_\tau = (d_{\tau,X}, d_{\tau,Y}, x_\tau, y_\tau, z_\tau). \tag{5}$$

The prediction horizon is set to T_f steps, and the optimization problem at time t is formulated as

$$\underset{v_0,v_1,\ldots,v_{T_f-1}}{\text{minimize}} \quad \sum_{k=1}^{T_{\text{step}}T_f} \sum_{(x,y)\in G} \text{PREC}(t+k,x,y)$$

$$\text{subject to} \quad w_{T_{\text{step}}\tau+1} = f(w_{T_{\text{step}}\tau}, v_\tau), \quad \tau = 0, 1, \ldots, T_f - 1, \tag{6}$$

$$w_{T_{\text{step}}\tau+l+1} = f_0(w_{T_{\text{step}}\tau+l}), \quad l = 1, 2, \ldots, T_{\text{step}} - 1,$$

$$d_{\tau,X} \in [\underline{d}, \bar{d}], \quad d_{\tau,Y} \in [\underline{d}, \bar{d}], \quad (x_\tau, y_\tau, z_\tau) \in U.$$

Among the intervention sequence $v_0, v_1, \ldots, v_{T_f-1}$ obtained by solving Optimization Problem (6), v_0 is applied as the actual intervention applied at time t. By executing this operation every T_{step} from $t = 0$ to T_e, the accumulated precipitation in the target region G from $t = 1$ to T_e, which is the control objective, is reduced.

2.3   Challenges of Optimizing an Intervention

This section discusses the challenges and difficulties associated with solving Eqs. 4 and 6. First, atmospheric phenomena are inherently nonlinear and highly

sensitive to initial conditions, causing even slight differences to amplify expo-
nentially [14]. As a result, the objective function often has a highly complex
landscape with non-convexity and numerous local optima. As many optimization
methods assume a smooth objective function, convergence becomes challenging.
Second, numerical weather prediction models are computationally expensive,
and incorporating high-resolution models like SCALE-RM into the optimization
loop significantly increases simulation costs. Furthermore, SCALE-RM solves
atmospheric equations through numerical time integration. However, due to the
complexity of this process, obtaining an explicit analytical gradient with respect
to state variables is challenging, making gradient-based methods inapplicable.
Additionally, constraints on momentum modifications d_X, d_Y, grid cell selection,
and model uncertainties further complicate finding an exact optimal solution.

Given these challenges, finding exact solutions under limited computational
resources is impractical. Instead, approximate methods, heuristics, and local
optimization techniques are required. To address the difficulty of obtaining gradient information directly, this study employs black-box optimization methods.

3   Evaluation Setting

This section first provides an overview of black-box optimization methods, followed by a description of the simulation setting used to evaluate their effectiveness.

3.1   Black-Box Optimization

Black-box optimization methods are used to search for optimal input values
that maximize or minimize an objective function when the function itself and
its constraints are not explicitly given, and only input-output data is available.
These methods are widely applied in various fields, including engineering design,
hyperparameter tuning of machine learning models, and simulations of complex systems [15]. In such optimization problems, the objective function is often
nonlinear, discontinuous, noisy, or computationally expensive. Therefore, selecting an efficient search method is crucial. Representative black-box optimization
methods include Bayesian Optimization (BO), Random Search (RS), Particle
Swarm Optimization (PSO), and Genetic Algorithm (GA). We omit the detailed
explanation of the methods due to page limitations.

3.2   Simulation Setting

First, we describe the implementation details and hyperparameter settings for
each optimization method used in this study. BO was implemented using the
`gp_minimize` function or the `Optimizer` class from the Scikit-Optimize library,
adopting the default hyperparameter settings. For PSO, we employed the Linearly Decreasing Inertia Weight Method (LDIWM) [16], a widely used parameter
adjustment technique. PSO requires defining the hyperparameters c_1, c_2, and w,

which determine particle movement. In LDIWM, these hyperparameters are set as $c_1 = c_2 = 2.0$, and w is linearly decreased as follows: $w = w_{\max} - \frac{n(w_{\max} - w_{\min})}{n_{\max}}$ where n is the current iteration count, and $n_{\max}$ is the maximum number of iterations. Following [16], we set $w_{\max} = 0.9$ and $w_{\min} = 0.4$. For GA, we used real-valued encoding. The selection, crossover, and mutation methods were set as follows: tournament selection, blend crossover [17], and random replacement with real values. The corresponding hyperparameters were: a tournament size of 3, a crossover rate of 0.8, a blending factor α of 0.5, and a mutation rate of 0.05. Since the control problems in this study involve a search space with integer values, when applying PSO or GA, the search space is defined in real numbers, and the obtained parameters are rounded to the nearest integers.

Next, we describe the simulation settings designed to ensure a fair evaluation of the control effectiveness of the black-box optimization methods. In weather intervention optimization, it is crucial to identify effective interventions with a minimal number of function evaluations. Therefore, simulations were conducted with various maximum function evaluation limits—specifically, 15, 30, 50, 100, 150, 200, 250, and 300. Since both PSO and GA require the specification of a population size, we set the population size based on the evaluation limit: a population size of 5 was used when the maximum function evaluations were 15, and a population size of 10 was employed for all other cases. Furthermore, to account for the inherent randomness of black-box optimization methods, we conducted simulations using 10 different random seeds, applying each optimization method to each seed. The results were then analyzed and discussed.

4   Warm Bubble Experiment

This section presents the simulation setup, results, and discussion in the context of the warm bubble experiment.

4.1   Experimental Settings

The warm bubble experiment is a widely used benchmark test for evaluating the performance of NWP models [10]. This experimental setup replicates atmospheric convection processes under idealized conditions. As demonstrated by Ohtsuka et al. [11], who validated control methods using this setup, the warm bubble experiment also serves as a benchmark for weather control studies.

In this study, simulations were conducted using SCALE-RM version 5.5.1. In this experiment, the horizontal grid spacing was set to 500 m in both the east-west and north-south directions, with 1 grid point in the east-west direction and 40 grid points in the north-south direction. The vertical domain consisted of 97 layers, with the model top at 20 km. The initial condition featured a warm bubble placed near the surface at the center grid in the y-direction. The warm bubble had a horizontal radius of 4 km, a vertical radius of 3 km, and a center temperature 3 K higher than the surrounding environment. This warm bubble triggered cumulus cloud formation and convection, leading to precipitation. This

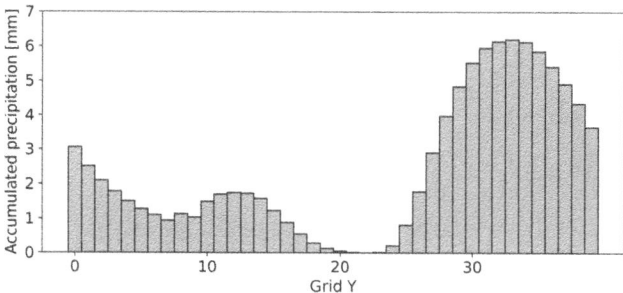

Fig. 1. Accumulated precipitation over one hour without any control. Grid Y represents the grid-cell index in the north-south (Y) direction; the length of one grid cell in Y is 500 m.

atmospheric motion is observed over a duration of one hour from the start of the experiment. Figure 1 presents the accumulated precipitation at each location over one hour without intervention. For each control problem, the objective is to determine an intervention that minimizes the total area of this bar graph, subject to the given constraints.

The two weather control problems described in Sect. 2 are formulated in a form applicable to the warm bubble experiment. The control objective is to reduce the total accumulated precipitation over the entire computational domain from time $t = 1$ [s] to $t = 3600$ [s]. Furthermore, since this experiment employs a north-south vertical plane as a two-dimensional setup, interventions to the atmospheric state are applied only to MOMY. In this experiment, the lower and upper bounds of d_Y, denoted by $\underline{d}$ and $\bar{d}$, were set to -30 and 30, respectively. Intervention requires predefined intensity limits. However, since this study does not assume a specific intervention method, the appropriate bounds are not uniquely determined. Therefore, based on preliminary experiments, we selected bounds sufficient for evaluating the control performance of each method.

Initial Value Intervention. In this control problem, we attempt to reduce accumulated precipitation by applying intervention only at the initial time step. At the initial time, the intervention applied to MOMY at a specific grid cell $(x, y, z) = (0, y, z)$ is defined as $u_0 = (d_Y, y, z)$. The corresponding optimization problem is then formulated as follows:

$$\underset{u_0}{\text{minimize}} \quad \sum_{t=1}^{3600} \sum_{y=0}^{39} \text{PREC}(t,0,y)$$

$$\begin{aligned}
\text{subject to} \quad & w_{t+1} = f_0(w_t), \quad t = 1,2,\ldots,3599 \\
& w_1 = f(w_0, u_0), \\
& 0 \le y \le 39, \quad y \in \mathbb{Z}, \\
& 0 \le z \le 96, \quad z \in \mathbb{Z}, \\
& -30 \le d_Y \le 30.
\end{aligned} \tag{7}$$

MPC. In this control problem, we attempt to reduce accumulated precipitation through sequential interventions applied every $T_{\text{step}} = 600$ seconds from the initial time. This interval balances the number of control patterns and computational efficiency, allowing effective evaluation of black-box optimization methods. Let the intervention applied at a future location $(x, y, z) = (0, y_\tau, z_\tau)$ at τ steps ahead of the current time be represented as $v_\tau = (d_{\tau,Y}, y_\tau, z_\tau)$. With a prediction horizon of T_f steps, the MPC optimization problem at a given time t can be formulated as follows:

$$\underset{v_0, v_1, \ldots, v_{T_f-1}}{\text{minimize}} \quad \sum_{k=0}^{600T_f} \sum_{y=0}^{39} \text{PREC}(t+k, 0, y)$$

$$\begin{aligned}
\text{subject to} \quad & w_{600\tau+1} = f(w_{T_{\text{step}}\tau}, v_\tau), \quad \tau = 0, 1, \ldots, T_f - 1, \\
& w_{600\tau+l+1} = f_0(w_{600\tau+l}), \quad l = 1, 2, \ldots, 599, \\
& 0 \le y_\tau \le 39, \quad y_\tau \in \mathbb{Z}, \\
& 0 \le z_\tau \le 96, \quad z_\tau \in \mathbb{Z}, \\
& -30 \le d_{\tau,Y} \le 30.
\end{aligned} \tag{8}$$

In this experiment, the prediction horizon at the initial time step is set to $T_f = 6$ and decreases by one step as control progresses. From the sequence of interventions $v_0, v_1, \ldots, v_{T_f-1}$ obtained by solving the optimization problem (8), only v_0 is applied as the actual intervention at time t. Since the total experiment duration is 3600 s, the optimization problem (8) is solved a total of six times.

4.2 Results and Discussion

This study aims to compare the effectiveness of black-box optimization methods as computational approaches for weather intervention. Since simulations require substantial computational resources, identifying appropriate parameters with a minimal number of function evaluations is crucial. To this end, for each of the two control problems, the control effectiveness of the optimal solutions obtained under different function evaluation limits is computed and presented in Figs. 2a and 2c. Furthermore, to evaluate the impact of the inherent randomness in black-box optimization methods on control performance, we conducted 10 simulations

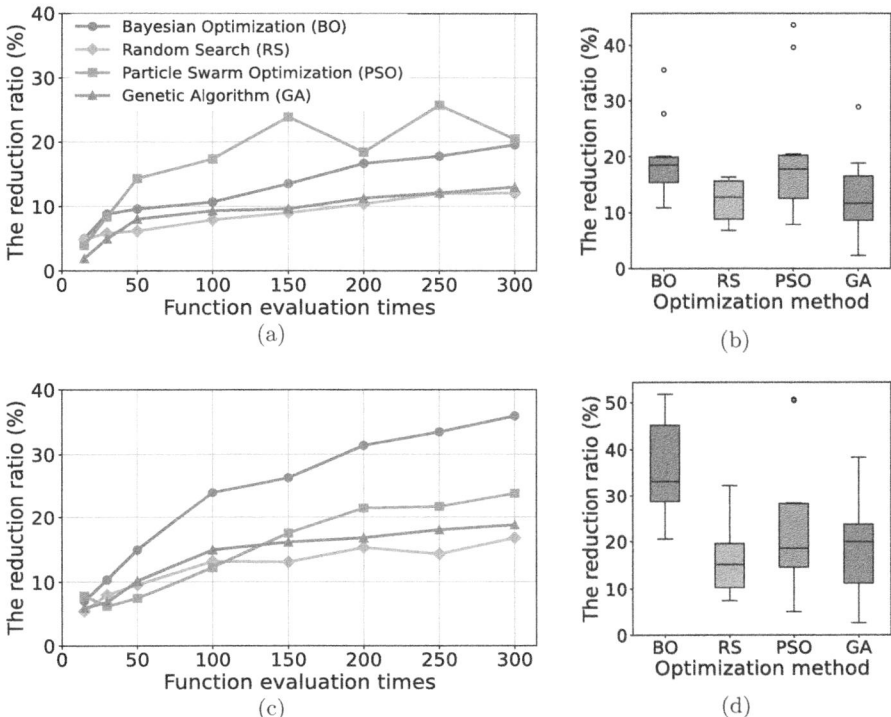

Fig. 2. Results of the warm bubble experiment. The upper panel shows the results of initial value intervention, while the lower panel shows those of MPC. (a) and (c) represent the average control effect at each function evaluation step, while (b) and (d) present the results with a function evaluation limit of 300.

with the number of function evaluations fixed at an upper limit of 300. The results of these simulations are presented in Figs. 2b and 2d.

Finally, the optimal solutions and corresponding optimal values explored by the black-box optimization methods for the control problem involving initial value interventions are presented in Fig. 3.

First, as shown in Figs. 2a and 2c, the reduction rate of accumulated precipitation generally improved with an increasing number of function evaluations, except when PSO was applied to the initial value intervention problem. For the initial value intervention problem, PSO achieved higher reduction rates, whereas for the MPC problem, BO performed better. This suggests that differences in control problems, such as the dimensionality of the search space and the use of feedback control, significantly impact control effectiveness.

Next, as shown in Figs. 2b and 2d, BO consistently achieved the highest reduction rate among all methods, even in the worst-case scenarios. This trend was particularly pronounced in the MPC problem, where, with the maxi-

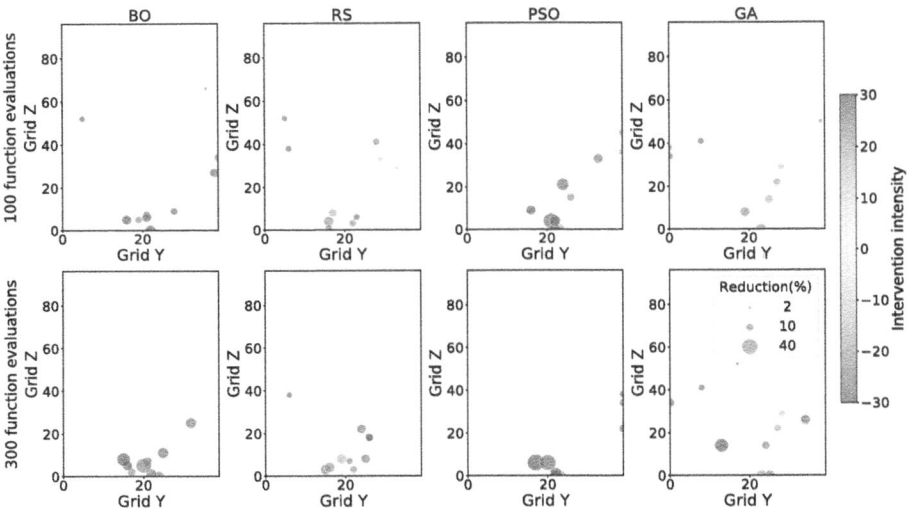

Fig. 3. The optimal solutions and corresponding values were obtained for each maximum number of function evaluations in the warm bubble experiment. The optimal solutions obtained from each simulation are represented by the position and color of the points. The size of each point indicates the reduction rate of accumulated precipitation achieved through control. Grid Y and Grid Z denote the grid-point indices in the meridional (north-south) and vertical directions, respectively. One Grid Y spacing is 500 m; Grid Z spacings vary with height (smaller near the surface, larger aloft). The upper panel shows results for a maximum of 100 function evaluations, while the lower panel shows those for 300 function evaluations.

mum number of function evaluations fixed at 300, the worst-case reduction rate achieved by BO exceeded the median reduction rates of other methods. This result can be attributed to BO's use of Gaussian process regression, which allows the regression model to approximate the objective function and effectively balance exploration and exploitation. Consequently, the random variability of initial sampling points had minimal impact on the final control performance. For PSO, outliers with exceptionally high reduction rates were observed in both control problems. This is likely due to PSO's convergence characteristics—once an exceptionally favorable search point is found, the swarm rapidly converges toward that search point. This suggests that PSO's control performance is strongly influenced by its inherent randomness.

Finally, Fig. 3 shows that grid cells with high control effectiveness, represented by large points, are concentrated in the region where $Z \leq 20$. Additionally, control effectiveness tends to increase as $|d_Y|$ becomes larger.

5   Real Atmosphere Experiment

This section presents the simulation setup, results, and discussion in the context of the real atmosphere experiment.

5.1   Experimental Settings

The real atmosphere experiment is a numerical simulation framework that replicates and predicts real-world weather phenomena based on actual atmospheric conditions and surface parameters. This framework requires substantial computational resources due to large-scale data processing and complex physical calculations. However, the higher computational cost enables more realistic behavior compared to idealized experiments. Therefore, this framework serves as a crucial tool for advancing real-world weather control.

In this study, simulations were conducted using version 5.5.1 or 5.5.3 of the SCALE-RM model. In these simulations, the horizontal grid spacing was set to 20 km in both the east-west and north-south directions, with 90 grid points in each direction. The model included 36 vertical layers, with an upper boundary approximately 28 km above the surface. The computational domain was centered around approximately 34 °N and 135 °E, encompassing Japan and the Korean Peninsula. Atmospheric data, including wind speed and temperature, as well as elevation and sea surface data at 00:00 UTC on 15 July 2007, were used as initial conditions.

The two weather control problems described in Sect. 2 are formulated in a form applicable to the real atmosphere experiment. First, a baseline simulation

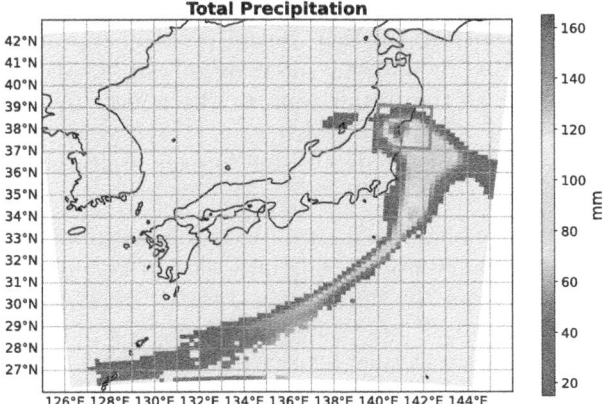

Fig. 4. The 6-h accumulated precipitation without any interventions is presented, with only grid cells receiving 20 mm or more of precipitation being colored. The red-framed region in the figure indicates the target area where the accumulated precipitation is to be minimized in this section.

was conducted without intervention to evaluate the 6-hour accumulated precipitation at each location in the computational domain. The results, shown in Fig. 4, indicate significant precipitation within the red-framed region. Therefore, the objective of the intervention is set to minimization of the total precipitation intensity [kg/m^2/s] over the target region $G = \{(x,y) \in \mathbb{Z}^2 \mid 65 \leq x \leq 74, 60 \leq y \leq 69\}$ during the period from $t = 1$ [s] to $t = 21600$ [s].

In a preliminary experiment, we attempted to control the system using the same settings as those in Sect. 4, by applying intervention only to a single grid cell. However, the reduction in accumulated precipitation was negligible. Here, we expand the number of grid cells to which intervention is applied. The set of grid cells where intervention is applied is defined as

$$R_{\ell,m,n} = \{(x,y,z) \in \mathbb{Z}^3 \mid \ell \leq x \leq \ell + 4, m \leq y \leq m + 4, n \leq z \leq n + 4\} \quad (9)$$

where ℓ, m, n represent the western, southern, and bottom boundaries of the region where intervention is applied, respectively. The same intervention is applied to all grid cells within this region. However, based on the results in Sect. 4, interventions in the upper atmosphere have limited effects. Therefore, interventions are applied near the surface by fixing $n = 0$. Furthermore, Fig. 4 indicates that interventions in the western part of the computational domain have a limited impact on precipitation reduction in the target region. Based on this observation, the feasible intervention is restricted to $\ell \in \{45, 46, \ldots, 85\}$ and $m \in \{0, 1, \ldots, 85\}$.

In the simulation, the intervention bounds were initially set to $\underline{d} = -30$ and $\bar{d} = 30$, as in the warm bubble experiment. However, this sometimes led to a loss of physical consistency, causing simulation failure. To prevent this issue, the bounds were adjusted to $\underline{d} = -20$ and $\bar{d} = 20$.

Initial Value Intervention. In this control problem, we attempt to reduce accumulated precipitation by allowing intervention only at the initial time. An intervention is applied at the initial time to the region $R_{\ell,m} = \{(x,y,z) \in \mathbb{Z}^3 \mid \ell \leq x \leq \ell + 4, m \leq y \leq m + 4, 0 \leq z \leq 4\}$, which is represented as $u_0 = (d_X, d_Y, R_{l,m})$. We omit the details due to page limitation.

MPC. In this control problem, we attempt to reduce accumulated precipitation by allowing sequential interventions. The intervention is applied at intervals of $T_{\text{step}} = 3600$, starting from the initial time. Let the intervention applied to a region $R_{\tau,\ell,m} = \{(x,y,z) \in \mathbb{Z}^3 \mid \ell_\tau \leq x \leq \ell_\tau + 4, m_\tau \leq y \leq m_\tau + 4, 0 \leq z \leq 4\}$ at τ steps ahead of the current time be represented as $v_\tau = (d_{X,\tau}, d_{Y,\tau}, R_{\tau,\ell,m})$. In this experiment, the prediction horizon T_f at the initial time step is set to 6. We omit the details due to page limitation.

5.2   Results and Discussion

In this section, simulation experiments similar to those in Sect. 4 are conducted. The control effect achieved with the optimal solutions was calculated for each

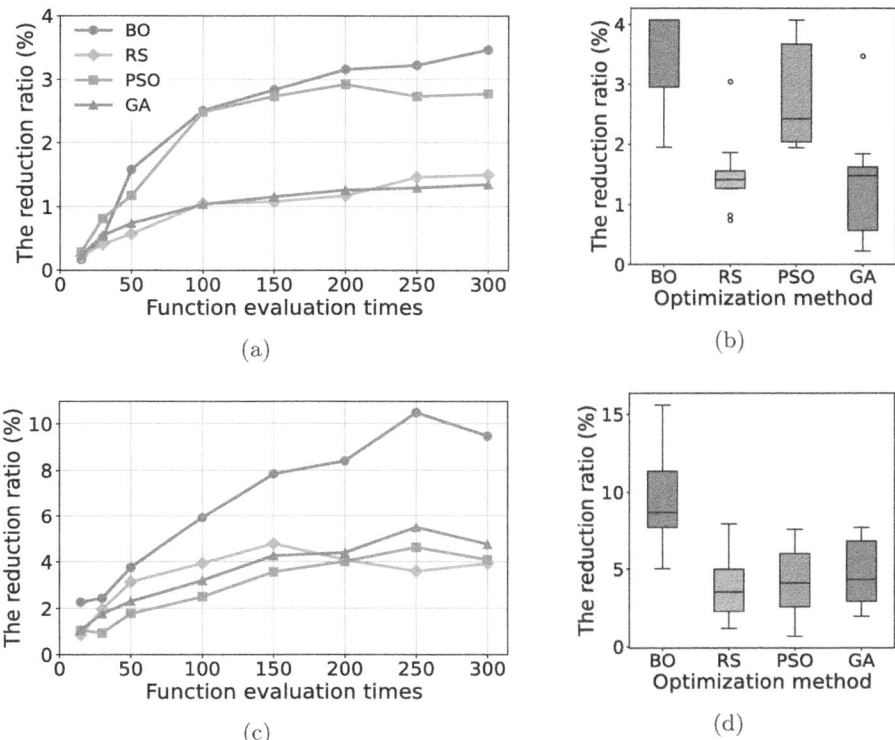

Fig. 5. Results of the real atmosphere experiment. The upper panel shows results with initial-value intervention, while the lower panel shows those with MPC. (a) and (c) represent the average control effect at each function evaluation times, while (b) and (d) present results with a function evaluation limit of 300.

upper limit on the number of function evaluations and the results are illustrated in Figs. 5a and 5b. Furthermore, the results of 10 simulations conducted with the number of function evaluations fixed at an upper limit of 300 are presented in Figs. 5b and 5d.

First, as shown in Figs. 5a and 5c, the reduction rate of accumulated precipitation generally improved with an increasing number of function evaluations, except when PSO was applied to the initial value intervention problem, similar to the warm bubble experiment. Furthermore, in both control problems, BO achieved the highest reduction rates among the four methods. Next, as shown in Figs. 5b and 5d, BO consistently achieved higher minimum reduction rates than other methods, aligning with the results of the warm bubble experiment. Notably, in the initial value intervention problem, when the number of function evaluations was fixed at 300, the median reduction rate equaled the maximum reduction rate, indicating exceptionally high convergence to the optimal solution with BO. A significant difference from the warm bubble experiment is that BO was also the most effective method for the initial value intervention problem. In the real atmosphere experiment, the search space dimensionality increased by one, which may have influenced the results. This is further supported by the observation that in the MPC problem, which has an even higher-dimensional search space, the control effectiveness of BO becomes even more pronounced.

6   Conclusion

In this study, we designed a weather intervention computation method based on black-box optimization and evaluated its effectiveness through simulations using the NWP model. The results suggest that, among the black-box optimization methods tested, BO demonstrates notably high effectiveness. It should be noted that these findings are based on specific experimental settings and may not necessarily generalize to other scenarios. Nevertheless, our results indicate the potential of black-box optimization as a viable approach for weather intervention computation and offer valuable insights into the distinct characteristics of each method.

However, several challenges remain. First, hyperparameters for PSO and GA were selected based on prior studies and preliminary testing, but their suitability for the specific tasks requires further validation. Second, the experimental setups may not fully capture the complexity of real atmospheric systems. Incorporating data assimilation and ensemble simulations could address this limitation. Third, the findings are based on a specific scenario and may not generalize to other weather control objectives or regions. Further testing under diverse conditions is needed. Lastly, the feasibility of implementing large-scale interventions in real-world conditions remains uncertain. Future research should explore realistic control strategies and assess their viability through detailed simulations.

Acknowledgments. This work was supported by JST Moonshot R&D Program Grant Number JPMJMS2284.

Disclosure of Interests. The authors have no competing interests to declare that are relevant to the content of this article.

References

1. Hoeppe, P.: Trends in weather related disasters – consequences for insurers and society. Weather Clim. Extremes **11**, 70–79 (2016)
2. Hoffman, R.N.: Controlling the global weather. Bull. Am. Meteor. Soc. **83**(2), 241–248 (2002)
3. Dong, X., et al.: Increase of precipitation by cloud seeding observed from a case study in November 2020 over Shijiazhuang, China. Atmos. Res. **262**, 105766 (2021)
4. Soldatenko, S., Yusupov, R.: An optimal control perspective on weather and climate modification. Mathematics **9**, 305 (2021)
5. Bennett, A.F., Chua, B.S., Leslie, L.M.: Generalized inversion of a global numerical weather prediction model. Meteorol. Atmos. Phys. **60**(1), 165–178 (1996)

6. Jones, D.R., Schonlau, M., Welch, W.J.: Efficient global optimization of expensive black-box functions. J. Global Optim. **13**, 455–492 (1998)
7. Sato, Y., Nishizawa, S., Yashiro, H., Miyamoto, Y., Kajikawa, Y., Tomita, H.: Impacts of cloud microphysics on trade wind cumulus: which cloud microphysics processes contribute to the diversity in a large eddy simulation? Prog. Earth Planet Sci. **2**(1), 1–16 (2015). https://doi.org/10.1186/s40645-015-0053-6
8. Nishizawa, S., Yashiro, H., Sato, Y., Miyamoto, Y., Tomita, H.: Influence of grid aspect ratio on planetary boundary layer turbulence in large-eddy simulations. Geosci. Model Dev. **8**, 3393–3419 (2015)
9. Honda, T., et al.: Advantage of 30-s-updating numerical weather prediction with a phased-array weather radar over operational nowcast for a convective precipitation system. Geophys. Res. Lett. **49**, e2021GL096927 (2022)
10. Zhang, F., Snyder, C., Sun, J.: Impacts of initial estimate and observation availability on convective-scale data assimilation with an ensemble Kalman filter. Mon. Weather Rev. **132**(5), 1238–1253 (2004)
11. Ohtsuka, T., Okazaki, A., Ogura, M., Kotsuki, S.: Convex optimization of initial perturbations toward quantitative weather control. SOLA **advpub**, 2025-020 (2025)
12. Jacobson, M.Z., Archer, C.L., Kempton, W.: Taming hurricanes with arrays of offshore wind turbines. Nat. Clim. Change **4**(3), 195–200 (2014)
13. Mayne, D.Q., Rawlings, J.B., Rao, C.V., Scokaert, P.: Constrained model predictive control: stability and optimality. Automatica **36**(6), 789–814 (2000)
14. Lorenz, E.N.: The predictability of a flow which possesses many scales of motion. Tellus **21**(3), 289–307 (1969)
15. Audet, C.: A survey on direct search methods for blackbox optimization and their applications. In: Pardalos, P.M., Rassias, T.M. (eds.) Mathematics Without Boundaries, pp. 31–56. Springer, New York (2014). https://doi.org/10.1007/978-1-4939-1124-0_2
16. Shi, Y.: Particle swarm optimization. IEEE Connect. **2**(1), 8–13 (2004)
17. Eshelman, L.J., Schaffer, J.D.: Real-coded genetic algorithms and interval-schemata. Found. Genetic Algor. **2**, 187–202 (1993)

Dual Adaptive Windows Toward Concept-Drift in Online Network Intrusion Detection

Xiaowei Hu[1,2], Duohe Ma[1,2], Wen Wang[1,2(✉)], and Feng Liu[1,2]

[1] Institute of Information Engineering, CAS, Beijing, China
{huxiaowei,maduohe,wangwen,liufeng}@iie.ac.cn
[2] School of Cyber Security, University of CAS, Beijing, China

Abstract. Network intrusion detection is a commonly used and critical defense mechanism in the field of cybersecurity for identifying abnormal traffic online. However, the phenomenon of concept drift leads to a decrease in the accuracy of online intrusion detection systems in recognizing malicious traffic. Traditional machine learning-based intrusion detection systems are unable to adapt to the changes in data distribution of dynamic data streams. To address this issue, we propose DWOIDS, an online intrusion detection system based on dual adaptive windows and a Hoeffding tree classifier. When concept drift occurs in network data streams, it employs dual adaptive windows to monitor the prediction error of the classifier, continuously refining the classifier's accuracy in identifying malicious traffic. We conducted experimental evaluations on multiple datasets. Our proposed method demonstrated superior classification performance when compared to the state-of-the-art.

Keywords: Concept-Drift · Intrusion Detection · Adaptive Windows

1 Introduction

In the increasingly interconnected digital ecosystem, the internet has facilitated a surge in networked devices, with network adversaries infiltrating these devices through data streams to commit data misuse. Consequently, the demand for robust online Network Intrusion Detection Systems (NIDS) has intensified. Online NIDS [1] are vital for maintaining security by continuously monitoring network traffic in real-time to detect malicious and suspicious activities. In recent years, research on Intrusion Detection Systems (IDS) has primarily been divided into two categories: rule-based IDS [8,25] and IDS based on Machine Learning (ML) [3,23] or Deep Learning (DL) [5,9,22]. Rule-based IDS relies on a predefined set of rules and identifies malicious features in network traffic through pattern matching. This approach offers the advantages of high detection accuracy and rapid detection speed. However, its static rules are unable to adapt to unknown threats and require regular updates. To overcome this limitation, researchers have recently begun to integrate ML and DL technologies to develop

M. H. Lees et al. (Eds.): ICCS 2025, LNCS 15903, pp. 210–224, 2025.
https://doi.org/10.1007/978-3-031-97626-1_15

IDS with greater adaptability. Gao et al. [12] integrated various classifiers such as decision trees, random forests, KNN, and DNN, and dynamically adjusted their weights to address declining classification accuracy. Chiche et al. [7] proposed a method combining machine learning and knowledge systems to enhance the adaptability and scalability of intrusion detection. These methods involve pre-training models and optimizing parameters using gradient descent to effectively fit threat traffic. Empirical studies have demonstrated that these methods possess a certain degree of effectiveness in detecting unknown threats. However, the presence of concept drift leads to a decline in the recognition accuracy of intrusion detection models trained on static datasets over time. While retraining these models with current data is a feasible solution to maintain their optimal performance, it incurs significant resources and time overhead.

The phenomenon of concept drift [18] refers to the non-stationarity of data stream distributions over time, which in the realm of cybersecurity manifests as unpredictable changes in the distribution of network data streams. Previous work typically assumes that data streams within a network originate from a single source and exhibit stable distributions. Consequently, when faced with dynamically shifting data distributions, the trained ML or DL intrusion detection models may fail to adapt to these changes, leading to a degradation in their predictive performance. Although the research on concept drift in IDS [20] has garnered attention, it remains in its infancy. The prevalent approaches involve directly transferring or integrating traditional concept drift detection techniques. For instance, the ADWIN [6] algorithm maintains a variable sliding window to statistically analyze the data distribution changes between two sub-windows, using confidence levels to ascertain whether the distribution change exceeds a predefined threshold. Similarly, the DDM [11] method detects concept drift by calculating whether the model's prediction error rate surpasses a threshold. Meanwhile, HDDM [10] computes the moving average of the model's predictions and employs the Hoeffding inequality to determine the threshold boundary for drift.

The challenges faced by existing IDS can be summarized into four key points:

- Traditional IDS, trained on fixed rules or static datasets, are vulnerable to future unknown attack patterns.
- The data distribution of online network traffic evolves over time, and ML or DL models trained on static datasets may struggle to adapt to such dynamic changes.
- The cost of retraining models to replace outdated ones in the event of concept drift is excessively high, posing practical challenges.
- Traditional concept drift detection methods have not been specifically optimized for online network traffic, thereby limiting their detection performance to some extent.

To overcome these challenges, we propose DWOIDS, an online intrusion detection system framework based on dual adaptive windows and the Hoeffding Tree classifier. This framework utilizes the Hoeffding Tree as the base classifier for real-time prediction of malicious behaviors in online network traffic. To effectively mitigate concept drift, we innovatively calculate the momentum of prediction errors within fast and slow windows, and estimate the degree of concept drift through the moving average difference of these momenta. Upon detection of concept drift, the framework automatically replaces the base classifier with a new classifier learned under the context of concept drift, thereby significantly reducing the overhead of model replacement and enhancing the framework's adaptive capabilities and detection efficiency.

The main contributions of this paper are as follows:

- We propose an online intrusion detection framework based on dual adaptive windows and the Hoeffding tree classifier. This framework effectively addresses the issue of concept drift in real-time network flows, thereby significantly improving the adaptability and detection efficiency of the intrusion detection system.
- To accurately identify concept drift in network flows, we design the DAWMA algorithm. This algorithm utilizes momentum and its moving average difference as estimators, effectively enhancing the intrusion detection performance by recognizing concept drift.
- We comprehensively evaluate our proposed framework on two public datasets, CIC-IDS2017 [21] and NSL-KDD [24], and compare it with state-of-the-art methods. The experimental results demonstrate that our framework excels in handling intrusion detection tasks with concept drift.

The structure of this paper is organized as follows: Sect. 2 systematically reviews the relevant research on concept drift in the field of IDS. Section 3 provides a detailed presentation of the innovative framework we propose. Section 4 discusses the experiments on the proposed framework to validate its effectiveness and superiority. Finally, Sect. 5 offers a comprehensive summary of the entire paper.

2 Related Work

In the context of online IDS, the application of concept drift is generally approached in two ways. One method involves incorporating the incoming network data stream into an adaptive window and then statistically analyzing the distributional changes between two sub-windows. If the change exceeds a predefined threshold, it is deemed that concept drift has occurred. Another method involves employing a classifier to initially predict network data streams and determine their real-time accuracy. Following this, the moving average of the accuracy or other statistical measures is evaluated. When the evaluated changes exceed a certain threshold, it is concluded that concept drift has taken place.

2.1 Data Distribution-Based Concept Drift Mitigation in IDS

Yang et al. [27] introduced the CADE system for detecting and explaining concept drift samples in security applications. CADE maps training data to a low-dimensional space using contrastive learning and learns a distance function between samples to identify drift samples deviating from the training distribution. However, It struggles with sparse data, high dimensionality, and limited performance on in-class evolution scenarios. Jain et al. [16] used a sliding window and employed K-Means clustering to reduce the data volume and update the training set. Subsequently, KL divergence is employed to measure data distribution differences and an SVM classifier is used for anomaly detection, and the model is retrained based on statistical tests. It achieves high accuracy but depends on manual parameter tuning and may not scale well with large or real-time data streams. Rajeswari et al. [19] introduced an efficient intrusion detection system based on concept drift in data streams and Support Vector Machines (SVM). The system enhances data quality through data cleansing and normalization techniques and facilitates pattern extraction and data change detection by incorporating a timestamp attribute for data streaming. The ADWIN algorithm is utilized to detect concept drift in the data stream, splitting the window into two parts at the cut-off point. An SVM classifier is then employed to categorize the data into normal and attack classes.

2.2   Prediction Statistics-Based Concept Drift Mitigation in IDS

Andresini et al. [2] introduced INSOMNIA, a semi-supervised intrusion detection framework designed to address the issue of concept drift in network traffic features over time. INSOMNIA employs a DNN classifier integrated with active learning, label estimation, and XAI to enhance adaptability. Drift is detected by assessing DNN accuracy against pseudo-labels from an NC classifier, triggering model updates via the NC classifier to align with new traffic patterns. While the approach demonstrates robustness to concept drift, it struggles with detecting low-prevalence attacks and requires further work to generalize across diverse attack types. Yang et al. [26] addressed the issue of concept drift in Internet of Things (IoT) data streams by proposing an integrated framework named PWPAE. PWPAE combines two popular drift detection methods (ADWIN and DDM) with two state-of-the-art drift adaptation methods (ARF and SRP) to construct base learners. Subsequently, the PWPAE is used to weight the base learners based on their real-time performance, and to integrate them into a robust anomaly detection ensemble model, thereby enhancing the drift adaptation performance. While PWPAE outperforms existing approaches in drift handling, broader applicability and efficiency in resource-constrained settings require further study.

Jain and Kaur [15] introduced a hybrid ensemble technique based on concept drift detection for distributed anomaly detection in network data streams. This technique integrates random forests and logistic regression as the first-level classifiers, with support vector machines serving as the second-level classifiers. To address concept drift, they employed a sliding window-based K-Means clustering technique to reduce the data volume and update the training set. Hnamte and Hussain [14] proposed a hybrid deep learning model named DCNNBiLSTM for network intrusion detection. This model leverages the strengths of both Convolutional Neural Networks (CNN) and Bidirectional Long Short-Term Memory networks (BiLSTM), and is optimized through a Deep Neural Network (DNN). The DCNNBiLSTM model first employs CNN to extract local features from the input data, followed by the use of BiLSTM to capture long and short-term dependencies in sequential data, with the final classification being performed by the DNN. However, the proposed model's high parameter complexity might exhibit overfitting tendencies. Seth et al. [20] proposed an intrusion detection method based on Adaptive Random Forest (ARF). ARF adapts in real-time to the evolving network environment and attack patterns, and prioritizes new data through an instance weighting mechanism, thereby enhancing the intrusion detection capability. Hoeffding's inequality and the moving average test provide statistical support to the system, aiding in the timely identification of concept drift and distinguishing between benign network changes and potential intrusions. However, Computational overhead from drift detection and adaptive updates may hinder scalability.

3   Methodology

3.1   Overall Architecture

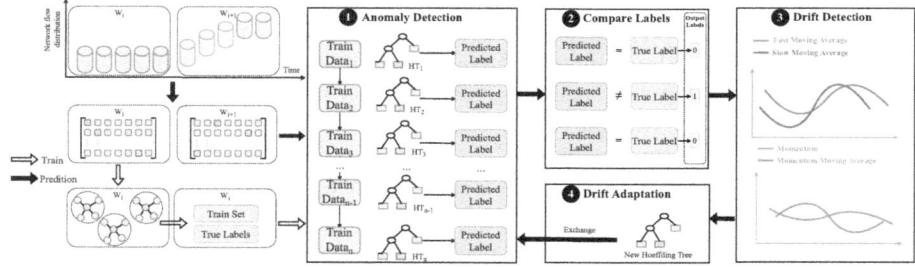

Fig. 1. Overview of DWOIDS.

Our proposed DWOIDS uses a dual adaptive window moving average technique (DAWMA) to detect concept drift and employs Hoeffding Trees as the base classifier. As shown in the Fig. 1, this method first learns the characteristics of network traffic in the W_i window, and then performs predictions in the W_{i+1} window. The proposed framework initially preprocesses the data stream through one-hot encoding, min-max normalization, and Birch [28] clustering to extract representative training data streams. Following this, it employs Hoeffding trees to learn and predict network traffic in the current environment, thereby generating corresponding prediction labels. The binary prediction results are then added to both a fast and a slow window, with the difference between the mean values of these windows being termed as the dual window mean. Subsequently, we introduce an estimator based on the difference between the momentum moving averages of fast and slow windows. Furthermore, we trigger a drift signal if the sign of the estimator at the current time step is opposite to that of the preceding time step. Lastly, the classifier that was learned in the context of the detected concept drift replaces the base classifier. This section offers a comprehensive description of the modules within the framework.

3.2 Data Processing

Due to the vast scale of network traffic data streams and in order to maintain the performance of the online IDS, we employ BIRCH clustering for the sampling of training data. We chose the BIRCH clustering algorithm because it automatically identifies high-quality cluster centers without pre-setting the number of clusters, making it highly effective for large-scale network data streams.

Algorithm 1: Birch Clustering Algorithm

    input : Network data stream setX $\{x_i, ...x_j\}$, label setY $\{y_i, ...y_j\}$,
              threshold T, branching factor B
    output: The final retained clusters

1 *Apply One-Hot encoding to the input sets;*
2 *Perform min-max normalization on the encoded data;*
3 *Initialize an empty CF Tree;*
4 **for** $i \leftarrow 1$ **to** $length(setX)$ **do**
5   *Find the closest leaf node L in the CF Tree to x_i;*
6   **if** *after adding x_i to node L, the diameter of $L < T$* **then**
7   adding x_i to L
8   **if** *the number of CFs in $L < B$* **then**
9   Create a new CF for x_i in L
10   **else** split L to L' and adding x_i to node L';
11 **if** *the number of CFs in the root of the CF Tree $\geq B$* **then**
12   compress the tree
13 *ClusterLabels $\leftarrow$ global clustering on CF Tree;*
14 *clusters $\leftarrow$ mapping of setX, setY and ClusterLabels;*
15 *nc $\leftarrow$ number of clusters;*
16 **for** $i \leftarrow 1$ **to** nc **do**
17   **if** *the ratio of malicious labels in the cluster$_i$ < 0.2* **then**
18   discard *cluster$_i$*
19 **return** clusters

In this paper, the network data stream is represented as setX $\{x_i, x_{i+1}, ..., x_j\}$, with i and j indicating different time points. And the label setY $\{y_i, y_{i+1}, ..., y_j\}$ comprises two states: 'normal' and 'malicious'. Specifically, we initially perform one-hot encoding on the data stream features to facilitate subsequent calculations. Afterwards, we apply min-max scaling to the encoded results to ensure data consistency and comparability. Algorithm 1 presents the specific steps of

BIRCH clustering. Through the precise application of the BIRCH clustering algorithm, we successfully delineated multiple clusters, each composed of similar data streams.

3.3   Detecting Concept Drift with Dual Adaptive Windows

We input the processed network data stream into the classifier for prediction, with details of the classifier provided in Sect. 3.4. The classifier outputs a set of predicted labels $\{p_i, ...p_j\}$. We introduce the prediction error (pe) as an quantify metric of classifier. As illustrated in Eq. 1, the prediction error is defined as follows: it takes the value of 0 when the predicted label matches the actual label. Conversely, the prediction error is set to 1.

$$pe_i = \begin{cases} 0, & p_i = y_i \\ 1, & p_i \neq y_i \end{cases} \tag{1}$$

To mitigate the limitations of relying solely on single-point prediction errors for detecting concept drift, we introduce the Dual-Adaptive Window Moving Average (DAWMA) algorithm. This approach aims to balance the sensitivity and accuracy of drift detection. The DAWMA algorithm employs a dual-sliding window mechanism, comprising a fast window and a slow window, both of which continuously receive the most recent pe_i values. Initially, the slow window leads the fast window by d units in length. As data continues to be input, the lengths of both windows gradually increase until concept drift is detected. During this process, we denote the lengths of the fast and slow windows as fl and sl, respectively, and their respective average values as fa and sa. The dual-window mechanism effectively balances the needs of short-term and long-term strategies. Furthermore, the windows in this algorithm are designed to dynamically adapt: if no concept drift is detected over a long period, the window size will automatically increase to capture broader data trends; conversely, if concept drift occurs frequently in the short term, the window will maintain a smaller size to enhance sensitivity to rapid changes.

$$M = (fa_k - sa_k) - \frac{\sum\limits_{i=k-n}^{k-1} (fa_i - sa_i)}{n} \tag{2}$$

As illustrated in the Formula 2, the DAWMA algorithm introduces the momentum of the slow window and the fast window to quantify the intensity of changes in the prediction error. Subsequently, the moving average difference M of these momenta is calculated to effectively filter noise and short-term data fluctuations. Furthermore, if the sign of the current M is opposite to that of the previous M, it indicates a trend of strengthening or weakening in the momentum of the prediction error, at which point we determine that a concept drift has occurred.

3.4   Classifier Based on Hoeffding Tree

Hoeffding Tree is an incremental decision tree algorithm for data stream classification. It processes large-scale data streams in real-time, learning and predicting on the fly without storing all historical data. In this study, we assume that the network data stream $\{x_i, x_{i+1}, ..., x_j\}$ is mutually independent and bounded ($a_i \leq x_i \leq b_i$). Based on this assumption, for any constant $\theta > 0$, the expression represented by Eq. 3 conforms to the Hoeffding's bound.

$$P\left(\sum_{i=1}^{n} x_i - E\left[\sum_{i=1}^{n} x_i\right] \geq \theta\right) \leq \exp\left(-\frac{2\theta^2}{\sum_{i=1}^{n}(b_i - a_i)^2}\right) \tag{3}$$

This study constructs a decision tree model based on Hoeffding's inequality. Initially, the Hoeffding Tree consists of a single root node, which is responsible for storing the class distribution information of the initial data. As new network data streams arrive, these data start from the root node and are directed to the corresponding child nodes based on the splitting attributes and feature values of each node. This process is recursive and continues until the data reach a leaf node. Upon arrival at a leaf node, the node updates its class distribution information to reflect the incorporation of new data. During the splitting decision of the decision tree, we use information gain as a metric, which is defined as the difference between entropy and conditional entropy. Then, we utilize Hoeffding's inequality to determine whether the attribute with the currently observed maximum information gain exceeds the Hoeffding bound, thereby deciding whether to perform node splitting. Since Hoeffding Tree is an incremental learning algorithm, once concept drift is detected, new concepts are gradually integrated into the Hoeffding Tree, thereby demonstrating the algorithm's robustness.

4   Experiments Evaluation

This section delves into the experimental evaluation process, wherein we utilized the CIC-IDS2017 [21] and NSL-KDD [24] datasets for testing. Under the condition of consistent dataset distribution and environmental setup, we reproduced and conducted a comparative analysis with other research works, thereby robustly validating the effectiveness of our proposed method.

4.1   Dataset and Processing

The CIC-IDS2017 dataset, published by the Canadian Institute for Cybersecurity, captures real network traffic from July 3, 2017, to July 7, 2017, and includes 15 common attack scenarios such as brute force FTP, brute force SSH, DoS, Heartbleed, web attacks, intrusions, Botnet, and DDoS. To enhance data quality, we utilized the improved version of the dataset by Liu et al. [17], which rectifies label errors and feature extraction inaccuracies in the original dataset and removes meaningless artefacts.Ultimately, we acquired a total of 2,090,564 network traces, comprising 1,657,069 benign traces and 433,495 malicious traces.

The specific distribution of the malicious traces is presented in Table 1. Due to the significant imbalance between benign and malicious tracks, it is necessary to consider the base-rate fallacy [4].

Table 1. Statistical Analysis of Malicious Traces

Attack Types	Count	Proportion
Infiltration	38	0.0088%
Brute Force	151	0.0348%
SQL Injection	12	0.0028%
XSS	27	0.0062%
DoS GoldenEye	7,567	1.7456%
DoS Hulk	158,469	36.5561%
DoS Slowhttptest	1,742	0.4019%
DoS Slowlori	4,001	0.9230%
Heartbleed	11	0.0025%
FTP Patator	3,973	0.9165%
SSH Patator	2,980	0.6874%
Botnet	738	0.1702%
DDoS	94,763	21.8602%
Portscan	159,023	36.6839%
SUM	433,495	100.0000%

The NSL-KDD dataset is an enhanced version of the KDD Cup 1999 dataset, specifically designed for the evaluation of network intrusion detection systems. It comprises 148,517 records, each consisting of 41 traffic features and a classification label. The dataset encompasses five distinct types of data, namely Normal (normal traffic), DOS (Denial of Service attacks), Probe (probing attacks), R2L (Remote to Local attacks), and U2R (User to Root attacks). Among these, benign traffic constitutes 53% of the dataset, while malicious traffic accounts for the remaining 47%. The dataset contains 125,973 training samples and 22,544 testing samples.

4.2   Experimental Setup

Our experiments were conducted in an environment featuring Ubuntu 22.04 as the operating system, an x86 hardware architecture, a 32-core AMD CPU, and 1TB of RAM. We implemented the proposed framework in Python and set the BIRCH clustering threshold to 0.5. In addition, to ensure a fair comparison, we obtained the source code for all baseline works and keep the best configurations for all hyperparameters. Furthermore, we conducted experiments using the same dataset distribution and proportions to replicate their results.

In our experiments with the CIC-IDS2017 dataset, we adopted the same approach as previous work [2,16] by using windows containing 50,000 instances to highlight the distribution characteristics of concept drift. For the NSL-KDD training set containing 125,973 samples, we configured the window size as 7,000, while for the test set with 22,544 samples, a window size of 2,000 was implemented. However, unlike prior studies, we discard the last window if it's smaller than the set size. Specifically, we trained a model in the current Window Id W_i (i starts from 0) and then used this trained classifier to predict in the subsequent Window Id W_{i+1}.

4.3   Evaluation Metrics

In this experiment, we employ accuracy, false positive rate (FPR), precision, recall, and F1 score as the metrics for performance evaluation. Specifically, true positive instances are denoted as (TP), false positive instances as (FP), true negative instances as (TN), and false negative instances as (FN). The mathematical expressions for these evaluation metrics are presented in Eqs. 4–8.

$$\text{Accuracy} = \frac{TP + TN}{TP + FP + TN + FN} \tag{4}$$

$$\text{FPR} = \frac{FP}{FP + FN} \tag{5}$$

$$\text{Precision} = \frac{TP}{TP + FP} \tag{6}$$

$$\text{Recall} = \frac{TP}{TP + FN} \tag{7}$$

$$\text{F1 Score} = 2 \cdot \frac{\text{Precision} \cdot \text{Recall}}{\text{Precision} + \text{Recall}} \tag{8}$$

4.4   Evaluation Results and Discussions

The aim of this study is to comprehensively evaluate the performance of our proposed DWOIDS in network intrusion detection with concept drift. To this end, we have designed the following two experiments:

– Classification Experiment: In this experiment, we do not apply special treatment to concept drift, aiming to evaluate the prediction accuracy of the base classifier (based on the Hoeffding tree) and address the issue of classifier prediction bias.
– Intrusion Detection with Concept Drift Experiment: We will conduct experiments using our proposed comprehensive framework, and compare it with baseline methods to demonstrate the superior performance of our proposed framework.

Table 2. The Window where the Classifier Fails

Dataset	Classifier	Failed Window Id	Failed Count
CIC-IDS2017	only Hoeffding tree	$W_1, W_6, W_7, W_{12}, W_{14}, W_{15}, W_{17},$ $W_{18}, W_{19}, W_{22}, W_{23}, W_{29}, W_{32},$ $W_{34}, W_{35}, W_{36}, W_{37}, W_{38}$	18
	add DAWMA	$W_{14}, W_{18}, W_{35}, W_{36}, W_{38}$	5
NSL-KDD	only Hoeffding tree	$W_3, W_4, W_6, W_{11}, W_{12}$	5
	add DAWMA	/	0

Classification Experiment. This study assesses the impact of concept drift handling on prediction performance using the CIC-IDS2017 and NSL-KDD datasets. The 2,090,564 CIC-IDS2017 samples are divided into 41 windows, while the 125,973 NSL-KDD samples are split into 17 windows. The experiment comprised two phases: In the first phase, the base Hoeffding tree classifier was applied to predict data streams within each window, with performance evaluated through accuracy and false positive rate (FPR). The second phase introduced the Dual-Adaptive Window Moving Average (DAWMA) method to address concept drift, followed by recalculation of these metrics post-optimization. The experimental results are presented in Fig. 2. Then, we compute the average "add DAWMA Accuracy" (average Acc) and "add DAWMA FPR" (average FPR). We establish a failure criterion: The base classifier is deemed to have failed in addressing concept drift if either the accuracy metric (acc_i) in window (W_i) satisfies the threshold condition defined in Eq. 9, or the false positive rate (fpr_i) reaches the critical value specified by Eq. 10. Table 2 presents the Window Ids where the base classifier fails before and after adding DAWMA. Empirical analysis reveals a substantial decrease in the number of windows experiencing base classifier failure due to concept drift following the incorporation of DAWMA, thereby enhancing the classifier's robustness.

$$acc_i < \text{average Acc} \times 0.9 \qquad (9)$$

$$fpr_i \times 0.2 > \text{average FPR} \qquad (10)$$

This experiment highlights three crucial findings: Firstly, concept drift indeed exists in network data streams, posing a significant challenge for traditional classifiers which often result in classifier failure. Secondly, the introduction of our proposed DAWMA strategy enables rapid recovery and stable performance of the classifier, thereby underscoring the necessity of accounting for concept drift in the context of network stream online intrusion detection. Thirdly, in the event of concept drift in the current window, our proposed framework swiftly replaces the outdated classifier with a newly trained one, which then effectively monitors the subsequent window. Experimental results demonstrate the effectiveness of our approach.

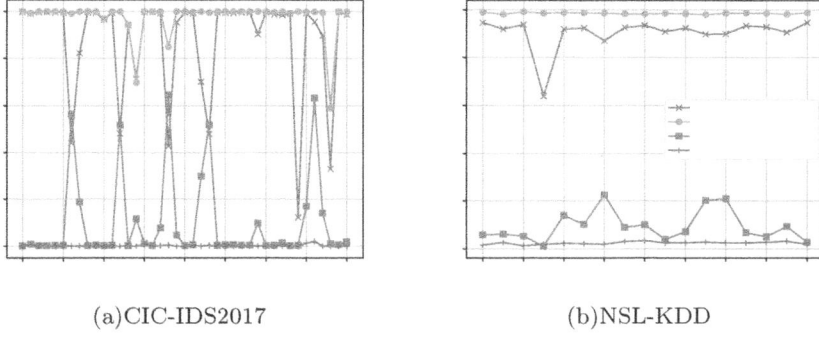

(a)CIC-IDS2017   (b)NSL-KDD

Fig. 2. CIC-IDS2017 and NSL-KDD in Concept Drift Handling.

Intrusion Detection with Concept Drift Experiment. In this study, We reproduce all baseline methods and perform detailed comparative analysis. The selected baseline methods are categorized into two groups: those that do not deal with concept drift, like Naive Bayes and LightGBM, and those that do, such as ARF+ADWIN [13], INSOMNIA [2].

For the CIC-IDS2017 dataset, we selected 1,396,914 traces from the last three days for validation with a window size of 50,000. For the NSL-KDD dataset, we used 22,544 test set traces for the experiment, with a window size of 2,000. The experiment was designed to train models in the current window and use the trained models to make predictions in the next window. Ultimately, we compared the accuracy, precision, recall, and F1 score of each model.

Specifically for the INSOMNIA method, which requires pre-training a deep neural network (DNN) model and making predictions on traces within the test window during testing, while also using fine-tuning strategies to adjust model performance in real-time, we used 693,650 traces from the first two days of CIC-IDS2017 and 125,973 training set traces from NSL-KDD to train the model for evaluation. Subsequently, we validated the model performance using the same approach as other methods and conducted a comparative analysis within the same test window interval to ensure the fairness of the experiment and the comparability of the results.

Table 3 shows the average results of all experiments conducted on the CIC-IDS2017 and NSL-KDD datasets. Our analysis reveals that methods accounting for concept drift significantly outperform those that do not when detecting real-time data streams. This is attributed to the former to adjust their original prediction strategies, either through fine-tuning or model replacement, following the detection of concept drift. This observation supports our contention that the distribution of network data streams evolves over time, and traditional static Intrusion Detection Systems (IDS) are less capable of adapting to concept drift. Notably, INSOMNIA encounters difficulties in detecting low-frequency and covert attacks (such as Infiltration attacks), and its update mechanism fails to effectively utilize the knowledge of attacks that occur only within a single win-

Table 3. Comparison of Algorithms for Online Intrusions Detection

Method	Dataset	Accuracy	Precision	Recall	F1 Score
Naive Bayes*	NSL-KDD	0.8563	0.8199	0.8730	0.8345
	CIC-IDS2017	0.8787	0.9212	0.9067	0.8758
LightGBM*	NSL-KDD	0.9418	**0.9829**	0.8802	0.9284
	CIC-IDS2017	0.9401	0.8657	0.8020	0.7982
ARF+ADWIN	NSL-KDD	0.9340	0.9426	0.9028	0.9210
	CIC-IDS2017	0.9383	0.9362	**0.9972**	0.9532
INSOMNIA	NSL-KDD	0.5455	0.4666	0.8178	0.5940
	CIC-IDS2017	0.8656	0.3941	0.4857	0.3758
We Proposed DWOIDS	NSL-KDD	**0.9463**	0.9524	**0.9214**	**0.9364**
	CIC-IDS2017	**0.9608**	**0.9480**	0.9969	**0.9653**

* indicates that this method does not address concept drift.

dow (such as Brute Force attacks), resulting in degraded detection performance in certain windows and thereby affecting the overall average performance metrics. Furthermore, our proposed framework achieved the best performance in accuracy, recall, and F1 score on the NSL-KDD dataset. Similarly, on the CIC-IDS2017 dataset, the framework also attained optimal results in accuracy, precision, and F1 score. Experimental results demonstrate that our proposed method not only outperforms traditional machine learning and deep learning algorithms but also exhibits significant advantages over other algorithms in handling concept drift in intrusion detection.

5   Conclusion and Future Work

This paper presents DWOIDS, an innovative online network intrusion detection system framework based on dual adaptive windows, designed to effectively address the challenge of concept drift in the field of network intrusion detection. This framework relies on a base classifier based on the Hoeffding tree to accurately identify malicious components in network traffic. Experiments on NSL-KDD and CIC-IDS2017 datasets demonstrate the necessity of addressing concept drift in online intrusion detection, and our proposed framework outperforms the baseline model. This enables real-time adaptation in critical scenarios, where concept drift frequently disrupts anomaly detection.

Currently, our validation is limited to the binary classification problem of online network streams using benchmark datasets. To bridge the academia-industry gap, we plan to deploy DWOIDS in real-world environments in future work, focusing on evaluating its robustness to multi-class concept drift against emerging network attack traffic.

Acknowledgement. This paper was supported by National Natural Science Foundation of China(No.62472418).

References

1. Ahmad, Z., Shahid Khan, A., Wai Shiang, C., Abdullah, J., Ahmad, F.: Network intrusion detection system: a systematic study of machine learning and deep learning approaches. Trans. Emerg. Telecommun. Technol. **32**(1), e4150 (2021)
2. Andresini, G., Pendlebury, F., Pierazzi, F., Loglisci, C., Appice, A., Cavallaro, L.: Insomnia: towards concept-drift robustness in network intrusion detection. In: Proceedings of the 14th ACM Workshop on Artificial Intelligence and Security, pp. 111–122 (2021)
3. Arp, D., et al.: Dos and don'ts of machine learning in computer security. In: 31st USENIX Security Symposium (USENIX Security 22), pp. 3971–3988 (2022)
4. Axelsson, S.: The base-rate fallacy and the difficulty of intrusion detection. ACM Trans. Inf. Syst. Secur. (TISSEC) **3**(3), 186–205 (2000)
5. Bao, H., et al.: Payload level graph attention network for web attack traffic detection. In: International Conference on Computational Science, pp. 394–407. Springer, Heidelberg (2023). https://doi.org/10.1007/978-3-031-36030-5_32
6. Bifet, A., Gavalda, R.: Learning from time-changing data with adaptive windowing. In: Proceedings of the 2007 SIAM International Conference on Data Mining, pp. 443–448. SIAM (2007)
7. Chiche, A., Meshesha, M.: Towards a scalable and adaptive learning approach for network intrusion detection. J. Comput. Netw. Commun. **2021**(1), 8845540 (2021)
8. Feng, X., et al.: Understanding and securing device vulnerabilities through automated bug report analysis. In: SEC'19: Proceedings of the 28th USENIX Conference on Security Symposium (2019)
9. Ferrag, M.A., Maglaras, L., Moschoyiannis, S., Janicke, H.: Deep learning for cyber security intrusion detection: approaches, datasets, and comparative study. J. Inf. Secur. Appl. **50**, 102419 (2020)
10. Frias-Blanco, I., del Campo-Ávila, J., Ramos-Jimenez, G., Morales-Bueno, R., Ortiz-Diaz, A., Caballero-Mota, Y.: Online and non-parametric drift detection methods based on hoeffding's bounds. IEEE Trans. Knowl. Data Eng. **27**(3), 810–823 (2014)
11. Gama, J., Medas, P., Castillo, G., Rodrigues, P.: Learning with drift detection. In: Bazzan, A., Labidi, S. (eds.) SBIA 2004. LNCS (LNAI), vol. 3171, pp. 286–295. Springer, Heidelberg (2004). https://doi.org/10.1007/978-3-540-28645-5_29
12. Gao, X., Shan, C., Hu, C., Niu, Z., Liu, Z.: An adaptive ensemble machine learning model for intrusion detection. IEEE Access **7**, 82512–82521 (2019)
13. Gomes, H.M., et al.: Adaptive random forests for evolving data stream classification. Mach. Learn. **106**, 1469–1495 (2017)
14. Hnamte, V., Hussain, J.: Dcnnbilstm: an efficient hybrid deep learning-based intrusion detection system. Telemat. Inf. Rep. **10**, 100053 (2023)
15. Jain, M., Kaur, G.: Distributed anomaly detection using concept drift detection based hybrid ensemble techniques in streamed network data. Clust. Comput. **24**(3), 2099–2114 (2021). https://doi.org/10.1007/s10586-021-03249-9
16. Jain, M., Kaur, G., Saxena, V.: A k-means clustering and svm based hybrid concept drift detection technique for network anomaly detection. Expert Syst. Appl. **193**, 116510 (2022)
17. Liu, L., Engelen, G., Lynar, T., Essam, D., Joosen, W.: Error prevalence in nids datasets: A case study on cic-ids-2017 and cse-cic-ids-2018. In: 2022 IEEE Conference on Communications and Network Security (CNS), pp. 254–262. IEEE (2022)

18. Lu, J., Liu, A., Dong, F., Gu, F., Gama, J., Zhang, G.: Learning under concept drift: a review. IEEE Trans. Knowl. Data Eng. **31**(12), 2346–2363 (2018)
19. Rajeswari, P.V.N., Shashi, M., Rao, T.K., Rajya Lakshmi, M., Kiran, L.V.: Effective intrusion detection system using concept drifting data stream and support vector machine. Concurr. Comput. Pract. Exp. **34**(21), e7118 (2022)
20. Seth, S., Chahal, K.K., Singh, G.: Concept drift–based intrusion detection for evolving data stream classification in ids: approaches and comparative study. Comput. J., bxae023 (2024)
21. Sharafaldin, I., Lashkari, A.H., Ghorbani, A.A., et al.: Toward generating a new intrusion detection dataset and intrusion traffic characterization. ICISSp **1**, 108–116 (2018)
22. Shen, Y., Mariconti, E., Vervier, P.A., Stringhini, G.: Tiresias: predicting security events through deep learning. In: Proceedings of the 2018 ACM SIGSAC Conference on Computer and Communications Security, pp. 592–605 (2018)
23. Sommer, R., Paxson, V.: Outside the closed world: on using machine learning for network intrusion detection. In: 2010 IEEE Symposium on Security and Privacy, pp. 305–316. IEEE (2010)
24. Tavallaee, M., Bagheri, E., Lu, W., Ghorbani, A.A.: A detailed analysis of the kdd cup 99 data set. In: 2009 IEEE Symposium on Computational Intelligence for Security and Defense Applications, pp. 1–6. IEEE (2009)
25. Wang, X., Bao, H., Li, W., Chen, H., Wang, W., Liu, F.: A framework for intelligent generation of intrusion detection rules based on grad-cam. In: International Conference on Computational Science, pp. 147–161. Springer, Heidelberg (2024). https://doi.org/10.1007/978-3-031-63783-4_12
26. Yang, L., Manias, D.M., Shami, A.: Pwpae: an ensemble framework for concept drift adaptation in iot data streams. In: 2021 IEEE Global Communications Conference (GLOBECOM), pp. 01–06. IEEE (2021)
27. Yang, L., et al.: {CADE}: detecting and explaining concept drift samples for security applications. In: 30th USENIX Security Symposium (USENIX Security 21), pp. 2327–2344 (2021)
28. Zhang, T., Ramakrishnan, R., Livny, M.: Birch: an efficient data clustering method for very large databases. ACM SIGMOD Rec. **25**(2), 103–114 (1996)

Estimating Airborne Transmission Risk for Indoor Space: Coupling Agent-Based Model and Computational Fluid Dynamics

Boon Leng Ang[1], Jaeyoung Kwak[1(✉)], Chin Chun Ooi[2], Zhengwei Ge[2], Hongying Li[2], Michael H. Lees[3], and Wentong Cai[1]

[1] Nanyang Technological University, Singapore 639798, Singapore
bang013@e.ntu.edu.sg, {jaeyoung.kwak,hongying.li,aswtcai}@ntu.edu.sg
[2] Institute of High Performance Computing, Singapore 138632, Singapore
{ooicc,ge_zhengwei,lih}@cfar.a-star.edu.sg
[3] University of Amsterdam, 1098XH Amsterdam, The Netherlands
m.h.lees@uva.nl

Abstract. The emergence of coronavirus disease (COVID-19) in late 2019 sparked a global pandemic, profoundly impacting societies and economies worldwide. To mitigate its spread, governments have implemented various preventive measures, prompting extensive research into transmission risk assessment. To evaluate the transmission risk systematically, we developed a framework integrating agent-based modeling (ABM) and computational fluid dynamics (CFD), and applied the framework to a preschool COVID-19 cluster in Singapore as a case study. Individual movement and behaviors are simulated with ABM, and CFD is employed to compute virus particle flow which is critical for transmission risk. In the case study, we categorized the infected individual's movement into three types based on the initial destinations and evaluated its impact on the transmission risk. Simulation results show that the average risk level is nearly the same for all three movement types and it changes across time depending on the degree of infected individual's active movement.

Keywords: Computational Fluid Dynamics (CFD) · Agent-Based Modeling (ABM) · Transmission Risk

1 Introduction

For several decades, the study of human crowds has been crucial in real-world applications, such as planning effective evacuation [4], studying disease transmission [8], and simulating virtual crowds for computer graphics [24]. This is especially relevant to the 2019 novel coronavirus disease (COVID-19) global pandemic that began in December 2019 [15]. As of January 2024, the pandemic has led to 774 million infections and taken the lives of 7 million people, severely and permanently affecting the livelihood and work of people across the world [19].

To curb the spread of the coronavirus, many nations have taken a variety of preventive measures such as vaccination campaigns, enforcing mask-wearing, and practicing social distancing. Extensive research has been conducted to evaluate the effectiveness of such measures. One stream of research focuses on the movement of individuals in line with agent-based model (ABM) and evaluating the transmission risk based on the inter-personal distance [21]. A diverse range of scenarios have been studied, such as supermarkets [27], train stations [13], and university campuses [1]. This type of approach allows us to make a straightforward estimation, but air flow is not considered although it is critical for airborne disease transmissions.

Another stream of research is based on computational fluid dynamics (CFD) simulations, which aim to compute the flow of droplets and aerosols in the air and the subsequent amount inhaled by individuals [18,20,29]. Although this approach can reflect details of virus particle dispersion in the air, most studies assume that individuals are stationary.

A few studies have proposed integrating ABM approach with CFD simulation. For instance, Vuorinen *et al.* [28] applied Monte-Carlo modeling for simulating the movement of susceptible and infected individuals, and CFD simulations to estimate the spread of aerosols and droplets in built environment such as a library and pub. In another study, Mendez *et al.* [16] simulated droplet trajectories in line with CFD simulations, aggregated the viral concentration at individual level, and then coupled the estimated concentration with pedestrian trajectories collected from different public places like train stations, markets, and street cafés. Those studies focused on transport of aerosol and droplets, for instance, aerodynamic effects like air flow and human motion, and droplet and aerosol dispersion and their behavior in indoor airflow.

In this work, we present an extension of the integration of ABM and CFD for the estimation of airborne transmission risk for indoor space. While the previous work [16,28] studied hypothetical scenarios, we applied our approach for various individual movement patterns for a preschool COVID-19 cluster in Singapore as a case study. Based on the available information of dimensions and air flow conditions of the study venue, we performed a series of crowd simulations to generate possible trajectories of individuals, computed location-specific virus particle concentration, and then integrated both components to systematically evaluate the airborne disease transmission risk.

The remainder of this paper is organized as follows: Sect. 2 gives a literature review on related work. Section 3 presents the simulation layout and the underlying models for individual movement and behaviors, and transmission risk evaluation. Section 4 presents the numerical experiments and discusses the results. We also have a discussion on the results as well as the limitations of the case study. Section 5 summarizes our work and provides possible research directions for the future.

2   Related Work

2.1   Crowd Simulations

There are differences in granularity when it comes to choosing the appropriate crowd simulation models. A few approaches including flow-based model and agent-based model have been developed to understand and replicate crowd behaviors [22]. In the flow-based model, individuals are portrayed as part of homogeneous crowds sharing common behaviors and destinations. This model is suitable for studying aggregate behavior in large gatherings showing homogeneous behaviors, for instance, walking toward the same destination [7]. On the other hand, the agent-based model represents individuals as members of heterogeneous crowds, each possessing intellectual capacity, unique traits, and the ability to react to others and their surroundings. This model is suitable for simulating pedestrian flow in a busy facility where we can see complex crowd dynamics in terms of various walking directions and constantly changing pedestrian volumes [6].

In many human crowd studies, agent-based simulation has emerged as the dominant approach for modeling crowds due to its ability to simulate detailed and complex environments. This is achieved by representing individuals as intelligent agents, thereby allowing for a more accurate depiction of the diverse aspects of human behaviors in the real world [14]. In a scoping review by Sun *et al.* [23], agent-based model has gained popularity in studying the effectiveness of policy intervention for COVID-19.

2.2   Infection Risk Models

With regards to the research into infection risk assessment of respiratory disease transmission, two approaches have been widely applied: the Wells-Riley model and dose-response model [25]. Wells-Riley is a simpler model predicting the risk of infection based on the concentration of infectious particles in the air and the duration of exposure. On the other hand, the dose-response model examines the risk of a response such as infection and illness severity due to a quantity (dose) of infectious particles.

In line with the dose-response model, Bale *et al.* [2] applied breathing zones to quantify the amount of virus particles an individual is exposed to. A breathing zone is defined to be a region of 3D space where particles are likely to be inhaled by an agent. The shape and dimensions of this region is theoretically arbitrary, but it is typically situated in front of an agent's nose and mouth. The authors assumed breathing zone dimensions of $10 \times 10 \times 15$ cm^3 which was further divided into 16 equally sized mesh blocks. Bale *et al.* [2] applied the concept of breathing zone for the estimation of individuals' virus particle exposure level.

3   Model

3.1   Scenario

For the case study, we modeled an outbreak at a preschool in Singapore, based on information reported by local news outlets [17,26]. One day, a staff of the preschool center was organizing a training session along with 30 staff members from other preschool centers. A few days later, it was found that the training session organizer was infected with COVID-19 when she was leading the session. According to the local news outlets, all the training session attendees had contact with the infected person and at least 16 of them were infected.

As not many details of the outbreak are available to us, we have made several assumptions on agent behaviors to develop the scenario. We first assumed a 15-min break in the middle of the training session. Before the break, the session organizer was standing in front of other attendees who were sitting at the table. In addition, the session organizer was leading the session and giving a presentation while other attendees were listening to her. Once the break began, all the attendees were standing and walking around the preschool center. This break time will subsequently be the focus of this study because the amount of interaction among the attendees would be significant in terms of investigating the level of virus particle exposure.

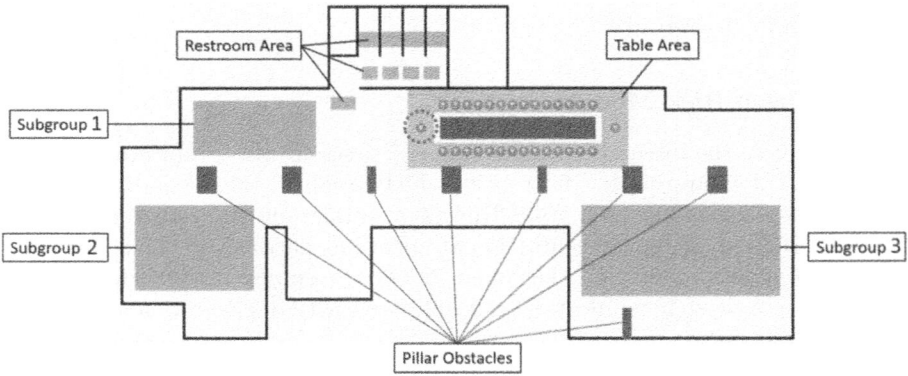

Fig. 1. The sketch of preschool scenario. The green circles represent the attendees. The contagious individual (session organizer) is indicated by a red dotted circle. (Color figure online)

Figure 1 shows the sketch of preschool layout, which is created based on existing preschools in Singapore. As can be seen from Fig. 1, the scenario can be separated into three different sections: table area, restroom area, and subgroup areas. In the table area, all 30 attendees are seated around the rectangular table during the training session. Specifically, 1 person each on the east and west side, 14 persons each on the north and south side. The session organizer, who is the first infected person among the attendees, is indicated by a red dotted circle in Fig. 1. When the 15-min break begins, everyone will leave their seats to visit other areas. The restroom area is a unisex facility consisting of an entrance, 4 cubicles, and 4 sinks. Before entering, people will form a queue at the restroom entrance. Upon entering, they will enter one of the 4 available cubicles with equal probability. After that, they will use the sink in front of the cubicle, before exiting the restroom. There are 3 subgroup areas in the preschool where individuals can gather and have casual conversations during the break.

3.2   Pedestrian Model

Our pedestrian movement model is implemented in MomenTUMv2 [10] based on the concept of hierarchical behavior modeling which describes the pedestrian behavior in terms of three interconnected layers: the strategic layer, the tactical layer and the operational layer [3,5].

The strategic layer is related to the destination choice. We used origin-destination (OD) matrix, which specifies the probability of a pedestrian visiting one area from another area. When the break begins, the attendees can either use the restroom, join a subgroup, or stay at their table seats. Additionally, the attendees that have used the washroom will either return to the table or join a subgroup. Over time, individuals will gradually leave the subgroups and return to the table. Figure 2 summarizes the movement flow of the pedestrians with a flowchart and Fig. 3 shows all the possible routes.

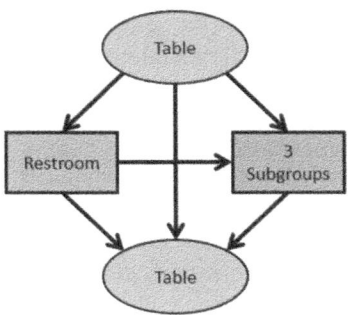

Fig. 2. Flowchart of attendee movement during the break.

The tactical layer is about how pedestrians approach their destinations. As the venue size of the preschool is small, we assume that the pedestrians have complete knowledge of all destinations. Hence, we utilized models that are computationally less intensive: Djikstra's algorithm to find the shortest route to the destinations and shifted random participating model to simulate pedestrian behavior of finding a position in a subgroup [11].

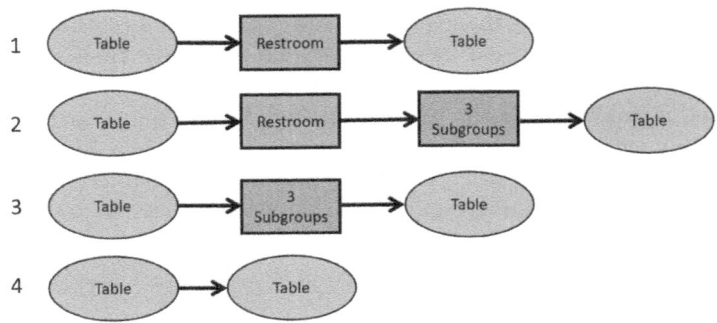

Fig. 3. All possible routes that an attendee can take.

The operational layer is associated with step-by-step movements during walking, queuing, and standing. We applied the social force model to simulate the walking behavior of pedestrians. As for the standing behavior, we utilized the model developed by Johansson et al. [9], so that we can ensure that the pedestrians in a queue stay responsive between standing and moving along the washroom queue. For the standing behavior, we implemented the Fixed standing model which prevents pedestrians in a subgroup from shuffling excessively when they are conversing at the spacious areas [12].

3.3   Exposure Level Estimation Model

We first computed the location-specific concentration of virus particles emitted by the contagious individual by means of computational fluid dynamics (CFD) simulations. Based on the work of Ooi et al. [18], we numerically solved the Navier-Stokes equation for conservation of mass and momentum, and the energy equation using a computational fluid dynamics (CFD) software (ANSYS FLUENT version 21.2). For simplicity, we assumed that the virus particle concentration level is in steady-state during the break in that the virus particles concentration level increased during the training session before the break. The virus particles concentration level does not change in the course of time during the break, thus we mainly consider the position of the contagious individual and other (i.e., susceptible) attendees, and the virus particle concentration around the susceptible individuals reflects the amount of inhaled virus particles. The mesh resolution of CFD simulations is around 10 cm.

Based on the study of Bale et al. [2], we assumed a $50 \times 50 \times 50$ cm^3 cubic region as the breathing zone. For the preschool scenario, the breathing zone was further simplified to be 50×50 cm^2 rectangular area from a top-down 2D view. This simplification was done because it was assumed that all attendees were roughly the same height and standing during the 15-min break, and hence their breathing zones would be located on the same 2D plane. The breathing zone was divided into a 10×10 cm^2 grid in line with the spatial resolution of virus particle concentration estimation. The coordinates of the 25 grid square centers

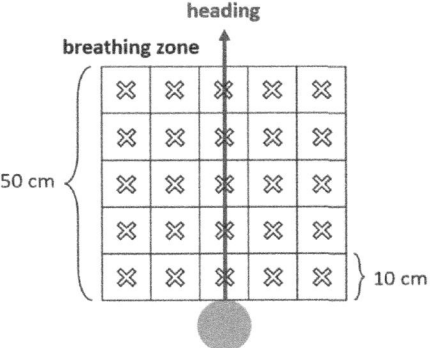

Fig. 4. 2D view of an attendee's breathing zone (not drawn to scale). A breathing zone contains 25 grid squares and the cross ($\times$) symbols indicate the center of grid squares. The breathing zone is attached to an attendee which is represented as a red circle. (Color figure online)

were used to calculate the representative value of virus particle concentration for a breathing zone. Additionally, the breathing zone is located in front of an agent, where the midpoint of the breathing zone bottom side touches the attendee's boundary. Figure 4 illustrates the setup of an attendee's breathing zone.

Using the pedestrian trajectory generated by MomenTUMv2 software, for each time frame, we obtained the coordinates of the contagious and susceptible individuals, and the headings of the susceptible individuals along with the coordinates of the 25 grid points in their breathing zone (50×50 cm^2 rectangular area in Fig. 4). For each susceptible individual, we computed the representative particle concentration value of the breathing zone by taking average of virus particle concentration at each grid point within the breathing zone. We then estimated the exposure level for each individual by summing the representative value of virus particle concentration for all time frames. It is noted that we computed the virus particle concentration as a proxy for an attendee's level of exposure to virus particles. This measure is related to the infection risk in that the larger the concentration, the higher the virus particle exposure level and infection risk.

The code that we used in this paper is available on github: https://github.com/jaeyoung82/iccs2025-ABM-CFD.

4   Simulation Results

In this section, we utilize the presented framework to perform a series of numerical experiments to evaluate how the initial destination of the sole contagious individual can affect the virus particle exposure level of other individuals in the scenario. In doing that, we performed three numerical experiments:

- Experiment R: The contagious individual goes to the restroom (R) first once the break starts, optionally joins a subgroup to talk with others, and then goes back to the table before the break ends. Refer to routes 1 and 2 shown in Fig. 3.

- Experiment G: The contagious individual joins one of three subgroups (G) to talk with others once the break starts, and then goes back to the table before the break ends. Refer to route 3 shown in Fig. 3.
- Experiment T: The contagious individual stays at the table (T) during the break, not walking around in the room. Refer to route 4 shown in Fig. 3.

For each experiment, we randomly assigned six individuals for each initial destination: restroom, subgroups 1, 2, 3, and table. We conducted 30 runs for each experiment (R, G, T) with random initial destinations of the susceptible individuals. For experiment G, we performed 10 runs for each subgroup (subgroups 1, 2, and 3). It is noted that three most probable initial destinations of the infected individual were selected as her actual movement trajectory is unknown.

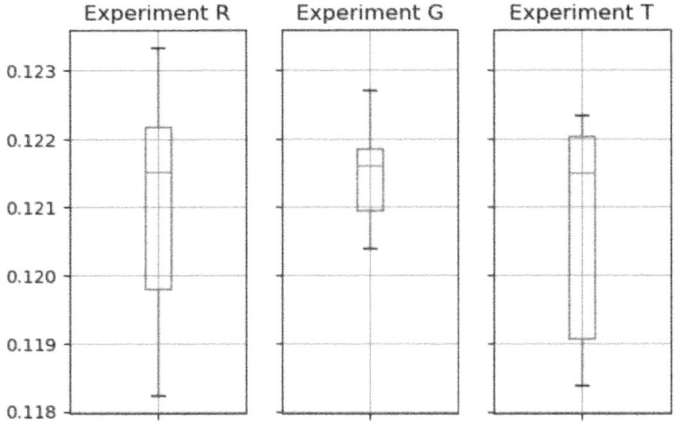

Fig. 5. Boxplots of individual's virus particle exposure level for each experiment. Note that the y-axis does not start at zero, focusing on the effective range between the minimum and maximum values.

Figure 5 shows boxplots of individual's virus particle exposure level for each experiment, reflecting the distribution of estimated transmission level. The average value of the individual's exposure level is nearly the same for all three experiments. This appears to be because the virus particle concentration level is assumed to be in steady-state. In addition, it can be seen that experiments R and T yield considerable variation, but the variation is less notable for experiment G. One can also notice that the level of virus particle exposure is positive for every susceptible individual regardless of the experiment type. This is because the virus particles were spreading in the scenario during the training session before the break, so the virus particle concentration level is considerable everywhere in the scenario.

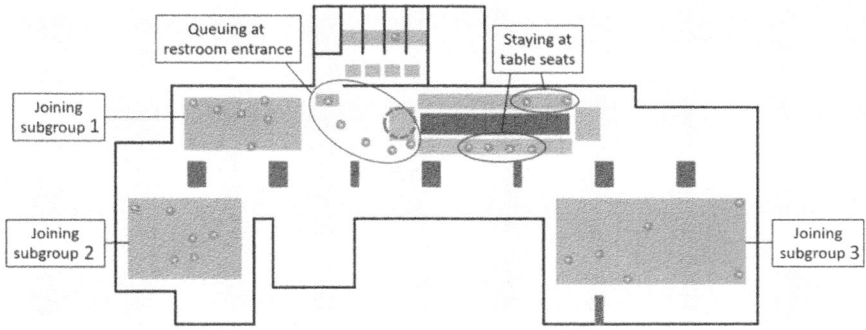

Fig. 6. A sample screenshot showing spatial distribution of training session attendees. The red dotted circle indicates the position of the contagious individual where she would stay in experiment T. (Color figure online)

This can be understood in terms of how much the contagious individual moves actively. In experiment R, it is possible for the contagious individual to walk the longest route by possibly visiting both the restroom and subgroup areas. In addition, attendees queuing at the restroom can encounter the contagious individual when she enters and leaves the restroom. For experiment T, the contagious individual stays at the table, but attendees in subgroups 1 and 2 need to bypass her and attendees queuing at the restroom might be exposed to her, see an example from Fig. 6. In contrast, for experiment G, the contagious individual goes to a subgroup almost immediately and tends to meets less individuals compared to other experiments.

The impact of distance between the contagious and susceptible individuals on the virus particle exposure level can be inferred from Spearman's correlation coefficient. As can be seen from Table 1, experiment T has strong correlation and experiment R has moderate correlation, while experiment G shows weak correlation between the two variables. It can be suggested that only considering the distance between contagious and susceptible individuals might not be enough to estimate the transmission risk, thus the inclusion of CFD is substantial in the risk estimation. In addition, for experiment G, susceptible individuals' virus particle exposure level is seemingly attributed to the virus particles in the air rather than the one emitted from the infected individual, thus the variation of virus particle exposure level is less significant among the susceptible individuals.

Furthermore, we plotted time series graphs to examine how the transmission risk changes in the course of time, see Fig. 7. It can be observed that all three graphs dip down to varying degree at start of the break, remain stable from minutes 3 to 10, and then quickly increase between minutes 12 and 14. It appears that the appearance of the plateau is attributed to the spatial distribution of attendees in that the attendees do not actively walk around the preschool center. The rapid increase from minutes 12 to 14 is seemingly due to the attendees' behavior of returning to the table. In addition, the curve of

Table 1. Correlation between the estimated transmission risk level and the distance between infected and susceptible individuals

Experiment	Correlation
R	−0.5414
G	0.0269
T	−0.9181

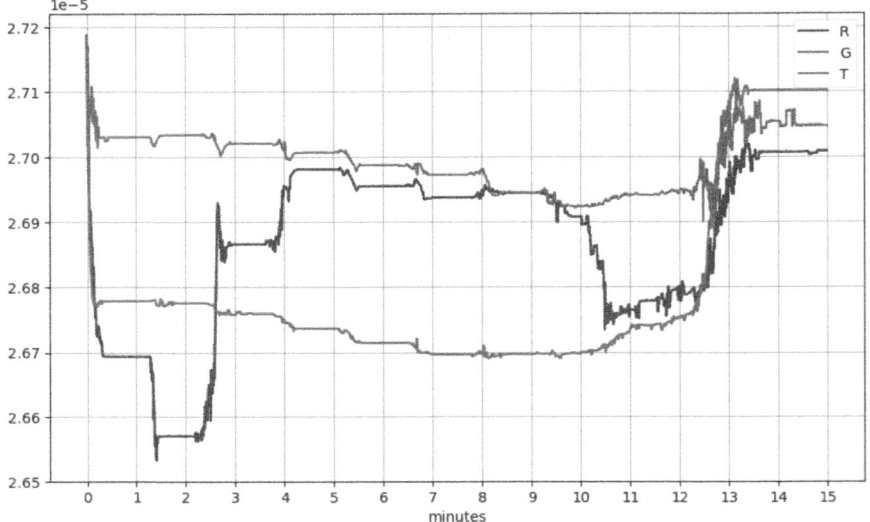

Fig. 7. Time series of transmission risk estimated for all three experiments. Note that the y-axis does not start at zero, zooming in the range of interest.

experiment R increases sharply between minutes 2 and 3. It is likely that the the queuing attendees encounter the contagious individual when she leaves the restroom while walking to the table or a subgroup.

5   Conclusion

We present a framework integrating agent-based modeling (ABM) and computational fluid dynamics (CFD) to systematically evaluate the airborne disease transmission risk. We modeled the movement of individuals in line with ABM and estimated virus particle concentration level based on CFD simulations. As a case study, the framework was applied to a preschool COVID-19 cluster in Singapore. We categorized the infected individual's movement into three types based on the initial destination and evaluated its impact on the transmission risk level. Simulation results show that the average risk level is nearly the same for all three movement types but it changes across time depending on the degree of infected individual's active movement.

Our study demonstrated potential in coupling ABM and CFD to estimate the airborne disease transmission risk, and the presented framework can be applied to different scenarios where the air flow condition is critical for the disease transmission risk. Although the presented results are produced based on a simple scenario, our study demonstrated potential in coupling ABM and CFD to estimate the airborne disease transmission risk for a real-world case in Singapore. The results are still useful for evaluating the airborne disease transmission risk and comparing possible movement of the contagious person in the scenario.

A limitation of the study is that we did not explicitly consider various respiratory activities (such as speaking, coughing, and singing) for simplicity. Computation of the aerosol and droplet dispersion reflecting such respiratory activities will allow us to better study the transmission risk as respiratory activities are highly relevant to the infectiousness and rapid spread of airborne diseases [2]. While this study used a 2D CFD model with point-to-point movement trajectories of individuals to make transmission risk evaluation simple, a 3D CFD model would provide greater model fidelity. In addition, a few simple experiments were designed and tested due to the lack of detailed information for the outbreak. The presented ABM can be further improved with more information of the attendees' behavior, for instance, their activity patterns before and after the training session, and during the whole session. Furthermore, the presented framework can be extended to estimate the critical amount of virus particle exposure level if it is known which attendees were infected. Another possible future work is to examine the impact of non-pharmaceutical control measures, for instance, social distancing, ventilation, and wearing a mask.

Acknowledgments. This research is supported by the Singapore Ministry of Health's National Medical Research Council under its National Epidemic Preparedness and Response R&D Funding Initiative (MOH-001041) Programme for Research in Epidemic Preparedness And REsponse (PREPARE).

Disclosure of Interests. The authors have no competing interests to declare that are relevant to the content of this article.

References

1. Alvarez Castro, D., Ford, A.: 3d agent-based model of pedestrian movements for simulating covid-19 transmission in university students. ISPRS Int. J. Geo Inf. **10**, 509 (2021)
2. Bale, R., Iida, A., Yamakawa, M., Li, C., Tsubokura, M.: Quantifying the covid19 infection risk due to droplet/aerosol inhalation. Sci. Rep. **12**, 11186 (2022)
3. Daamen, W.: Modelling passenger flows in public transport facilities. Phd dissertation, Delft University of Technology, Delft, the Netherlands (2004). https://repository.tudelft.nl/islandora/object/uuid:e65fb66c-1e55-4e63-8c49-5199d40f60e1. Accessed 11 Apr 2024

4. Helbing, D., Farkas, I., Vicsek, T.: Simulating dynamical features of escape panic. Nature **407**, 487–490 (2000)

5. Hoogendoorn, S., Bovy, P.: Pedestrian route-choice and activity scheduling theory and models. Transp. Res. Part B: Methodol. **38**, 169–190 (2004)

6. Hoy, G., Morrow, E., Shalaby, A.: Use of agent-based crowd simulation to investigate the performance of large-scale intermodal facilities: case study of union station in Toronto, Ontario, Canada. Transp. Res. Rec. **2540**, 20–29 (2016)

7. Hughes, R.: A continuum theory for the flow of pedestrians. Transp. Res. Part B: Methodol. **10**, 2205255 (2002)

8. Johansson, A., Batty, M., Hayashi, K., Al Bar, O., Marcozzi, D., Memish, Z.: Crowd and environmental management during mass gatherings. Lancet. Infect. Dis **12**, 150–156 (2012)

9. Johansson, F., Peterson, A., Tapani, A.: Waiting pedestrians in the social force model. XXPhys. A **419**, 95–107 (2015)

10. Kielar, P., Biedermann, D., Borrmann, A.: Momentumv2: A modular, extensible, and generic agent-based pedestrian behavior simulation framework. Technische Universität München, Munich, Germany, Tech. rep. (2016)

11. Kielar, P., Borrmann, A.: Coupling spatial task solving models to simulate complex pedestrian behaviour patterns. In: In Proceedings of the 8th Conference on Pedestrian and Evacuation Dynamics, pp. 229–235 (2016)

12. Kielar, P., Borrmann, A.: Modeling pedestrians' interest in locations: a concept to improve simulations of pedestrian destination choice. Simul. Model. Pract. Theory **61**, 47–62 (2016)

13. Lee, J., Marinov, M.: Analysis of rail passenger flow in a rail station concourse prior to and during the covid-19 pandemic using event-based simulation models and scenarios. Urban Rail Transit **8**, 99–120 (2022)

14. Luo, L., Chai, C., Ma, J., Zhou, S., Cai, W.: Proactivecrowd: modelling proactive steering behaviours for agent-based crowd simulation. Comput. Graph. Forum **37**, 375–388 (2018)

15. Maxmen, A.: Who report into covid pandemic origins zeroes in on animal markets, not labs. Nature **592**, 173–174 (2021)

16. Mendez, S., Garcia, W., Nicolas, A.: From microscopic droplets to macroscopic crowds: crossing the scales in models of short-range respiratory disease transmission, with application to covid-19. Adv. Sci. **10**, 2205255 (2023)

17. Mothership: Covid-19: Sparkletots preschool cluster increases to 20 cases, forming 3rd largest local cluster (2020). https://mothership.sg/2020/03/covid-19-sparkletots-preschool-coi/. Accessed 22 Apr 2025

18. Ooi, C., et al.: Risk assessment of airborne covid-19 exposure in social settings. Phys. Fluids **33**, 087118 (2021)

19. Our World in Data: Coronavirus pandemic (covid-19) (2020). https://ourworldindata.org/. https://ourworldindata.org/coronavirus

20. Peng, S., Chen, Q., Liu, E.: The role of computational fluid dynamics tools on investigation of pathogen transmission: prevention and control. Sci. Total Environ. **746**, 142090 (2021)

21. Ronchi, E., Lovreglio, R.: Exposed: an occupant exposure model for confined spaces to retrofit crowd models during a pandemic. Saf. Sci. **130**, 104834 (2020)

22. Saeed, R., Recupero, D., Remagnino, P.: Modelling group dynamics for crowd simulations. Pers. Ubiquit. Comput. **26**, 1299–1319 (2022)

23. Sun, Z., Bai, R., Bai, Z.: The application of simulation methods during the covid-19 pandemic: a scoping review. J. Biomed. Inform. **148**, 104543 (2023)

24. Sung, M., Gleicher, M., Chenney, S.: Scalable behaviors for crowd simulation. Comput. Graph. Forum **23**, 519–528 (2004)
25. Sze To, G., Chao, C.: Review and comparison between the wells–riley and dose-response approaches to risk assessment of infectious respiratory diseases. Indoor Air **20**, 2–16 (2010)
26. Today Online: 49-year-old preschool staff among 373 new covid-19 cases (2020). https://www.todayonline.com/singapore/new-covid-19-cases-may-28.   Accessed 22 Apr 2025
27. Tsukanov, A., Senjkevich, A., Fedorov, M., Brilliantov, N.: How risky is it to visit a supermarket during the pandemic? PLoS ONE **16**, e0253835 (2021)
28. Vuorinen, V., et al.: Modelling aerosol transport and virus exposure with numerical simulations in relation to sars-cov-2 transmission by inhalation indoors. Saf. Sci. **1303**, 104866 (2020)
29. Zhang, Z., Han, T., Yoo, K., Capecelatro, J., Boehman, A., Maki, K.: Disease transmission through expiratory aerosols on an urban bus. Phys. Fluids **33**, 015116 (2021)

A Dynamic Model of Customers Behavior: Integrating Econophysics and Physics-Informed Neural Networks

Kirill Zakharov$^{(\boxtimes)}$ ⓘ, Anton Kovantsev ⓘ, and Alexander Boukhanovsky ⓘ

ITMO University, Kronverksky Pr. 49, Saint Petersburg 197101, Russia
kazakharov@itmo.ru

Abstract. The transactional activity of a bank's customers contains a wealth of information regarding their behavior. By examining the transactions of either a specific group or all clients of the bank, it is possible to gain insight into the macroeconomic environment and utilize this to anticipate various outcomes. In this paper, we proposed a dynamic behavioral model for bank clients based on econophysics principles. Additionally, we identified the non-homogeneous function in the dynamic equation using physics-informed neural network. We also interpreted this non-homogeneity through the lens of news reports from social media platforms and news agencies. The model demonstrated accurate results in a numerical simulation of the restoration of the initial dependency. We also showed the potential for creating scenarios in which news events impact the behavior of bank customers.

Keywords: econophysics · Lagrangian mechanics · physics-informed neural networks · customers behaviour · bank transactions

1 Introduction

In banks or other financial organisations that handle customer transactions, it is essential to understand customer behavior in order to anticipate certain scenarios and enhance the user experience. Transactions contain a substantial amount of data, enabling the provision of insights into the historical patterns of customer behavior and potential purchasing preferences within specific product or service categories.

The primary applications of transactional data encompass: forecasting expenditures [1] and cashback categories [2], providing personalized recommendations [3], detecting fraudulent activities [4], implementing privacy-preserving mechanisms [5], analyzing advertising impact [6], and managing risks [7], among others. All these applications are widely used in modern banking and finance organizations to enhance organizational profitability and analyze potential investment opportunities.

In our article, the objective was to describe the dynamics of customer spending. Specifically, we were interested in a particular feature that denotes the

M. H. Lees et al. (Eds.): ICCS 2025, LNCS 15903, pp. 238–252, 2025.
https://doi.org/10.1007/978-3-031-97626-1_17

amount of money utilized in the current transaction. This amount of spending is a continuous variable and, furthermore, strictly positive. By understanding the spending patterns of a bank's customers, it is possible to gain insight into the state of the economy, including the presence of potential crisis situations, holidays or other significant events. For instance, the paper [8] explores the recognition of the behavior and experiences of individuals during crisis situations. The work [9] specifically focuses on bank customers and their transactional activity.

Additionally, it is important to understand how bank customers react to external economic factors such as politics, macroeconomic indicators and news. This understanding provides government influence to essential factors that can increase or decrease the purchasing power through the imposition of restrictions, prohibitions, or other regulatory measures.

The description of economic entities and agents is approached through various methods, including the use of various mathematical theories. For example, agent-based modelling, cellular automata, gravitational models. Therefore, studies that utilize a priori economic knowledge and specific physical techniques to construct models of consumption and production [10], social group interactions [11], and other social aspects are of particular interest. However, due to the vast number of interactions between economic agents and other external factors, it is not possible to fully describe behavioural patterns.

We used the tools of dynamical systems theory to identify the behavioural patterns. Specifically, we are concerned with second-order differential equations, as they are relevant to an approach originating from econophysics. This method combines advanced physical principles with economic phenomena through the interpretation of economic indicators using physical analogies.

In econophysics the economic entities, such as production volumes, are expressed in terms of generalized coordinates, and kinetic and potential energies are determined for them. By incorporating the necessary components into the potential energy, production processes can be understood in various ways. For example, potential energy can be defined in such a way that it represents an opportunity for the company to produce more output. This accumulated effect can be realized by increasing the company's utility function [10].

Transactions typically comprise a series of both continuous and discrete factors. These factors may include, for instance, the age of the customers, the amount spent, the transaction category, the terminal address, the customer identification number, the date and time of the transaction.

The process of receiving transactions is a dynamic process, although the receipt of transactions occurs at discrete points in time and at irregular intervals. However, it is more convenient to view transactions as a continuous stream and consider discrete points of time as defined by historical data. For instance, categorical features can be encoded to numerical values, as demonstrated in [12].

Numerous research papers are focused on predicting customer behavior using machine learning techniques. Understanding customer behavior can lead to enhanced company processes and improved customer experiences, as evidenced by studies such as [13] and [14]. Additionally, works related to customer purchases and digital marketing strategies, such as [15] and [16], demonstrate the potential of predicting customer behavior based on transaction data using numerical and categorical attributes. However, these models primarily focus on identifying patterns in the data without fully explaining the underlying causes.

Therefore, despite some work done in the area of modelling the behaviour of bank customers, models are either based on a predetermined behavioural model or on a specific set of features, with further identification of non-linear relationships between these features and the target variable. The solution may involve identifying the underlying forces, the external field, which help to explain the discrepancies between empirical observations and theoretical predictions.

Our contribution:

i) We developed a dynamic model that describes customer expenditures using Lagrangian mechanics. This derivation is based on the kinetic and potential energy associated with the customers behavior.

ii) We identified the non-homogeneous function in the dynamic equation utilizing a physics-informed neural network.

iii) We interpreted the non-homogeneous function by utilizing information from social media and news agencies. We also created scenarios based on news in order to alter the non-homogeneity function and, therefore, change the dynamics of the system.

2   Methods

2.1   Problem Formulation

We consider customer behaviour through their bank transactions. Each client has their own views and unique experiences that determine their actions. However, their behavior may exhibit some periodic patterns and seasonal variations, as well as some characteristics that deviate from conventional patterns due to some external factors.

We classified the transaction data into three categories by MCCs (Merchant category codes). The first category consists of all codes related to personal development, *self-realization*. They may include educational pursuits, namely university studies, vocational training courses, musical academies, or specialized master classes. The second category pertains to *socialization* activities. These encompass outings to restaurants, museums, or exhibitions. It can also involve spendings for cinemas, shopping malls, and travel-related activities [17].

The final category of goods and services encompasses everything that individuals require for *survival*. Specifically, pharmacies, hospitals, food and beverages, clothing, utility payments, etc.

Figure 1 shows the client expenses. We have presented the initial data and also constructed a smoothing trend using a period of 14 days and a one-sided filter. The Figure shows, some distinct peaks for the New Year event, corresponding to an increase in the purchasing power. Furthermore, there has been an increase in the all three categories, indicating that people are actively involved in events, buying gifts, and purchasing food for the holidays.

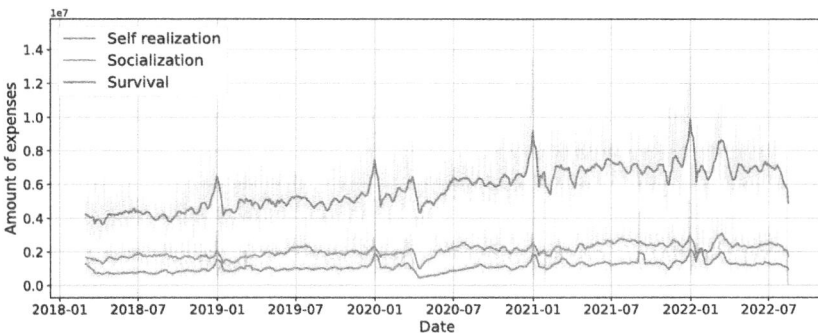

Fig. 1. Customer expenses. The X-axis shows the date, and the Y-axis indicates the total amount spent. The initial data are indicated in gray, while their smoothed trends are represented in color. (Color figure online)

It is remarkable that the fairly stable amount of spending in the categories of self-realization and socialization, in contrast to survival, where stable growth occurs, with the exception of crisis periods. We also observed a substantial reduction in expenditures during the first six months of 2020, which coincides with the onset of the COVID-19 pandemic. Furthermore, there was a significant shock in the early periods of 2022 due to a special military operation in Russia. Individuals experienced heightened uncertainty and increased their food purchases. In addition, there were instances where individuals spent money to relocate.

In this study, our objective was to address the problem of identifying the dynamics of customer banking behavior, as influenced by macroeconomic factors and seasonal variations. Specifically, our analysis focused on the amount of spending by customers in various currencies as an indicator of their behavior.

2.2 Econophysics and Lagrangian Formalism

Economics shares numerous similarities with physics, enabling the application of physical terminology and established methods from physics to solve economic problems. Energy is a fundamental concept in physics that enables bodies to move and accumulate potential energy that can be realized at a later time. For instance, just as a ball held above the ground possesses potential energy, enterprises, factories, and economic actors have the capacity to accumulate resources that can later convert into income or losses. This implies that when the ball is released, its potential energy transforms into kinetic energy, until the system reaches a state of minimum energy.

However, the question arises regarding the selection of a unit of energy measurement in the economy [18]. For instance, in work [10], money serves as the unit of measurement, representing currency. Consequently, the energy exchange is interpreted as a commodity-money relationship between two economic agents. Without external energy, a physical system that has attained its lowest possible energy state cannot transition to a state of higher energy.

This concept corresponds to an investment in the company's economy. When investments are low or loan rates are high, the company is unable to develop. Consequently, the only resource it can utilize is its accumulated capital, which can be likened to potential energy.

In our work, we also considered money as energy, but we adopted an abstract concept as units of measurement, such as γ. Different tasks may require varying values of γ. It can be represented as money, quantity, and other economic units [18]. We assumed that γ serves as an arbitrary unit to obtain the appropriate coefficients in equations.

We considered the time series of the length T, which corresponds to the amount of transactions accumulated by all clients. The X defines our system, i.e., there is one degree of freedom, where X is the generalized coordinate. The unit of measurement for the amount X is an arbitrary currency β (in our cases rubles). In fact, we included all the transformations we performed on the original time series in this variable. Specifically, during training the neural network, we normalized the values to the range from -1 to 1.

The generalized velocity $\dot{X}$ represents the rate of change in the amount of transactions over one period Ω, with units of $\frac{\beta}{\Omega}$. The duration of change can be equivalent to a day, a week, or any other number of days, depending on the trend smoothing period employed.

The kinetic energy is the quadratic form of the generalized coordinates, that is, in the one-dimensional case is equal to $T = \frac{1}{2}m\dot{X}^2$. The inertial mass of manufacturing encompasses all elements that suppress alterations in the manufacturing process, including inflexible technology and difficulties in adjusting manufacturing inputs [10]. This mass should have units of measurement such that the quantity $m\dot{X}^2$ is measured in γ, that is, $\frac{\gamma\Omega^2}{\beta^2}$.

One possible method to derive the dynamic equation is to employ the Lagrangian formalism. For most physical systems, it is feasible to determine its Lagrangian. The Lagrangian is a function that encapsulates the behavior of all components within the system. Once the Lagrangian is known, the dynamic equation can be derived through the Euler-Lagrange equation with dissipative function F,

$$\frac{d}{dt}\frac{\partial L}{\partial \dot{X}} = \frac{\partial L}{\partial X} - \frac{\partial F}{\partial \dot{X}}. \tag{1}$$

The primary challenge in constructing the model lies in determining the suitable potential energy function to derive the dynamic equation. To address this, we analyzed the system's behavior through the Taken's embeddings in the subsequent section.

2.3 Embeddings and Phase Trajectories

Let us assume that the dynamics of the system is governed by a multidimensional system under an ordinary differential equation $\dot{y} = v(x)$, where $x \in \mathcal{M}$ is the state, $\mathcal{M}$ is the n-dimensional smooth compact manifold and v is the vector field defined on $\mathcal{M}$, i.e., v is an element of the tangent bundle $T\mathcal{M}$. Under the initial condition $z \in \mathcal{M}$, the solution of the system is a flow $g^t : \mathcal{M} \to \mathcal{M}$. This map represents the one-parametric group of diffeomorphisms $\{g^t\}_{t \geq 0}$ such that $v(x) = \frac{d}{dt}|_{t=0}(g^t x)$.

We denoted the projections of this system as $\{\alpha(g^t)\}_{t \geq 0}$, where $\alpha : \mathcal{M} \to \mathbb{R}$ is the projection function which is so-called an observation function. In our case, these projections represent the volume of customer expenses. Our objective was to analyze the values of the observation function and understand the qualitative behavior of the system. This objective realized through the application of Taken's theorem.

We refered to the embedding as the map, which is defined as follows: the differentiable map $f : \mathcal{M} \to \mathcal{N}$ is called immersion if $d_x f : T_x \mathcal{M} \to T_{f(x)} \mathcal{N}$ is an injective map for all $x \in \mathcal{M}$; then embedding is the immersion which is also the homeomorphisms between $\mathcal{M}$ and $f(\mathcal{M}) \subset \mathcal{N}$. We denoted time delay as τ.

Theorem 1 (Taken's theorem). *Suppose $\tau \geq 0$ and $m \geq 2n+1$. Let g^t satisfy the conditions:*

(i) There is no periodic orbits of the period τ and 2τ.
(ii) There is a finite number of periodic orbits of the period $k\tau, k \in \mathbb{N}, k \geq 3$.
(iii) If $v(x) = 0$, then eigenvalues of $d_x g^t$ are all distinct and not equal to one.

Then $\mathcal{F}_{\alpha,g,\tau} : \mathcal{M} \to \mathbb{R}^m$, such that $\mathcal{F}_{\alpha,g,\tau}(x) := (\alpha(x), \alpha(g^\tau x), \ldots, \alpha(g^{(m-1)\tau} x))$ $\in \mathbb{R}^m$ is an embedding.

Thus, by establishing a specific phase flow, it is possible to construct an embedding that is devoid of self-intersections and uniquely corresponds to the initial dynamics of the system. Utilizing this embedding, we interpreted the behavior of a system that we do not directly observed, provided we restricted our observations to specific projections.

If the time series X has the one-dimensional observations and T data points, then the embedding's values are equal to

$$
\begin{pmatrix}
X(t) & X(t+\tau) & \ldots & X(t+(m-1)\tau) \\
X(t+1) & X(t+\tau+1) & \ldots & X(t+(m-1)\tau+1) \\
\vdots & \vdots & \ddots & \vdots \\
X(t+(m-1)) & X(t+\tau+(m-1)) & \ldots & X(t+(m-1)\tau+(m-1))
\end{pmatrix}, \quad (2)
$$

where m is the embedding dimension, τ is the time delay and $t = 1, \ldots, T - (m-1)\tau - (m-1)$.

We assumed the time delay $\tau = 1$. In our tasks, initially the data is already discrete. We have selected days for transaction date measurements by grouping the hours and minutes during which processing of the source data takes place, and then smoothing the results. In other words, we have chosen this specific sampling period based on the requirements of the task at hand. We determined the optimal embedding dimension using false nearest neighbors [19].

Since the embedding dimension does not display the entire trajectory of the system, we employed the dimensionality reduction method, t-distributed stochastic neighbor embedding [20] (t-SNE), to display the phase trajectory of the system on a plane.

Analyzing Fig. 2, we noted the cyclic pattern in the transactions trajectory. The colors in the graph correspond to the years, commencing with 2018 (brown) and concluding with 2022 (blue). The graph's points align with the annual cycles, as evident in instances such as January 2020, 2021, and 2022. The Figure shows a deviation in the period for the month of April in 2020 due to the COVID-19 pandemic.

It is also noteworthy that the cycles of March in 2018, 2019, 2020, and 2021 exhibit similar patterns of behavior, with points exhibiting these patterns having close orbits. In contrast, the month of March 2022 stands out due to the outbreak of the Russian-Ukraine conflict. The trajectory's movements are directed towards the point marked for April 1, 2022. During these months, the system encounters external influences and deviates from its typical phase trajectories.

Fig. 2. Projection of the phase portrait of a system onto a plane using the dimensionality reduction method t-SNE.

This structure of orbits resembles periodic fluctuations. One of the most basic models in physics, which describes periodic oscillations, is the mathematical pendulum model. However, in addition to the seasonal fluctuations, there are also some irregularities that occur in the crisis state of the economy. Therefore, we have taken into account a pendulum that is affected by an external field, which varies over time. This external field can be understood in different ways, namely store closures, tax increases, restrictions on public gatherings, negative news events, and so forth. Finally, it also makes sense to assume a damping effect in our system, since if we do not affect the system at all, it will go into some kind of equilibrium economic state, which we do not observe in the real world due to a large number of random factors and geopolitical and macroeconomic decisions.

2.4   Dynamic Equation

In the absence of external forces and frictional forces, the pendulum will persist in a specific oscillatory pattern. However, under the persistent influence of frictional forces, the pendulum will gradually decline and reach a state of equilibrium. Subsequently, upon the introduction of even a minor external perturbation, the pendulum will deviate and, due to the frictional forces, gradually decelerate until it returns to its usual state and, thereby, achieves a state of equilibrium.

In our interpretation, the pendulum is perpetually affected by various factors. Crisis situations correspond to significant disruptions, which induce the pendulum to adopt a novel oscillatory mode. However, due to the inherent forces of friction, it inevitably returns to a stable equilibrium, where fluctuations are solely sustained by seasonal perturbations.

Thus, it is feasible to analyze the potential energy in a manner analogous to a pendulum.

$$U(t, X, \dot{X}) = \frac{kX^2}{2} - F(\dot{X}) - \psi(t)X, \tag{3}$$

where $F(\dot{X}) = \frac{\alpha\dot{X}^2}{2}$ is related to a dissipate function, $\psi(t)$ represents the external forces (non-homogeneous function in the right hand side of the dynamic equation), k is the «spring» constant (the measure of the stiffness of the system).

In our case, the «spring's» stiffness can be understood as the extent with which bank customers' transaction patterns return to their pre-shocked state after experiencing a shock. The higher the «spring» coefficient, the faster the customers return to their original spending levels.

After analyzing the seasonal patterns of X, we identified the two seasonal components with periods of 12 and 31. The first component corresponds to monthly fluctuations, while the second component corresponds to daily fluctuations. We decided to model these components through the periodic functions $\cos(\frac{2\pi}{12}i)$ and $\cos(\frac{2\pi}{31}j)$, where $i \in \{1, \ldots, 12\}, j \in \{1, \ldots, 31\}$. The amplitudes of the oscillation are equals to c_m and c_d, correspondingly. Given the continuous nature of time, we recalculated the frequencies in the cosine argument to ensure that the argument remains a function of time continuously, where $t \in [0, 1]$. We denoted new frequencies as ω_d and ω_m for day and month periods.

Finally, the potential energy is equal to

$$U(t, X, \dot{X}) = \frac{kX^2}{2} - \left[\phi(t) + c_n f(t)\right]X - \frac{\alpha\dot{X}^2}{2}, \tag{4}$$

where $\phi(t) = c_d \cos(\omega_d t) + c_m \cos(\omega_m t)$.

The Lagrangian of the system equals to

$$L = \frac{1}{2}m\dot{X}^2 + \psi(t)X - \frac{kX^2}{2} - \frac{\alpha\dot{X}^2}{2}. \tag{5}$$

After taking derivatives in (5) and substituting them to (1), the resulting dynamic equation is

$$\ddot{X} = -\frac{k}{m}X + \frac{1}{m}\psi(t) - \frac{\alpha}{m}\dot{X}, \tag{6}$$

or if we redefine the terms as $\frac{k}{m} = b, \frac{\alpha}{m} = a,$

$$\ddot{X} + a\dot{X} + bX = \frac{1}{m}\psi(t). \tag{7}$$

Given that we calibrated the parameters a, b and determined the function ψ such that the solution of the dynamics equation best aligns with the initial data, the term m disregarded, as it incorporated into the function ψ. Alternatively, it is possible to set m to 1 by using a re-normalized units of measurement.

3 Results

We evaluated the quality of our model with two steps. Firstly, we assessed whether the solution of the dynamic equation with non-homogeneity, restored through the PINN, aligns with the initial data. Secondly, we investigated how the non-homogeneity function could be predicted using the news background data. For the network architecture we used the inverse physics-informed neural network with three feed forward layers and tanh activation, and with three losses: (i) data loss; (ii) initial conditions; (iii) equation loss to satisfy (7).

We utilized transaction data from debit cards of commercial bank clients in Russia for the period of January 2018 to July 2022. This data encompasses information about approximately $10,000$ active consumers' expenses, which were sampled daily. For news parsing, we utilized a regional news aggregator (https://sanktpeterburg.bezformata.com).

3.1 Empirical Non-homogeneity Function

By employing the appropriate external force signals and periodic components, we discerned the corresponding patterns of customers behavior. Figure 3 shows the periodic signal $\phi(t)$ and shocks $f(t)$ from the economy. The periodic component is achieved by combining two cosine functions with the periods of 12 and 31, which correspond to the year and month seasons. Using the Fourier analysis we also detected seasons with the periods of 7 and 14 days. However, fluctuations with this periodicity do not significantly affect the trend component of transaction amount. Therefore, to accelerate the calculations, they were not included in the analysis. The amplitudes c_d, c_m of cosine functions were determined from the data with the Fourier analysis.

Based on empirical considerations, we identified the function $f(t)$. Firstly, during the initial phase of the COVID-19 pandemic (first six months of 2020), purchasing power experienced a substantial decline due to the restrictions on offline shopping and the extended quarantine period. Secondly, during the New

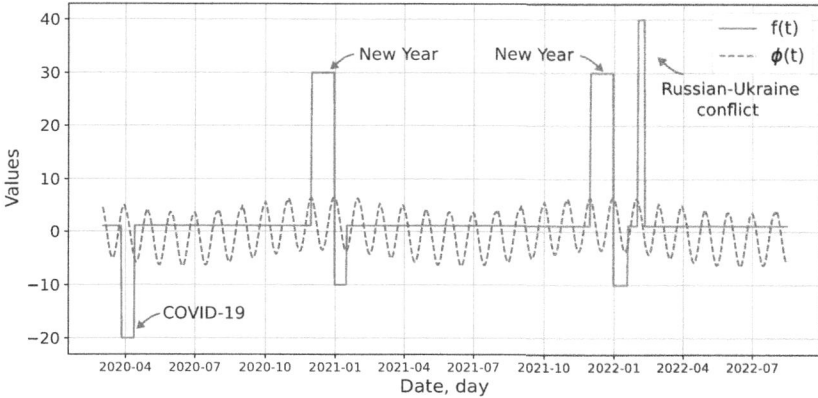

Fig. 3. Non-homogeneity function of the equation with periodic and empirical components. The absolute values on the Y-axis represent the relative influence of external factors. The higher the absolute value, the greater the influence.

Year period, individuals exhibited a significant increase in spending, aligning with the customary holiday season in December. However, after the New Year, the frequency of discounts decreased sharply, prompting individuals to adopt a saving mindset following their substantial spending. Therefore, spending patterns experienced a decline. Lastly, the period commencing in 2022 coincided with the onset of the Russian-Ukraine conflict, during which people actively engaged in purchasing essential commodities.

The numerical solution of the Eq. (7) with this empirical signals is demonstrated in the next subsection.

3.2   Solution of the Dynamic Equation

Figure 4 shows the equation solution with empirical signals (red color) obtained using the Runge-Kutta numerical scheme of order 4, related to calibrated parameters determined through the L-BFGS-B optimization method (limited and bound constrained quasi-Newton method of Broyden, Fletcher, Goldfarb, and Shanno [21]). The numerical solution closely approximated the normalized value of expenditures. The solution captured all the significant peaks from the economic shocks, as well as similar seasonal variations.

However, it is significantly more advantageous to obtain these signals directly from the data and endeavor to comprehend the factors that generate such signals. The green color in Fig. 4 denotes the PINN's solution, which is the approximation obtained using a neural network. It closely resembles the initial data (black color) because we solved the inverse task for PINN. However, the most interesting aspect is the non-homogeneity $\hat{\psi}$ achieved during the PINN's training process.

The solution obtained the Runge-Kutta numerical scheme with $\hat{\psi}$ is shown in blue. It is evident that the calibrated function precisely aligns with the dynamic

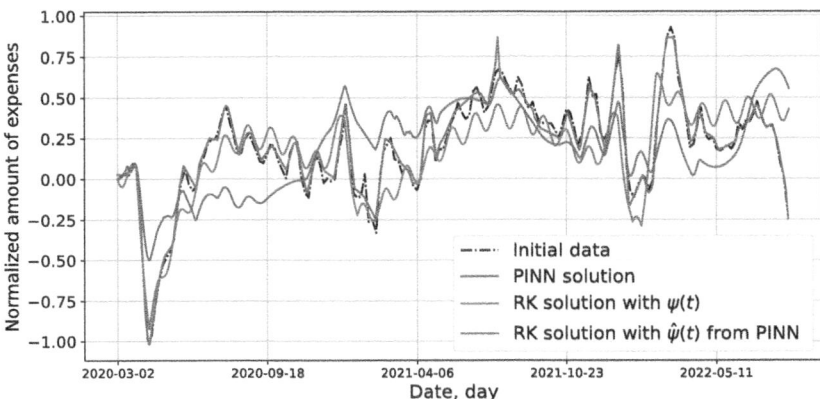

Fig. 4. Solutions obtained using the Runge-Kutta method for the empirical non-homogeneity function $\psi(t)$ and for the function obtained from PINN $\hat{\psi}(y)$.

equation, as the solution exhibits all the primary characteristics of the initial data. However, it contains less information regarding the monthly seasonal patterns, capturing more general trends. However, when considering customer behavior in reaction to economic shocks, we wish to capture general trends rather than specific local patterns.

From a practical standpoint, it is crucial not only to identify the non-homogeneity (external impact on the system) but also to explain the underlying causes. Without making purely empirical assumptions about the initial data, it is challenging to ascertain when and how the impact was applied to the system. Therefore, we proceeded to interpret this external influence in further detail.

3.3   The Non-homogeneity Interpretation by News

To comprehend the impact of news on customers behavior, it is crucial to identify the relevant signals from news sources. To achieve this, we employed data collection techniques from social media and reputable news agencies. Utilizing specialized filters, we identified significant events related to COVID-19 within the specified time frame, spanning from January 14, 2020, to June 30, 2021.

Firstly, we cleaned the news data from non-semantic words using regular expressions. Then, we selected keywords relevant to COVID-19 events, such as virus, vaccine, COVID, epidemic, sick, disease and others. Subsequently, for each day, we employed these keywords to identify sentences related to our topic. Afterward, we counted the number of sentences with the semantic meaning to our topic and divided it by the total number of sentences in our news database for the current date.

Figure 7 presents the computed average number of news articles related to COVID-19. It is noteworthy that news activity experienced an uptick in March

2020, followed by a decline in the summer of the same year, and further there were fluctuations around the value 0.1 (Fig. 5).

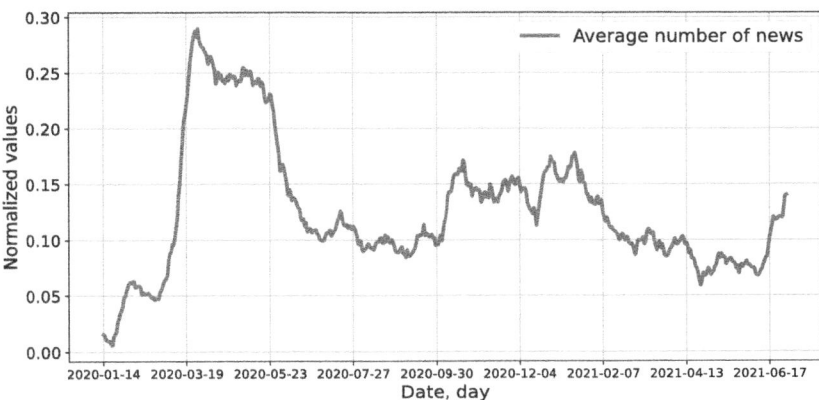

Fig. 5. The average quantity of news coverage over a specified time frame, spanning from January 14, 2020, to June 30, 2021.

To explain the non-homogeneous function within the dynamic equation, a regression model was employed, wherein the independent variables were the dates, expressed in terms of days, months, and years, as well as the mean value of news pertaining to a given topic, derived from the analysis of news datasets.

We employed the gradient boosting regression model from XGBoost library with a total of 100 trees and a maximum depth of 3. To ensure statistical significance, we randomly partitioned the data into training and testing sets with a ratio of 8 : 2. This process is repeated 200 times.

To assess the accuracy of the attributes in describing the non-homogeneity function, we utilized the coefficient of determination. This coefficient indicates the proportion of variance that is accounted for in the model's predictions. In our case its average values equals 0.8 with standard deviation 0.09 on the repeated train and test samples. This suggests a good descriptive capacity of the news, given that we have selected them based on fixed regular expressions.

Figure 6 shows the SHAP values. We used four indicators, and as can be seen, the most significant indicator was related to news. A large number of average news has a negative impact on the predicted external effect. The year can also be clearly divided into two parts. However, the indicators associated with the day and month of news have both positive and negative impacts, with some values being large and others small.

Figure 7 illustrates the scenarios that were generated by varying the news factor. Specifically, in the left-hand figure, we increased the coverage of news related to COVID-19 by 15% for the period spanning March to June 2021. Then, we predicted a non-homogeneous function and solved the differential equation (7) numerically. As can be seen, the number of expenditures increased during

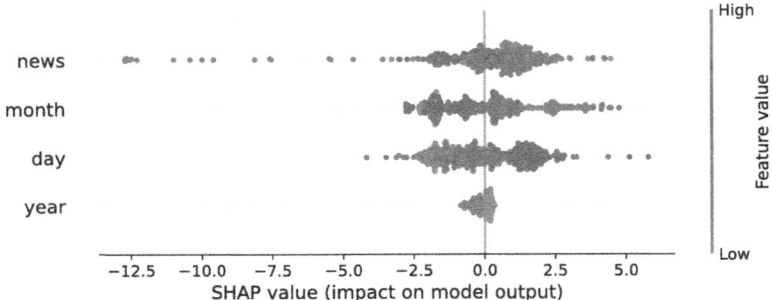

Fig. 6. Feature impact visualization. The feature names are plotted along the Y-axis. The X-axis indicates the impact on the model's prediction. Values above zero indicate a positive impact, while negative values indicate a negative impact. The color coding represents the magnitude of the feature's values.

this time period. In another scenario, we reduced news coverage of COVID-19 by 20%. The period selected corresponds to the onset of the pandemic. It can be observed that if news coverage is reduced, people would be less exposed to external influences, and therefore their spending would decrease less. At the same time, seasonal factors remain.

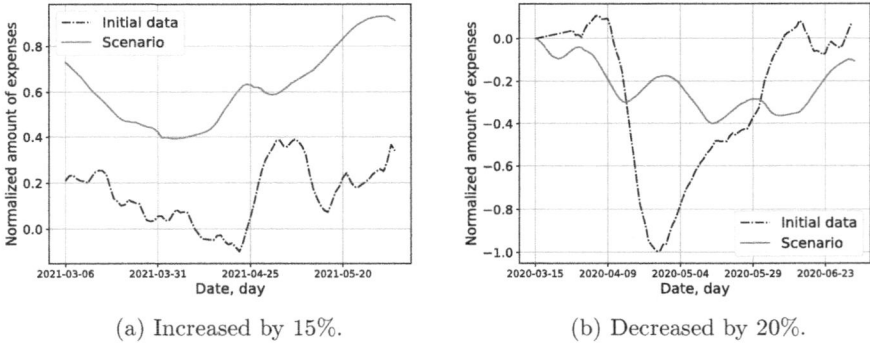

(a) Increased by 15%.   (b) Decreased by 20%.

Fig. 7. Scenario modeling based on the variation of the amount of news about COVID-19.

4   Discussion and Conclusion

In this study, we have addressed the issue of understanding the behavior of bank customers based on transactional data. Rather than relying solely on machine learning algorithms, we employed a hybrid approach that combines physics-informed neural networks and econophysics techniques. A key point in our work is the connection of the economic system through the concepts of Lagrangian mechanics. Analogies with physics have made it possible to analyze the behavior of customers and interpret the external factors with selected set of variables, namely news, day of the week, month, and year. As a result, news proved to be a significant factor, accurately reflecting the state of the national economy.

We analyzed phase trajectories of a system that describes the complete dynamics of human behavior, considering the projections of this system as customer expenditures and applying the Takens' embedding theorem.

At the same time, we believe that further improvements to the description of non-homogeneity can be achieved by considering not only the news context, but also macroeconomic indicators of the country in question. Additionally, it would be interesting to explore the use of LLM-based semantic filtering of news.

Moreover, we conducted scenario simulations, demonstrating how manipulating news inputs can influence consumer behavior. This finding is attributed to individuals' tendency to adapt their spending patterns in response to news events, aiming to optimize their personal utility. We believe that our research can be beneficial for managers at banks and other financial institutions to analyze and forecast the behavior of their customers by modeling various scenarios, including news reports.

References

1. Paramesha, M., Rane, N.L., Rane, J.: Artificial intelligence, machine learning, deep learning, and blockchain in financial and banking services: a comprehensive review. Partners Univers. Multidisciplinary Res. J. 1(2), 51–67 (2024)
2. Xu, L., Roy, A.: Cashback as cash forward: the serial mediating effect of time/effort and money savings. J. Bus. Res. 149, 30–37 (2022)
3. Vullam, N., Vellela, S.S., Reddy, V., Rao, M.V., SK, K.B., Roja, D.: Multi-agent personalized recommendation system in e-commerce based on user. In: 2023 2nd International Conference on Applied Artificial Intelligence and Computing (ICAAIC), pp. 1194–1199. IEEE (2023)
4. Al-Hashedi, K.G., Magalingam, P.: Financial fraud detection applying data mining techniques: a comprehensive review from 2009 to 2019. Comput. Sci. Rev. 40, 100402 (2021)
5. Zakharov, K., Stavinova, E.: Time-dependent differential privacy for enhanced data protection in synthetic transaction generation. In: Proceedings of the 2024 13th International Conference on Software and Computer Applications, pp. 112–117 (2024)
6. Wuisan, D.S., Handra, T.: Maximizing online marketing strategy with digital advertising. Startupreneur Bus. Digital (SABDA J.) 2(1), 22–30 (2023)
7. Khandani, A.E., Kim, A.J., Lo, A.W.: Consumer credit-risk models via machine-learning algorithms. J. Bank. Financ. 34(11), 2767–2787 (2010)
8. Saretzki, J., Pretsch, J., Pretsch, E., Grossmann, G.: Experience, behavior, and action in crisis situations: a literature review. Eur. J. Soc. Sci. Educ. Res. Art. 8 (2021)
9. Koshkareva, M., Kovantsev, A.: Crisis behaviour strategy recognition using transactional data. Procedia Comput. Sci. 229, 208–217 (2023)

10. Estola, M., Dannenberg, A.A.: Newtonian and Lagrangian mechanics of a production system. Hyperion Int. J. Econophys. New Econ. **9**(2), 7–26 (2016)
11. Sandler, U.: S-Lagrangian dynamics of many-body systems and behavior of social groups: dominance and hierarchy formation. XXPhys. A **486**, 218–241 (2017)
12. Zakharov, K., Stavinova, E., Lysenko, A.: Trgan: a time-dependent generative adversarial network for synthetic transactional data generation. In: Proceedings of the 2023 7th International Conference on Software and e-Business, pp. 1–8 (2023)
13. Kumari, L., Bhattacharjee, K., Sharma, N., Kumar, S., Kumari, A.: Machine learning models in customer behaviour prediction: A comparative analysis. In: 2024 7th International Conference on Contemporary Computing and Informatics (IC3I), vol. 7, pp. 957–959. IEEE (2024)
14. Harish, V., Benitta, D.A.: Construction of prediction model and forecasting trends based on consumer behaviour. In: 2024 OPJU International Technology Conference (OTCON) on Smart Computing for Innovation and Advancement in Industry 4.0, pp. 1–6. IEEE (2024)
15. Deniz, E., Bülbül, S.Ç.: Predicting customer purchase behavior using machine learning models. Inf. Technol. Econ. Bus. **1**(1), 1–6 (2024)
16. Sakthi, B., Sundar, D.: An effective customer behavior prediction system in digital marketing using ensemble serial cascaded network. In: 2024 International Conference on Intelligent Algorithms for Computational Intelligence Systems (IACIS), pp. 1–8. IEEE (2024)
17. Guleva, V.Y., Kovantsev, A.N., Chunaev, P.V., Gornova, G.V., et al.: Value-based modeling of economic decision making in conditions of unsteady environment. J. Sci. Tech. Inf. Technol. Mech. Opt. **147**(1), 121 (2023)
18. Zakharov, K., Kovantsev, A., Boukhanovsky, A.: Coupling of Lagrangian mechanics and physics-informed neural networks for the identification of migration dynamics. Smart Cities **8**(2), 42 (2025)
19. Kennel, M.B., Brown, R., Abarbanel, H.D.: Determining embedding dimension for phase-space reconstruction using a geometrical construction. Phys. Rev. A **45**(6), 3403 (1992)
20. Van der Maaten, L., Hinton, G.: Visualizing data using t-sne. J. Mach. Learn. Res. **9**(11) (2008)
21. Byrd, R.H., Lu, P., Nocedal, J., Zhu, C.: A limited memory algorithm for bound constrained optimization. SIAM J. Sci. Comput. **16**(5), 1190–1208 (1995)

Adaptive PCA-Based Outlier Detection for Multi-feature Time Series in Space Missions

Jonah Ekelund[1](✉), Savvas Raptis[2], Vicki Toy-Edens[2], Wenli Mo[2], Drew L. Turner[2], Ian J. Cohen[2], and Stefano Markidis[1]

[1] Computer Science Department, KTH Royal Institute of Technology, Stockholm, Sweden
`jonahek@kth.se`
[2] Applied Physics Laboratory, Johns Hopkins University, Laurel, MD, USA

Abstract. Analyzing multi-featured time series data is critical for space missions making efficient event detection, potentially onboard, essential for automatic analysis. However, limited onboard computational resources and data downlink constraints necessitate robust methods for identifying regions of interest in real time. This work presents an adaptive outlier detection algorithm based on the reconstruction error of Principal Component Analysis (PCA) for feature reduction, designed explicitly for space mission applications. The algorithm adapts dynamically to evolving data distributions by using Incremental PCA, enabling deployment without a predefined model for all possible conditions. A pre-scaling process normalizes each feature's magnitude while preserving relative variance within feature types. We demonstrate the algorithm's effectiveness in detecting space plasma events, such as distinct space environments, dayside and nightside transients phenomena, and transition layers through NASA's MMS mission observations. Additionally, we apply the method to NASA's THEMIS data, successfully identifying a dayside transient using onboard-available measurements.

Keywords: Outlier detection · Incremental PCA · Space mission data · Online learning

1 Introduction

Space missions, such as NASA's Magnetospheric Multiscale (MMS) Mission [1] and Time History of Events and Macroscale Interactions during Substorms (THEMIS) [2], generate large volumes of multi-featured time series data from in-situ measurements. These data require efficient methods for identifying and prioritizing scientifically relevant events. More broadly, large-scale data collection is essential across various scientific domains, including Earth-based weather

---

This work is supported by the European Commission, with Automatics in Space Exploration (ASAP), project no. 101082633.

M. H. Lees et al. (Eds.): ICCS 2025, LNCS 15903, pp. 253–267, 2025.
https://doi.org/10.1007/978-3-031-97626-1_18

monitoring [3] and Hyper Spectral Imaging (HSI) from Earth observation (EO) satellites [4]. In space missions, limited onboard computational resources and restricted downlink capacity necessitate real-time event detection to prioritise the most valuable data for transmission [1].

A challenge in analyzing space mission data is detecting outlier data points. These outliers may indicate both instrument failures and important physical phenomena occurring in the near-earth space environment, such as plasma environment crossings, magnetic reconnection events, or transient events. While some outliers may result from sensor anomalies or noise, others may reveal structures or processes not accounted for by existing models [5]. Traditional anomaly detection methods often assume static data distributions, which is generally not true for space plasma environments: dynamic changes in solar wind conditions and magnetospheric interactions can lead to evolving data characteristics.

One additional challenge for in-situ automatic data analysis is that onboard computational resources are largely constrained. These devices can only support lightweight and adaptive algorithms capable of processing high-dimensional data streams in real time [6]. Many existing methods require extensive prior knowledge of the data or rely on complex models that are impractical for deployment on resource-limited spacecraft. An effective approach should balance computational efficiency with the ability to detect scientifically meaningful outliers without extensive pre-training on specific datasets.

In this work, we present an outlier detection algorithm tailored for space missions. The method leverages reconstruction error from Principal Component Analysis (PCA) for feature reduction and adapts dynamically to new data distributions using Incremental PCA. This enables real-time outlier detection onboard spacecraft. Our focus is on evaluating the algorithm's principle for use in detecting outliers and demonstrating the algorithm's effectiveness in identifying plasma events such as bow shock crossings in MMS data and foreshock bubbles in THEMIS observations. The main contributions of this work are the following:

– Introduce an adaptive outlier detection algorithm based on the PCA reconstruction error, which can be applied to streaming data. The algorithm works without a pre-training step and dynamically adapts to the dominant features in new data.
– Extend the algorithm to handle different feature types with different magnitudes by coupling the feature scaling to a group with features of the same type. This approach preserves the relative variance within each feature group.
– Demonstrate the algorithm's effectiveness in finding boundary crossings and other scientifically interesting events in MMS data using multiple features.
– Show that the algorithm can detect scientifically relevant events in data available onboard the spacecraft using the THEMIS mission, potentially supporting real-time identification.

2   Background

In this work, we focus on the use-case of finding outliers in multi-featured space plasma time-series data using MMS [1], launched in 2015 and THEMIS, launched in 2007 [2], as examples. MMS was launched to investigate magnetic reconnection in the boundary regions of Earth's magnetosphere using unprecedented time scales. It consists of four spacecraft flying in formation through the dayside magnetopause and the nightside magnetotail regions [1].

The THEMIS mission was launched to investigate the trigger and evolution of substorms. The mission consists of five satellites, lining up to track particles along the magnetotail. While the primary mission goal was to perform measurements in the magnetotail, in the nightside region of Earth's magnetosphere, the spacecraft also obtained measurements from the dayside region [2]. In this work, we have used an interval of dayside data from THEMIS.

Both MMS and THEMIS have limited storage capabilities, constraining the amount of data that can be collected. Furthermore, the presence of limitations in the downlink capabilities means that the collected data of the highest possible value has to be prioritized. The collection limitation and prioritization were performed using pre-specified temporal region-of-interests (ROI) with corresponding locations in space, using onboard calculated indicator values and human-made selections. These selections are intended to prioritize the most valuable mission data, especially those that are significantly different from the general state of the regions and have a high likelihood of containing interesting events [1,2].

As spacecraft orbit Earth, they transition through multiple regions, including the magnetosphere, where Earth's magnetic field shapes the plasma environment and areas beyond its protective influence, where the solar wind and the Sun's magnetic fields dominate. They also traverse key boundaries such as the bow shock, where the solar wind slows and deflects and the magnetopause, which separates Earth's magnetic domain from the solar wind. Classifying which region the measurements belong to can help direct scientists toward phenomena of more interest [7]. Determining the regions and finding the crossings between them can help to optimize the data collection onboard the spacecraft.

2.1   Outlier Detection in Streaming Data

Outlier detection deals with finding samples, or groups of samples, that differ from the general structure of the remaining samples. Traditionally, most outlier detection methods deal with static data where the entirety of the data range is known [8]. If the data is high-dimensional, feature reduction techniques, such as Principal Component Analysis (PCA) [9] or Autoencoders [10], can be utilized to simplify the problem and reduce the search space [5].

However, streams of data add to the complexity of finding outliers. The temporal variability of the data means that what is considered an outlier can shift with time. The nominal values of a feature and feature importance may also shift in time [5]. This can render an initial data model incorrect and any outlier detection algorithm needs to be able to adapt to these shifts in the data.

PCA is a linear feature reduction method that surfaces the top N features, the Principal Components (PC) containing the most variance (i.e. information), from the original feature space [11]. Once in the reduced feature space, the samples can be separated into groups using different clustering techniques [10,12]. Then, outliers can be located based on the distance to these clusters' centroids or the density. Another option for finding outliers is to transform the samples back to the original feature space and evaluate how much the reconstructed features (R_f) differ from the original features (F_f). This reconstruction error (E_f):

$$E_f = R_f - F_f \tag{1}$$

is the loss of information from the feature reduction. Samples where features have large reconstruction errors will deviate from the PCA model and can, therefore, be considered outliers compared to the other samples [13]. Incremental PCA is a variant of PCA that can be used to build the PCA model incrementally [11]. This can be necessary if the data is too large to load all the samples into memory at once, or for the use-case presented in the paper, not all the data is available when the initial model is created.

3   Methodology

In this work, we present a method for detecting outliers in streaming multi-feature data based on evaluating the reconstruction error stemming from feature reduction using Incremental PCA [11]. The reconstruction error E_i, for a sample i, is calculated as the Euclidean norm of the reconstruction errors E_f for all the features f. By looking for large and rapid changes in this error, we can find samples that deviate from the model and the previous samples.

3.1   Outlier Detection

The outlier detection algorithm[1] operates in three different modes, Initialization, Check and Calibrate, as shown in Fig. 1. In the initialization mode, (Fig. 1 - top), the initial model is built based on the first samples retrieved by the algorithm. Following the initialization mode is the check mode, which is the main operating mode (Fig. 1 - bottom right). Here, samples are retrieved and evaluated for outliers. If multiple sequential samples are labeled as outliers, then a calibration using these outlier samples is initiated. In the calibration mode, the PCA model is updated based on the latest outlier samples.

There are five parameters controlling the execution of the algorithm: the calibration buffer size (S_c), the number of components (N) used in the PCA, the mean buffer size (S_m), the threshold (λ) and the outlier limit (L_o). S_c controls the size of the calibration buffer (c_buffer). The calibration buffer is used to store samples to use for calibration. When this buffer is full, either in the initialization mode or the check mode, the PCA is calculated using the samples

---
[1] https://github.com/Jonah-E/multi-feature-outlier-detection.

in this buffer. N controls how many components of the PCA are calculated and thereby the size of the reduced feature space. A lower value will, in general, mean a higher reconstruction error, but changing this parameter can also change what events are discovered by the algorithm. S_m controls the size of the mean buffer (m_buffer). The mean buffer is a fixed width circular buffer used to store the reconstruction error (E_i) of the latest S_m samples not labeled as outliers. This buffer is then used to calculate the current mean μ and standard deviation σ.

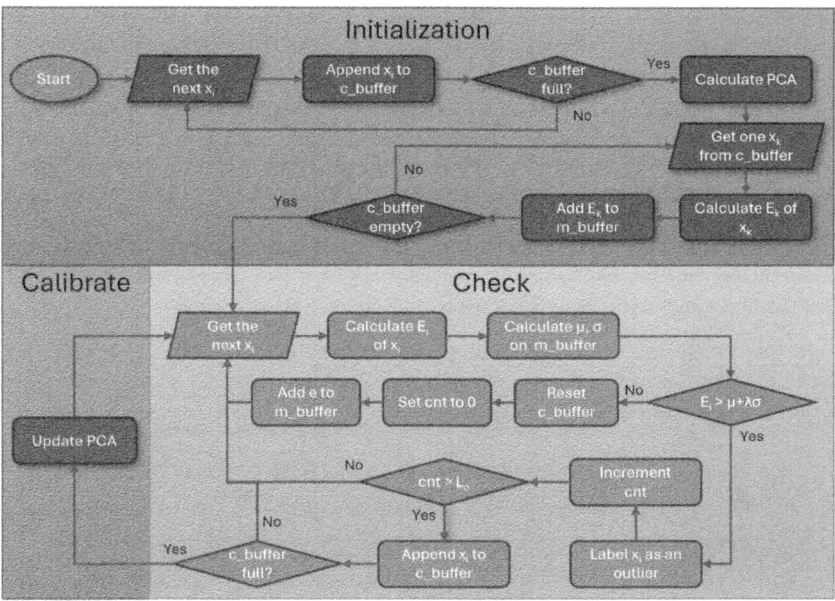

Fig. 1. Flowchart describing the working principles of the adaptive algorithm.

The μ and σ values are then used together with the threshold λ to calculate the maximum allowed reconstruction error, or the error threshold T_i, according to the equation

$$T_i = \mu + \lambda\sigma \tag{2}$$

A sample x_i with error E_i exceeding T_i will be labeled an outlier. The last parameter, L_o, controls how many samples x_i can be labeled as outliers before samples are added to the calibration buffer. Setting this to a value larger than zero will exclude the L_o first outlier samples from the calibration buffer.

Initialization mode: is the start of the algorithm. Here, the algorithm builds the initial model by retrieving the next sample x_i and then adding it to the calibration buffer. When the calibration buffer is full, the PCA is calculated using the samples in the calibration buffer. Following this, the reconstruction error (E_k) of the samples (x_k) in the calibration buffer is calculated and added to the mean buffer. Depending on the application of the algorithm, it could be beneficial to pre-calculate the PCAs on a curated dataset containing data from expected regions. Then, the initialization step could be skipped.

Check Mode: The first step in the check mode is to retrieve the next sample x_i and calculate the E_i of this sample. Then μ and σ are calculated using the samples in the m_buffer and T_i can be calculated according to Eq. 2. If the E_i is larger than T_i, then the sample is labeled as an outlier and a counter (cnt) is incremented. If this counter exceeds L_o, then the sample x_i is also added to the calibration buffer. If the sample is not an outlier, the calibration buffer is cleared, the counter is reset to 0 and the E_i is added to the mean buffer. Only samples not labeled as outliers are added to the mean buffer to prevent outlier samples from inflating the error threshold. The last step is to evaluate if the calibration buffer is full. If it is full, multiple samples in a row have been labeled as outliers and the PCA model is likely no longer correct. Then, a calibration is initiated.

The calibration mode: is where the PCA is updated using the samples from the calibration buffer, bottom left in Fig. 1, before the algorithm reenters check mode and the next sample is processed.

Table 1. MMS intervals with primary ROI, dayside data intervals are originating from Ref. [14] and nightside intervals from Ref. [15].

Nr	Data Interval	ROI	Day or Nightside
0	2017-12-14 16:00 → 2017-12-14 05:00	-	Dayside
1	2017-12-17 16:00 → 2017-12-17 22:00	17:52 → 17:54	Dayside
2	2018-01-12 00:50 → 2018-01-12 06:00	01:50 → 01:52	Dayside
3	2018-12-14 02:00 → 2018-12-14 07:20	04:21 → 04:22	Dayside
		04:40 → 04:42	
4	2018-12-10 04:00 → 2018-12-10 11:00	05:12 → 05:25	Dayside
		06:27 → 06:31	
5	2019-01-05 16:00 → 2019-01-05 19:00	17:38 → 17:41	Dayside
6	2021-01-12 00:00 → 2021-01-12 06:00	01:18 → 01:21	Dayside
7	2021-02-13 10:00 → 2021-02-13 18:00	11:05 → 11:06	Dayside
8	2022-05-02 18:00 → 2022-05-02 22:00	18:23 → 18:25	Dayside
9	2022-11-24 02:00 → 2022-11-24 09:00	04:16 → 04:18	Dayside
10	2023-01-16 06:00 → 2023-01-16 11:00	08:21 → 08:24	Dayside
11	2017-07-23 12:00 → 2017-07-23 18:00	16:55 → 16:56	Nightside
12	2021-08-14 16:00 → 2021-08-15 06:00	01:23 → 01:25	Nightside

3.2   Data and Pre-processing

MMS Data: The primary data used in this paper is from MMS. Specifically, the omni-directional ion spectrum and ion velocities in GSE coordinates from the Fast Plasma Investigation (FPI) instrument [16] and the magnetic field (hereafter referred to as the B-field) from fluxgate magnetometers (FGM), part of the FIELDS instrument suite [17], from the MMS-1 spacecraft. Table 1 lists eleven dayside intervals of MMS data together with some regions of interest. These intervals are either from MMS passing into or out of the Earth's magnetosphere from the Solar wind. Transitions from the solar wind to the magnetosheath are called *bow shock* crossings, while transitions from the magnetosheath to the magnetosphere are called *magnetopause* crossings. The data from the FGM is collected at a significantly higher frequency, ~ 8 Hz to ~ 16 Hz, than the FPI data, ~ 0.2 Hz, we therefore down-sampled to the sample frequency of FPI. The data used is level-2 data, which is post-processed on Earth and would not be available onboard the spacecraft.

The dayside ROI in Table 1 are taken from Raptis *et al.* [14] and are transient phenomena upstream of the bow shock in the ion foreshock region, which are known to energize particles and cause space weather effects [18]. Therefore, finding these, even in level-2 data, is important for further scientific research. In addition to these transient events, the data intervals also contain other phenomena, such as bow shock and magnetopause crossings, which are operationally critical to identify automatically.

We have used two nightside data intervals from MMS-1, listed in Table 1, to evaluate how the algorithm generalizes to different regions. These are intervals from the nightside magnetosphere of Earth, containing two events analyzed by Richard *et al.* [15]. These are fast plasma flows associated with geomagnetic disturbances and can, therefore, cause space weather effects [19,20].

THEMIS spacecraft is used as an alternative source of space plasma data with different instruments compared to MMS. In particular, we use THEMIS-C probe measurements of the ion energy flux and ion velocity from the Electrostatic Analyzer (ESA), these are measurements with minimal processing, which are readily available for onboard algorithms [2,21]. The interval investigated is from 2008-07-15 and contains a foreshock transient [22].

Multiple Features: The PCA method finds the axis with the highest variance; therefore, for PCA to be effective, all features must be normalized to the same range. When PCA is used in data with one type of feature, for example, the omnidirectional ion spectrum used in this paper, the values of all features can be expected to have the same magnitude. However, if multi-feature types are included, these new features can have values with magnitudes that differ significantly. Then, the feature types with the highest magnitudes will dominate the PCA. This can be seen in the leftmost plot of Fig. 2, where Ion spectrum channels 19 to 21 have the highest variance, while the B-field variance is not visible. A common way to solve this issue is to scale the features, for example, MinMax scaling, which scales each feature to be between zero and one. However, this destroys the relative variance within a feature type. This can be seen

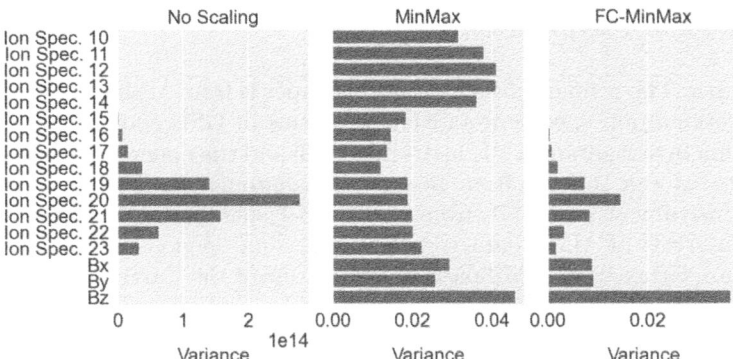

Fig. 2. Comparison of the variance of the different features for no scaling (Left), Min-Max scaling (Middle) and FC-MinMax scaling (Right). The features are MMS omni-directional ion spectrum, channel 10 to 23 and B-field from 2017-12-17, 20:00 to 21:50, while MMS is traversing the magnetosheath region.

in the middle plot of Fig. 2, where the highest variance for the ion spectrum data has now moved to channels 12 and 13. To solve this issue, we introduce a coupling between features of the same type when scaling. This Feature Coupled MinMax (FC-MinMax) scaling scales the coupled features to the same min and max values, which are calculated over the group of features of the same type.

The variance after using the FC-MinMax scaling can be seen in the rightmost plot of Fig. 2. Here, the relative variance within each group, ion spectrum and B-field are retained while scaling the features to have the same magnitude. We can now see that the B-field, together with the ion spectrum, will affect the PCA. In our tests, we have used data interval 0 from Table 1 to calculate the scaling for each feature group when testing the data from the MMS dayside intervals. For other data, we have calculated the scaling based on the specific data.

4   Results

Figure 3 shows the outlier detection algorithm performance when applied to a single time-series of MMS data. The features used are the ion omnidirectional spectrum, the B-field and the ion velocities. Each feature is scaled using the FC-MinMax scaling with the scaling factors calculated on data interval 0, see Table 1. The bottom three plots describe the functionality of the algorithm. Samples with the flag 'outlier' or 'calibration' are outlier samples. However, samples with the flag 'calibration' are used to update the PCA model.

The initialization stage can be seen furthest to the left in the plot; here, there are no outlier detections and the flag is set to 'Calibrating'. Following this, we can see how the reconstruction error and detection threshold are updated as more samples are ingested. At 01:50, the algorithm detects ROI from Table 1. The next region of interest in the data is the bow shock crossing at 03:22, where

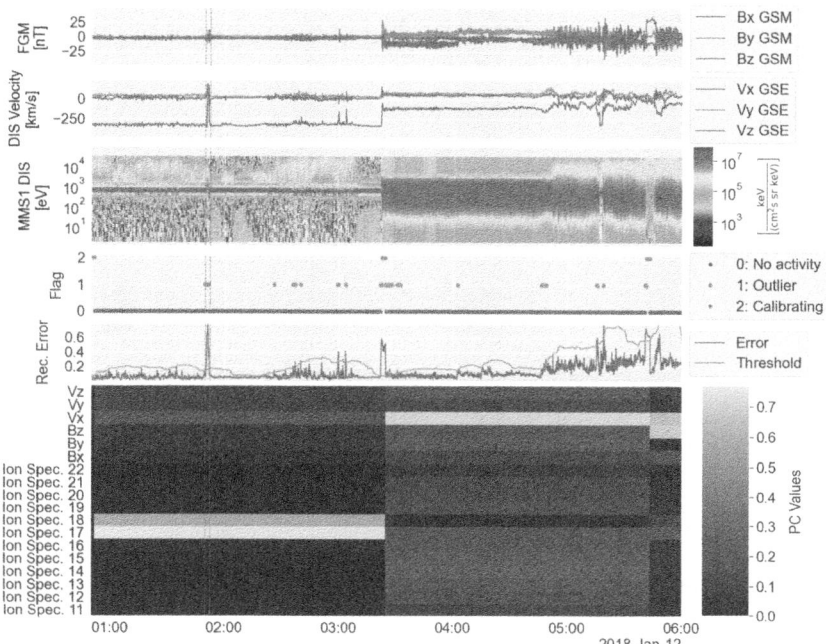

Fig. 3. Algorithm applied to multi-feature data from MMS dayside interval 2, the primary ROI is marked with dashed lines, $S_c = 25$, $S_m = 150$, $L_o = 10$ and $\lambda = 4$. The bottom plot is the principal component (PC) at each time; ion spectrum channels 0–10 and 23–31 have values close to zero and are not shown.

the spacecraft enters the magnetosheath. The algorithm detects the crossing by the sudden increase in the reconstruction error. However, as the error does not decrease below the threshold of 35 samples ($L_o + S_c$), the entry triggers a calibration for the new region, which is shown by the samples labeled 'Calibration'. After the calibration, the error decreases below the threshold. One more calibration is triggered at the short entry into the magnetosphere region at around 05:40 after a magnetopause crossing. Both of these crossings highlight different physical dynamics and their automatic finding can facilitate scientific research.

The bottom panel in Fig. 3 shows how each feature affects the principal component, inversely, this also shows which feature affects the reconstruction error the most. We can see there that in the solar wind region, the PCA model is dominated by the ion spectrum channels 17 and 18, while after entering the magnetosheath, multiple spectrum channels, together with the B-field and velocities, affect the model, where the Vx feature is now dominant.

Optimal Parameters: As mentioned in Sect. 3.1, five parameters control the algorithm. We have performed a rudimentary parameter optimization on the set of dayside data intervals from Table 1, maximizing the detection of the listed ROIs. For this data, with the feature types, ion spectrum, B-field and ion velocities, we have found that setting the parameters as $S_c = 25$, $N = 2$, $S_m = 150$,

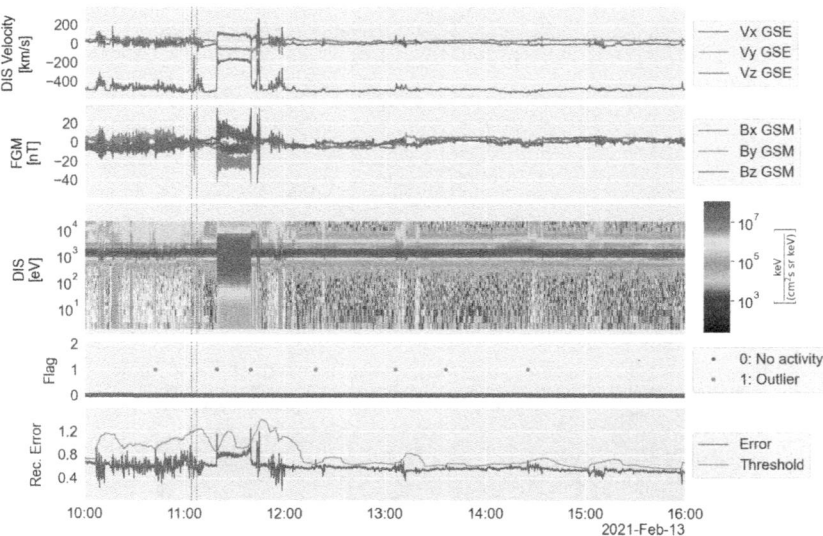

Fig. 4. Example of a ROI (dashed lines) the algorithm showed no detection.

$\lambda = 5$ and $L_o = 10$, balances finding the ROIs with false detections. With these parameters, the algorithm can find eleven out of the twelve dayside ROI specified in Table 1 when all the data intervals are fed in sequence to the algorithm without restarting it between intervals. However, the time shift between the intervals also introduces an artificial shift in the reconstruction error. When this shift is towards a larger error, the algorithm over-calibrates to the region at the start of the new interval, which can help it detect the primary ROIs. For a more realistic analysis, the mean buffer was reset to the first two samples when entering each new data interval. With these settings, most of the samples are marked as either 'No Activity' or 'Outlier', 98.7% and 1.1% respectively, only 0.2% are used to update the model.

The detection result for data interval 7 is plotted in Fig. 4. Here, the algorithm is missing the primary ROI. However, it does find the two crossings into and out of the magnetosheath at 11:19 and 11:39. The following quick crossing in and out of the magnetosheath, at 11:43 to 11:44, is, however, missed due to the sudden drop in reconstruction error from the first exit causing an increase in the standard deviation and thereby in the error threshold. If the ion spectrum is removed from the feature space, thereby only using the velocities and B-field, the marked ROI is found. All the crossings above are also found. However, multiple recalibrations are triggered at the entry to the magnetosheath.

MMS-1 Nightside Data: When applying the algorithm to the nightside data from Table 1, it can find the ROI at 2021-08-15 01:23 from the 12th data interval but not the ROI from the 11th interval. However, by increasing the number of PCAs to three and lowering the threshold to 3.5, the algorithm can find both

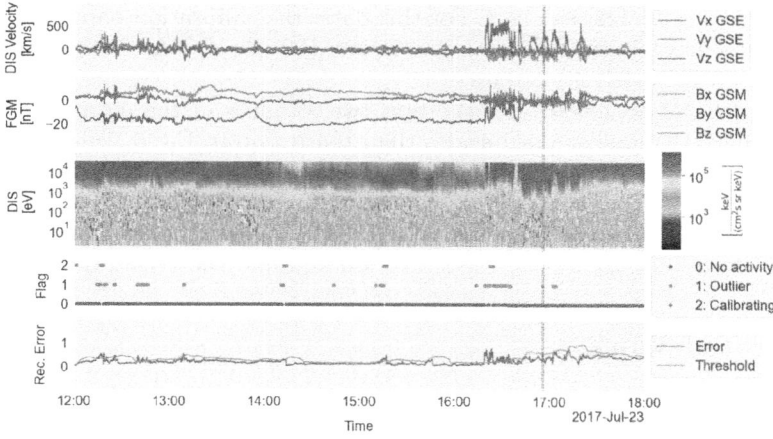

Fig. 5. The algorithm applied to MMS Nightside data interval 11 from Table 1.

regions. Here, the FC-MinMax scaling is calculated on the same data interval as the algorithm is applied. If scaling is instead calculated on dayside interval 0, the algorithm will over-calibrate to the ion velocities, as these have a larger value range in parts of the nightside data (Fig. 5).

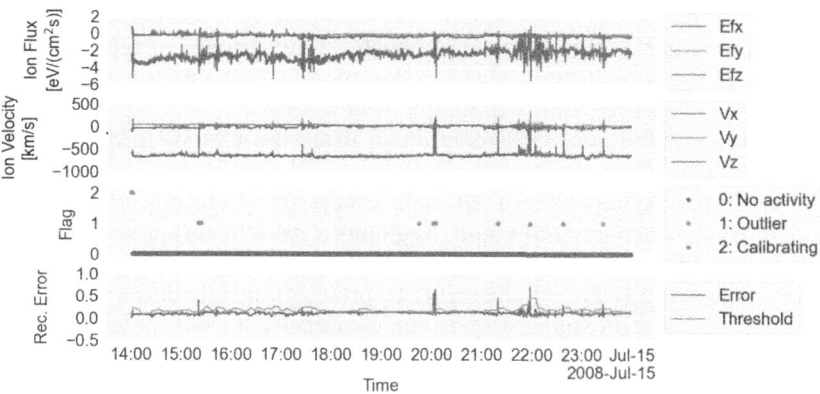

Fig. 6. The algorithm applied to an interval of dayside data from THEMIS C.

THEMIS C Dayside Data: Applying the algorithm to THEMIS C intervals, with a scaling calculated on the same interval, we find several possible interesting events; The longest of these is between 20:01:51 to 20:04:36 and 21:57:44 to 21:58:08. This second event starting at 21:57:44 has been noted in previous works as a foreshock bubble [22,23]. The start and stop times indicated by our algorithm also correspond well to the start of expansion to end of compression, shown by Liu *et al.* [22] in Fig. 2. For this data, the calibration batch and mean window size have been increased to 30 and 180 due to the higher sampling frequency of the THEMIS C data.

For the full day, 2008-07-15, the first twelve hours of the data have values with significantly smaller magnitudes than the region in Fig. 6. When applying the algorithm to this wider data interval, with a scaling calculated on the full interval, the algorithm detects the start of the new region but does not trigger a recalibration. Lowering the threshold to 3.5 causes the algorithm to more aggressively recalibrate during the transition and find the events in Fig. 6.

5   Related Work

Plasma Region and Bow Shock Identification: In the area of space plasma physics, bow shock crossings, which are coupled with the transition between the solar wind region and the magnetosheath are areas of active research. There have been numerous works developing methods for identifying plasma regions (e.g., solar wind and magnetosheath) and their transitions (e.g., bow shock). Both Olshevsky *et al.* [24] and Breuillard *et al.* [25] utilized neural-network-based approaches to identify different plasma regions. The bow shock transitions are either identified by where network output indicates a high probability for neighboring regions [24] or as specific labels [25]. These approaches are powerful and displayed high accuracies on the test sets. However, as the methods rely on having a training dataset, they cannot adapt to changing input data. While the algorithm presented in this paper can benefit from samples of relevant regions to create an initial model, the algorithm does not require a training set. Instead, it can be applied directly and will adapt to new data.

More recent work by Toy-Edens *et al.* [7] utilized a Gaussian Mixture Model to classify the MMS dayside region, which displayed high accuracy using a method more lightweight than a full neural network. Bakrania *et al.* [10] used a pipeline of autoencoder, PCA and Agglomerative Clustering to classify data from the ESA's Cluster spacecraft [26]. Innocenti *et al.* [12] utilizes PCA, self-organizing maps and K-means clustering to process multiple different quantities, such as B-field, particle velocity and particle density, to obtain a classification of the plasma region.

PCA-Based Outlier Detection. The two-step approach, presented by Finley *et al.* [9], analyzes a region of MMS data for interesting events by subdividing the region into smaller windows and performing a PCA on the resulting matrix of subregions. Then, a One-Classifier support vector machine (OCSVN) is used to find the windows containing outliers. This technique shows promising results; however, it is very computationally expensive, as the PCA and OCSVN have to be recomputed for each analyzed region. Zamry *et al.* [6] utilizes a PCA feature reduction coupled with a One-Class Support Vector Machine to detect anomalous sensor readings in data from a Wireless Sensor Network. Bhushan *et al.* [13] applied an Incremental PCA on streaming data with geospatial spread. They only consider finding outliers in one type of feature from one sensor type and the outliers are artificially introduced into the data. Furthermore, the PCA model is constantly updated, increasing the computational complexity. The algorithm presented in this paper limits the computational complexity by only updating the PCA model when necessary.

6   Discussion and Conclusion

In this work, we have presented a lightweight outlier detection algorithm for streaming multi-feature data. The algorithm can adapt to new data regions, enabling its deployment in applications where all the data regions are not known or fully understood. We utilize the Incremental PCA to have a lightweight algorithm that could be deployed on low-performing hardware. However, the basic principles in the algorithm can be applied using other unsupervised feature reduction techniques, where a reconstruction error can be calculated, for example, with autoencoders. Furthermore, while we have demonstrated the algorithm using space mission data, it can also be used to analyze other types of streaming data, such as measurements from a weather monitoring station. This will have to be evaluated for the specific data in question.

With optimal parameters, the algorithm can find most dayside ROIs, together with several crossings from solar wind and ion foreshock to the magnetosheath. For the nightside data, the number of components needed to be increased and the threshold slightly lowered. If we compare where our primary ROI is located in the dayside data to where it is located in the nightside data, we can understand why. In the dayside data, the ROIs are located in the Solar Wind/Ion foreshock regions, where the ion spectrum mainly consists of a tight beam of energy, as we can see in Fig. 3. This region is well-defined by only two parameters and outliers are visible in the reconstruction error. The nightside data is more varied around the marked ROI, leading to the ROI being masked by other errors. By increasing the number of components, the model can better describe the structure of the data, allowing the ROI to be found. This result indicates that the algorithm can also be used as a data mining tool to find interesting events in the dayside and nightside regions of the Earth's magnetosphere. This can help scientists find events for further studies. However, the different parts of the nightside can vary significantly and further studies will be necessary to understand how well the algorithm performs on data from the nightside region.

The algorithm can also be used on data from different spacecraft, as demonstrated by the use of THEMIS C data. In this case, the data used would be available onboard the spacecraft, indicating the algorithm's potential to be used onboard to inform the decision on which data to prioritize for downlink. In a real mission scenario, the algorithm would be applied to data which has already been subjected to a robust onboard pre-processing check for sensor failures. However, the algorithm needs to be more extensively tested using data available onboard, from multiple spacecraft, to show that the principle is robust enough for use in future space missions. Furthermore, the algorithm has to be evaluated on representative hardware before it can be used onboard during a space mission. Therefore, future work will focus on the creation and evaluation of a version of the algorithm optimized for low-performing hardware.

The FC-MinMax scaling used to scale the different feature types enables the use of multiple feature types with different magnitudes. It also opens up the possibility of adding a feature selection by weighing the different feature types differently. However, the necessity of this pre-scaling step is one of the two main weaknesses of the current algorithm, the second is how the threshold is defined. To further advance the algorithm, we will focus on this pre-scaling step and evaluate ways to integrate it into the algorithm itself. It would then be calibrated at the same time as the algorithm in the initialization stage. During a recalibration, the new data can be checked against the existing scaler to see if it needs to be updated. However, this could mean that the PCA model has to be recalculated and not just updated using the new samples.

For the second weakness, the main issue is that defining the threshold using the standard deviation can lead to large thresholds when there are large decreases in the reconstruction error, as seen in Fig. 4. This can lead to missed detections of interesting events. Furthermore, an incorrectly selected threshold can lead to excessive recalibration of the algorithm. To address these issues, the threshold could be dynamically updated either continuously or during calibration. Another option to limit excessive recalibration could be to introduce a forgetting factor to the IncrementalPCA. This would decrease the importance of earlier data and allow the new data to be more prominent in the model [11].

The algorithm presented in this paper is a promising approach to finding outliers in multi-feature data. It highlights areas of interest for further investigation and can inform decisions on which data should be prioritized for downlink. To further prioritize the data, the algorithm's output can be evaluated for the number of sequential outliers and the height of the error signal above the threshold.

References

1. Burch, J.L., et al.: Magnetospheric multiscale overview and science objectives. Space Sci. Rev. **199** (2016)
2. Angelopoulos, V.: The themis mission. In: Burch, J.L., et al. (eds.) The THEMIS Mission. Springer, New York (2009)
3. Perez, R.C. et al.: Chapter 5 - oceanographic buoys: providing ocean data to assess the accuracy of variables derived from satellite measurements. In: Nalli, N.R. (ed.) Field Measurements for Passive Environmental Remote Sensing. Elsevier (2023)
4. Qian, S.-E.: Hyperspectral satellites, evolution, and development history. IEEE J. Sel. Top. Appl. Earth Obs. Remote Sens **14** (2021)
5. Souiden, I., et al.: A survey of outlier detection in high dimensional data streams. Comput. Sci. Rev **44** (2022)
6. Zamry, N.M., et al.: Lightweight anomaly detection scheme using incremental principal component analysis and support vector machine. Sensors, **21** (2021)
7. Toy-Edens, V., et al.: Classifying 8 years of mms dayside plasma regions via unsupervised machine learning. J. Geophys. Res. Space Phys. **129** (2024)

8. Yeh, C.-C.M., et al.: Matrix profile i: all pairs similarity joins for time series: a unifying view that includes motifs, discords and shapelets. In: 2016 IEEE 16th International Conference on Data Mining (ICDM) (2016)

9. Finley, M.G., et al.: Generalized time-series analysis for in situ spacecraft observations: anomaly detection and data prioritization using principal components analysis and unsupervised clustering. Earth Space Sci. **11** (2024)

10. Bakrania, M.R., et al.: Using dimensionality reduction and clustering techniques to classify space plasma regimes. Front. Astron. Space Sci. **7** (2020)

11. Ross, D.A., et al.: Incremental learning for robust visual tracking. Int. J. Comput. Vis. **77** (2008)

12. Innocenti, M.E., et al.: Unsupervised classification of simulated magnetospheric regions. Ann. Geophys. **39** (2021)

13. Bhushan, A., et al.: Incremental principal component analysis based outlier detection methods for spatiotemporal data streams. ISPRS Ann. Photogrammetry, Remote Sens. Spat. Inf. Sci. (2015)

14. Raptis, S., et al.: Revealing an unexpectedly low electron injection threshold via reinforced shock acceleration. Nat. Commun. **16** (2025)

15. Richard, L., et al.: Are dipolarization fronts a typical feature of magnetotail plasma jets fronts? Geophys. Res. Lett. **49** (2022)

16. Pollock, C., et al.: Fast plasma investigation for magnetospheric multiscale. Space Sci. Rev. **199** (2016)

17. Torbert, R.B., et al.: The fields instrument suite on mms: scientific objectives, measurements, and data products. Space Sci. Rev. **199** (2016)

18. Kajdič, P., et al.: Transient upstream mesoscale structures: drivers of solarquiet space weather. Front. Astron. Space Sci. **11** (2024)

19. Angelopoulos, V., et al.: Bursty bulk flows in the inner central plasma sheet. J. Geophys. Res. Space Physics **97**(A4), 4027–4039 (1992)

20. Sitnov, M., et al.: Explosive magnetotail activity. Space Sci. Rev. **215** (2019)

21. Variable descriptions. https://themis.igpp.ucla.edu/var_desc.shtml. Accessed 23 Feb 2025

22. Liu, T.Z., et al.: Relativistic electrons generated at earth's quasi-parallel bow shock. Sci. Adv.**5** (2019)

23. Wilson, L.B., et al.: Relativistic electrons produced by foreshock disturbances observed upstream of earth's bow shock. Phys. Rev. Lett. **117** (2016)

24. Olshevsky, V., et al.: Automated classification of plasma regions using 3d particle energy distributions. J. Geophys. Res. Space Phys. **126** (2021)

25. Breuillard, H., et al.: Automatic classification of plasma regions in near- earth space with supervised machine learning: application to magnetospheric multi scale 2016–2019 observations. Front. Astron. Space Sci. **7** (2020)

26. Escoubet, C., et al.: Introduction the cluster mission. Ann. Geophys. Copernicus GmbH, **19** (2001)

Adaptive Physics Refinement for Anatomic Adhesive Dynamics Simulations

Aristotle Martin$^{(\boxtimes)}$ ⓘ, William Ladd ⓘ, Runxin Wu ⓘ,
and Amanda Randles ⓘ

Dept of Biomedical Engineering, Duke University, Durham, NC 27705, USA
{aristole.martin,william.ladd,wendy.wu,amanda.randles}@duke.edu

Abstract. Explicitly simulating the transport of circulating tumor cells (CTCs) across anatomical scales with submicron precision—necessary for capturing ligand-receptor interactions between CTCs and endothelial walls—remains infeasible even on modern supercomputers. In this work, we extend the hybrid CPU-GPU adaptive physics refinement (APR) method to couple a moving finely resolved region capturing adhesive dynamics between a cancer cell and nearby endothelium to a bulk fluid domain. We present algorithmic advancements that: enable the window to traverse vessel walls, resolve adhesive interactions within the moving window, and accelerate adhesive computations with GPUs. We provide an in-depth analysis of key implementation challenges, including trade-offs in data movement, memory footprint, and algorithmic complexity. Leveraging the advanced APR techniques introduced in this work, we simulate adhesive cancer cell transport within a large microfluidic device at a fraction of the computational cost of fully explicit models. This result highlights our method's ability to significantly expand the accessible problem sizes for adhesive transport simulations, enabling more complex and computationally demanding studies.

Keywords: Adaptive physics refinement · Adhesive dynamics · Fluid-structure interaction · Multiscale modeling · Heterogeneous computing

1 Introduction

Understanding the mechanisms driving cancer cell transport through the bloodstream requires models that can capture cellular behavior at the adhesion level while spanning long anatomical length scales. Circulating tumor cells (CTCs) interact with fluid forces and vascular walls through adhesive interactions, processes that are central to cancer metastasis, but remain poorly understood due to their complexity and multiscale nature. Computational modeling offers a unique opportunity to explore these dynamics by integrating cellular adhesion mechanisms with fluid transport across physiologically relevant scales. However, existing *in silico* models are limited in scope and are constrained by the computational cost of resolving submicrometer adhesive binding interactions over meterlength CTC trajectories. Current approaches predominantly focus on idealized

© The Author(s), under exclusive license to Springer Nature Switzerland AG 2025
M. H. Lees et al. (Eds.): ICCS 2025, LNCS 15903, pp. 268–282, 2025.
https://doi.org/10.1007/978-3-031-97626-1_19

microvessels [2,3,9,16,19], providing valuable but incomplete insights into the localized relationships between fluid flow, cell mechanics, and adhesion. A critical need remains for a unified computational framework of cancer transport capable of maintaining high fidelity at the adhesion scale while efficiently simulating the transport of CTCs across anatomic scales. Addressing this challenge could help uncover new insights into the biophysical mechanisms of cancer progression and inform therapeutic strategies targeting CTC adhesion and vascular interactions. This study (Fig. 1) represents a critical step in this direction by introducing an adaptive physics refinement-based adhesive dynamics (APR-AD) model. The key contributions of this work include:

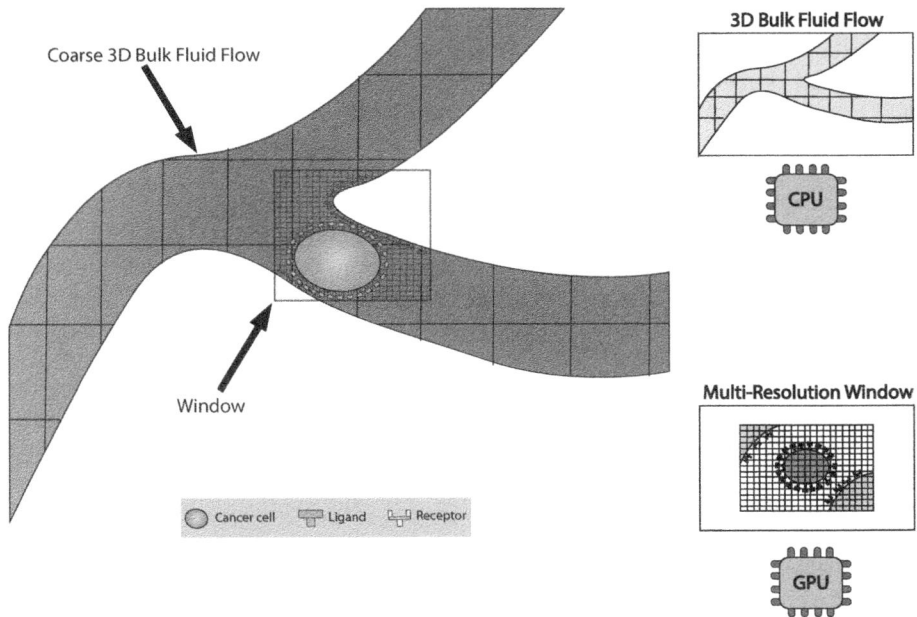

Fig. 1. Adaptive physics refinement-based adhesive dynamics (APR-AD) model overview depicting a cancer cell (blue) coated in ligands (green) undergoing adhesive interactions with wall receptors (yellow) within a finely resolved moving window that is coupled to a coarse bulk fluid-only domain. The bulk fluid simulation is performed on the CPUs, whereas the cellular calculations are done by GPUs. (Color figure online)

1. **Algorithmic advancements** that extend APR to span vessel walls.
2. **Integration of a detailed adhesive dynamics model**, enabling high-fidelity simulation of ligand-receptor binding events.
3. **Optimization for GPU architectures,** ensuring computational efficiency for simulations across large-scale domains.

By enabling high-resolution simulations of adhesive transport at anatomic scales, the APR-AD framework represents a significant step forward in multiscale modeling. This approach not only bridges existing gaps in computational cancer research, but also provides a scalable platform for addressing critical questions about the interplay between fluid mechanics, adhesion, and cancer cell transport that can probe fundamental drives of metastasis across physiologically relevant scales.

2 Application Overview

The present work employs HARVEY [12,13], a massively parallel, multiphysics code. The fluid flow is resolved by solving the Lattice Boltzmann Bhatnagar-Gross-Krook (LBGK) equation on a standard D3Q19 lattice [11]. Fluid-structure interactions between finite element (FEM) meshed cells and the background fluid are computed using the Immersed Boundary method (IBM) [1,7]. Adhesive interactions between cell surface ligands and receptor-lined endothelial walls are resolved using the stochastic adhesive dynamics (AD) model from [4,5].

An adaptive physics refinement (APR) algorithm developed in [10,14] is used by HARVEY to resolve tumor cell transport over large spatial distances. Other multiphysics approaches to modeling cancer transport processes include nanoparticle-based simulations [15]. Within the context of APR, high-resolution grids are used in the window to resolve fluid-structure interactions of the tracked cancer cell, while a coarser grid is employed outside the window to resolve the background fluid dynamics. The information exchange at the bulk-window boundary is handled through a multi-block approach detailed in [10]. An important quantity in the APR scheme is the ratio between the coarse and fine grid spacings, or the multi-resolution ratio n. Due to the cubic dependence of the total number of lattice sites N_s on grid spacing Δx ($N_s \sim \frac{1}{\Delta x^3}$, for three spatial dimensions), significant memory savings can result from higher values of n [6]. The APR method is optimized for heterogeneous workloads, with the bulk computations performed by the CPUs, and the fine window calculations handled by the GPUs (Fig. 1).

3 Algorithmic Advances to Capture Adhesive Interactions with APR

In this section, we describe the methodology for extending APR to include AD. Sections 3.1 and 3.2 detail the incorporation of walls and endothelial receptors, respectively, in APR. An overview of the window move algorithm is provided in Sect. 3.3. Finally, the development of a GPU-accelerated AD implementation is documented in Sect. 3.4.

3.1   Incorporation of Walls Into APR

A major advancement made in the present work is the additional ability of the moving window to traverse vessel walls. In previous versions of the APR algorithm, the window was assumed to be entirely submerged in fluid [10]. Allowing the APR window to span vessel walls was a necessary prerequisite to accurately resolving adhesive phenomena within the window. Several changes were made to the existing framework to accommodate walls.

When the window is first created, the identities of the grid points (i.e., fluid or wall) must be determined from the surrounding bulk. Previously, all window interior points were assumed to be fluid. In the updated setup, bulk tasks classify grid points as either fluid or wall points and then communicate the subset of points that intersect with the window (Fig. 2(A)). Within the window, categorization of misaligned points into either fluid or wall type relies on interpolation from nearby bulk points (Fig. 2(B)). Specifically, if a fine-scale point is completely surrounded by coarse fluid points, it is classified as fluid; otherwise, if a single neighboring coarse point is a wall, the corresponding window point is designated as a wall.

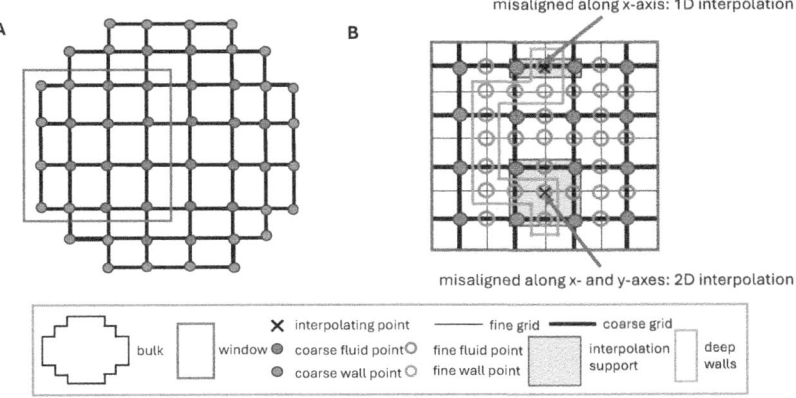

Fig. 2. Schematic illustration of the process for setting up the window points. (A): the bulk tasks communicate the subset of points intersecting with the window box. (B): the window tasks interpolate misaligned points from the bulk. For this example, the multi-resolution ratio $n = 2$. Two fine points are denoted with the "x" mark with their interpolation boxes shown in yellow to illustrate how their types are determined.

As a result, the fine-scale fluid point boundary remains aligned with the bulk fluid boundary, while additional "deep" wall points fill the gap between the last fine fluid point and neighboring coarse wall points. The thickness of this inner layer of wall points will depend on the resolution ratio n, but does not exceed one coarse lattice unit. These deep wall points are subsequently pruned from the vessel geometry to ensure consistency in the final representation.

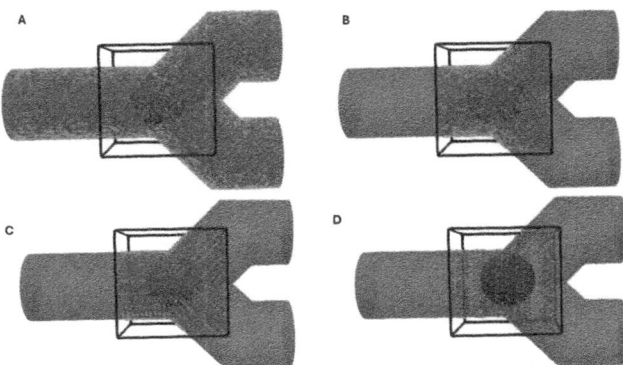

Fig. 3. Overview of different methods for placing wall receptors in the APR framework. (A) Method 1: bulk-driven placement, with random selection of receptors. (B) Method 2: window placement, with random selection of receptors. (C) Method 3: window placement with spatial patterning function Φ from Eq. 1, and constant threshold Γ. (D) Method 3, with variable threshold Γ from Eq. 2. The black outline of the APR window is depicted in each case, with the cancer cell in blue, and wall receptors in green.

3.2   Wall Receptor Placement

The initialization of endothelial wall receptors within the APR window requires careful consideration. To address this, we developed three distinct methods (Fig. 3). The first method initializes wall receptors globally (Fig. 3(A)). Bulk tasks generate receptor distributions over the full geometry and communicate their positions to the window tasks. While this approach ensures precise mapping of wall receptor distributions (e.g., based on mesh coloring, as in [9]) to the window, it comes with increased memory demands and communication overhead due to a large number of receptors.

In contrast, the second method (Fig. 3(B)) localizes receptor initialization to the window itself. During window creation, receptors are spawned randomly by the window root rank and then broadcast to the rest of the window tasks. By delegating receptor selection to a single root process, this approach guarantees deterministic receptor placement irrespective of window process count. The downside of this approach is that receptor spawning is serialized, causing other window tasks to remain idle during initialization. Nevertheless, this approach significantly reduces memory usage, as only the wall receptors that are needed by the window tasks are stored at any given time. It is particularly well-suited for scenarios where a relatively uniform wall receptor distribution is assumed.

The third method balances the flexibility of the first method with the memory efficiency of the second. Here, window tasks independently generate receptors in parallel using a spatial patterning function, Φ, which determines receptor placement based on the global position of the wall point. A receptor is placed if the function output exceeds a threshold Γ. A spatial patterning function yielding a checkerboard pattern is shown in Eq. (1).

$$\Phi(x, y, z) = \sin(2\pi f x)\sin(2\pi f y)\sin(2\pi f z) \tag{1}$$

The frequency f in Eq. (1) is set according to the desired wall receptor density. A simple stripe pattern (Fig. 3(D)) of wall receptors along the z direction can be obtained through specifying Γ as in Eq. (2).

$$\Gamma(z) = \frac{\mathrm{erf}(k_1\sin(z) + k_2) + 1}{2} \tag{2}$$

where erf is the error function, and k_1 and k_2 are constants selected to control the stripe spacing.

Ultimately, the choice of receptor placement method depends on the application. The first method offers maximal flexibility with arbitrary receptor distributions based on colored meshes. The second method (Fig. 3(B)) serves as an effective first pass approximation when uniform receptor distributions are sufficient. When a specific pattern is required that can be described mathematically, the third method (Fig. 3(C–D)) provides a flexible, function-driven approach.

3.3   Moving the Window

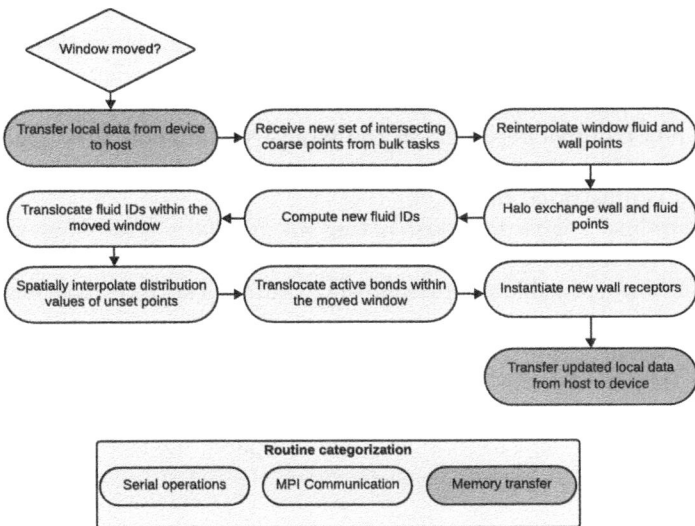

Fig. 4. Overview of the window move algorithm with incorporation of walls, from the window task perspective.

When the tracked cell nears the edges of the APR region, a window move is triggered, initiating a cascade of operations outlined in Fig. 4. Some steps mirror those taken for updating a fully submerged window: data are transferred from the device to the host, which then re-orients fluid points within the window following the move and spatially interpolates distribution values for incoming fluid points near the leading edge of the window [10]. However, to accommodate vessel walls, several routines are significantly modified, and additional steps are introduced. When the window relocates, new fluid and wall point positions are determined through re-interpolation of intersecting bulk points (Fig. 2). This operation is followed by a communication exchange to update wall and fluid points at the window task boundaries. In the case of a fully submerged window, this step is unnecessary, as the coordinates of halo fluid points remain unchanged, preserving the communication structure. However, for a partially submerged window, refreshing the fluid communication tables is essential, as the number of fluid points can change due to the presence of passing wall points. Consequently, the indices of fluid points, referred to as "fluid IDs", must then be re-computed causing the new fluid IDs to become out of sync with the original data. To efficiently reconcile these discrepancies, a map data structure is used to translate fluid IDs and ensure accurate data movement.

After a window move, newly uncovered window points are updated through spatial interpolation from the surrounding bulk. In contrast to a stationary window, where interpolation occurs in two dimensions along interfacial planes, these new points require full 3D interpolation. For a fully submerged window, tricubic interpolation using the Catmull-Rom spline [17] is applied over a cubic support region spanning four coarse points in each spatial dimension, with the interpolating point positioned near the center of the box. However, volumetric interpolation becomes more complex in a partially submerged window, where the presence of walls disrupts uniform support regions. In this case, identifying the appropriate support points requires determining the largest rectangular subprism of coarse fluid points that fully encloses the interpolating point. The algorithm we devised for this task is illustrated in Fig. 5.

The procedure begins by constructing an auxiliary array that records the maximum prism height values based on the point classifications. The algorithm then identifies all possible subprisms that contain the interpolating point P. Candidate subprism formation starts at an anchor column, extending downward until the maximal height drops below the height at the anchor value. Once a base face is established, the prism is extruded along the depth direction as far as allowed by the height values. A final check ensures that the resulting subprism contains the point P, and its volume V_R is compared against the running maximum value. The resulting subprism becomes the 3D support used for spatial interpolation.

Once the window moves to encompass new wall points, new wall receptors are instantiated in accordance with the desired receptor placement scheme (Fig. 3). The shuffling of wall receptor indices that the window move introduces requires the re-mapping of preexisting ligand-receptor bonds to their new index locations.

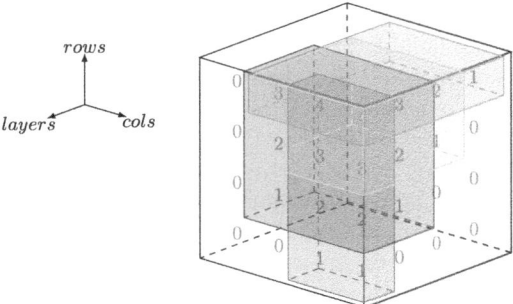

Fig. 5. Schematic illustration of subprism finder algorithm. The support matrix is shown, with nonzero values denoting the maximal height values of coarse fluid points, and zeros indicating the presence of coarse walls within the support. In this example, the algorithm is evaluating subprisms from the upper right position of the box. Four candidate rectangles are identified, with the interpolating location assumed to be near the center of the cube. The algorithm extrudes each of the rectangles identified as far as allowed based on the height values of deeper points to form a set of potential subprisms that contain the interpolating point.

3.4   GPU Acceleration of Adhesive Dynamics Calculations

The AD routines, originally designed to execute on multi-core CPUs, were re-designed as GPU kernels to be executed by the window tasks. Biological cells can express multiple species of ligand, that we collectively refer to as ligand sets. Each ligand set has a user-specified density that determines the number of ligands of that set placed on the cell surface. To model microvilli, ligands are permitted to interact with multiple wall receptors, though only one wall receptor can be engaged in a bond at any given time. The GPU parallelization strategy involves threading over ligands. To optimize memory access, data structures for adhesive cells, such as wall receptor positions, were reformatted from an array-of-structures layout to a structure-of-arrays format. At a high level, the AD process consists of three main components: (1) random number generation (RNG), (2) bond formation, and (3) bond rupture.

RNG is a prerequisite for both (2) and (3), as these routines rely on probabilistic interactions. We implemented RNG directly on the GPU using software-emulated linear feedback shift registers (LFSRs). An LFSR is a deterministic shift register that generates new values based on a linear function of its previous state. With an appropriate tap configuration, an LSFR of a given width can produce uniformly distributed values over a predefined period. For this work, we selected a 64-bit LFSR with the following feedback polynomial:

$$x^{63} + x^{61} + x^{60} + 1 \tag{3}$$

where the exponent terms represent the tapped bits. This tap configuration ensures a maximal cycle length, as documented in [18]. The LFSR was chosen for its efficient representation on the GPU and low computational overhead. To integrate LFSRs into the AD process, we allocated an array of LFSRs to store random values for each ligand. Each ligand set has what we denote as its "root" LFSR, which is seeded based on a combination of parameters that uniquely identify the cell and particular ligand set for the cell. The binary sequences of subsequent LFSRs within the ligand set are derived by shifting the root a number of times corresponding to the relative position of the ligand within the set, guaranteeing that no two ligands draw identical values at any given time. Beyond its simplicity, this approach also minimizes communication overhead. When the CTC transfers ownership between window tasks, only the root LFSR of each ligand set needs to be communicated. The receiving task can then locally reconstruct the full set of LFSR states, maintaining determinism in the simulation. The LFSR kernel is written to leverage shared memory and gives an approximate 20% time reduction over an RNG block-based approach.

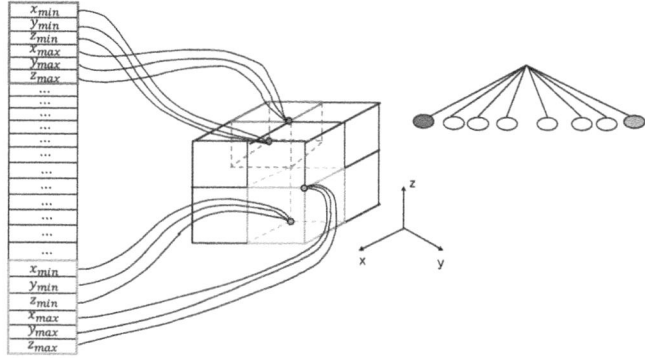

Fig. 6. Schematic representation of an octree in array form on the GPU. An octree with a single refinement level is shown for clarity. The octant array stores the minimum and maximum bounds of each octant box in an array-of-structures format.

Following the RNG phase, the bond formation step is executed. During each time step, adhesive ligands must identify the subset of wall receptors within range of bond formation. A naïve search through all wall receptors, for every cell ligand, leads to $O(N^2)$ time, making it computationally infeasible. To accelerate this process, we adopted an octree-based spatial partitioning scheme inspired by the CPU implementation [8] and optimized it for GPU execution (Fig. 6). For simplicity, a complete octree is used, ensuring a structured hierarchy. To guarantee a one-to-one mapping between cell ligands and octant indices, octant boxes are extended with a halo depth equal to the reactive distance, which is an intrinsic property of the ligand type. During bond formation, each ligand determines the index of its corresponding leaf node, which serves as a key to efficiently look up the start and end locations defining the segment of the wall receptors that fall into the corresponding octant (the wall receptors are sorted by octant index during tree creation, further improving lookup efficiency).

One of the challenges of parallelizing the adhesive bond formation step on the GPU is the constraint that each wall receptor can only engage in a bond with one ligand at a time. To resolve thread contention over shared receptors, a simple arbitration mechanism is implemented: only the thread with the smallest ID is permitted to form a bond with a given wall receptor, ensuring deterministic and conflict-free assignments.

The final phase of AD involves evaluating whether bonds are broken. This operation is the simplest to parallelize and is completed in a single kernel invocation.

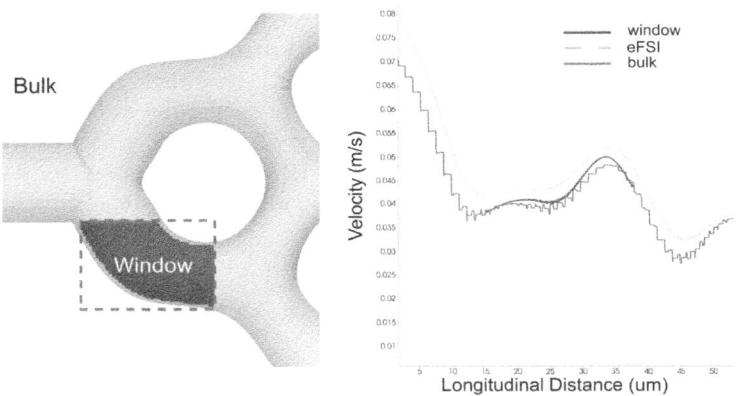

Fig. 7. APR walls validation in a stationary window. Left: region of the simulated microfluidic device, with the window location denoted by the red box and window fluid points indicated in dark blue. Right: Longitudinal velocity profiles for the bulk (green), window (blue), and eFSI (dotted yellow). (Color figure online)

4   Results

4.1   Validation of APR Walls

To validate the multi-block scheme with the addition of walls, we simulated a stationary window placed within a complex vessel representing a microfluidic device (Fig. 7). The bulk fluid resolution was set to 0.6 μm, while the window resolution was refined to 0.15 μm ($n = 4$). As a reference, we used the flow profile from an explicit fluid structure interaction (eFSI) model with a consistent resolution across the entire domain that matched the finer resolution of the window domain. The velocity profile error between the window and eFSI cases remained within 5% with minor discrepancies attributed to interpolation errors and fluid convergence effects.

4.2 Timing Composition Analysis

To evaluate the performance of the APR scheme, a timing composition analysis
was conducted, as shown in Fig. 8. Conceptually, the simulation can be decomposed into two distinct phases, representing periods where the cell is advecting
through a stationary window, and when the window itself is being moved to
follow the cell. The timing diagram reflects this distinction by showing the total
simulation loop time in Fig. 8(A), and the portion of the loop time spent in moving the window (denoted by "Window Move" in grey) is depicted in Fig. 8(B).
Notably, the bulk spends a considerable amount of time in "Window Move."
Fig. 8(B) indicates that the "Window Test" routine accounts for most of the window movement time. During the "Window Test", the root window rank checks
whether the window needs to be moved, and broadcasts the Boolean result to
the remaining window tasks, as well as the bulk. Altogether, the results suggest
that the bulk tasks spend a significant portion of their time waiting on the window tasks. This result is expected since the bulk has less work assigned to it
than the window (bulk fluid compared with IBM+AD), coupled together with
the need for the window tasks to perform n window time steps' worth of work
for each corresponding bulk time step.

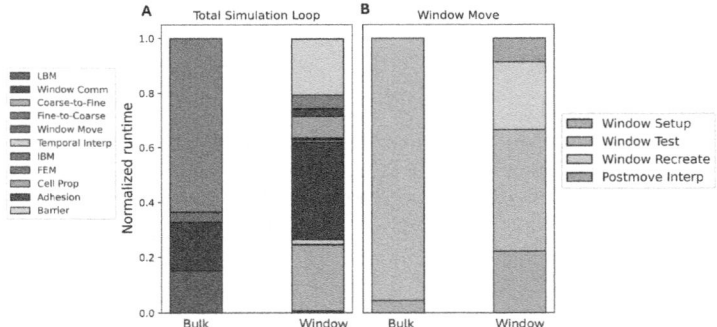

Fig. 8. Timing composition of APR simulation from Fig. 10. A: Total runtime breakdown for bulk (left) and window (right). B: Window move breakdown for bulk (left)
and window (right). Subroutines with negligible runtime were excluded from the timing
diagram.

The dominant runtime factor in the APR window is MPI communication
(dark blue) (Fig. 8(A)). This result reflects a shift in the primary bottleneck
from computation to communication, as core computational kernels (LBM, IBM,
FEM) have been offloaded onto the GPU. Unlike the bulk, which only handles
fluid data exchange, the window tasks must also manage cellular data communication, further increasing overhead. A significant portion of simulation time is
spent in the "Coarse-to-Fine" routine, which involves receiving fluid data from
bulk tasks at the multi-resolution interface and performing spatial interpolation.

Since this routine remains CPU-bound, it is considerably more time-consuming than purely GPU-executed kernels such as IBM. The large "barrier" time indicates the load imbalance within the window, a consequence of the naive domain decomposition scheme used for window tasks. In contrast, the bulk benefits from a bisection load balancer, resulting in more even workload distribution. As shown in Fig. 8(B), window tasks spend most of their time performing the "Window Test" routine, where the root process determines whether a window move is required and broadcasts the result. This finding highlights the cost of global communication in the window task workflow. Outside of "Window Test", most of the window movement time is spent in "Window Recreate", which involves interpolating new wall points and reconstructing window data, and "Window Setup", which updates bulk-window communication tables.

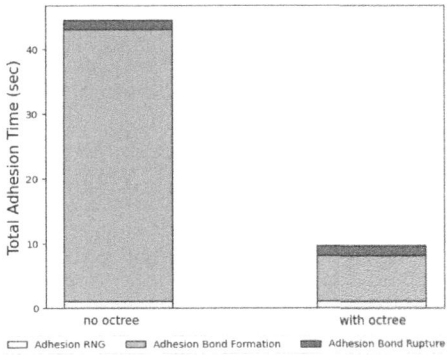

Fig. 9. Impact of GPU octree on adhesion time, during a 40,000 time step simulation in the microfluidic device (Fig. 10) on Aurora. Left: Total adhesion time without octree. Right: Total adhesion time with octree enabled.

4.3   Performance Optimizations

To better leverage the parallelism offered by the CPUs, a subset of bulk task functions were re-written using SIMD-style programming (i.e., SYCL ND-range kernels) analogous to the GPU. SIMD efficiency was enhanced by converting bulk data structures to structure-of-arrays (SoA). These code changes resulted in an approximate 25% reduction in overall runtime. Furthermore, implementing the search for wall receptors based on octrees for bond formation kernels significantly accelerated adhesive bond formation, which was previously the primary computational bottleneck in adhesive dynamics, as shown in Fig. 9. This optimization greatly reduced the runtime of bond formation calculations, improving overall simulation performance.

4.4   APR-AD Simulation in Complex Microvessel

The capabilities of our APR-AD framework are demonstrated using the microfluidic vessel geometry shown in Fig. 10. In this simulation, the APR window
tracked an adhesive cancer cell as it flowed through the lower branch of the
geometry. A bulk grid spacing of 0.6 μm was selected along with a resolution
ratio $n = 4$, yielding a window grid spacing of 0.15 μm. The simulation was performed on three nodes of the Aurora supercomputer (Argonne National Laboratory), yielding significant resource savings, as summarized in Table 1. To validate
the APR framework, we also conducted an explicit adhesive transport simulation
at the window resolution using 32 nodes of Aurora. A sinusoidal wall receptor
patterning function (Eq. (1)) was used in both the eFSI and APR simulations
for consistency. Both qualitative (Fig. 10(A)) and quantitative (Fig. 10(B)) comparisons between the explicit (eFSI) and APR simulations indicate strong agree-

Table 1. Comparison of resource requirements between APR-AD and explicit FSI
(eFSI) simulations within the microfluidic device shown in Fig. 10.

Model	Δx (μm)	fluid points	wall receptors	memory usage
APR-AD (window)	0.15	7.83×10^5	1.41×10^5	1.15 GB
APR-AD (bulk)	0.6	1.41×10^6	0	0.59 GB
eFSI	0.15	1.09×10^8	6.60×10^5	42.6 GB

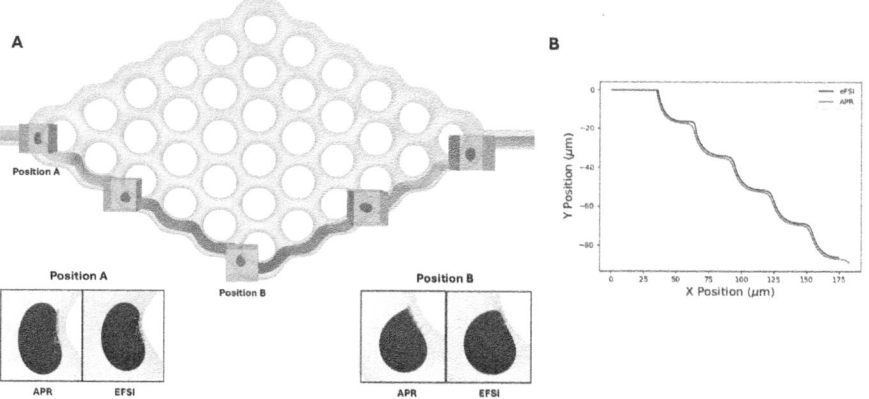

Fig. 10. APR simulation of a ligand-coated cancer cell (blue) traversing a microvessel
geometry. (A): The APR window (yellow) tracking the cancer cell is shown at different time points during the simulation, with bonds formed between the cell and the
endothelial receptors shown in green. The overall trajectory taken by the cell is indicated by its velocity pathline. The cell is shown at select positions for both the APR
and corresponding eFSI simulations. (B): Quantitative comparison of CTC trajectory
between eFSI (blue) and APR (magenta) in the left half of the geometry. (Color figure
online)

ment. Minor discrepancies observed in receptor-ligand binding stemmed primarily from differences in the RNG schemes between the CPU and GPU. Despite this, the overall cell trajectories (Fig. 10(B)) and CTC morphologies (Fig. 10(A)) remained consistent across both cases.

5   Conclusion

In this work, we demonstrated the capability of our novel APR-AD framework to resolve submicrometer ligand-receptor interactions of a cancer cell in a large region of practical utility. The APR-AD scheme offers significant computational savings compared to an explicit FSI approach (Table 1). By allowing the bulk to be simulated at a coarser resolution than the window, the number of fluid points and endothelial wall receptors are substantially reduced. In particular, the total memory footprint is reduced by more than an order of magnitude ($25X$, greatly expanding the feasible simulation volume for studies of adhesive dynamics.

The presented work provided a detailed examination of the underlying multiscale modeling techniques and key implementation considerations. The presented study serves as a proof-of-concept for a novel approach to investigating the adhesion cascade of circulating tumor cells at anatomical scales, paving the way for experimental validation studies and future applications in physiological environments. Future efforts will seek to address the limitations of the presented model by incorporating heterogeneous cell populations into the APR window and modeling the stochasticity of the adhesive receptor parameters.

Acknowledgements. The authors thank Wentao Ma, Daniel Puleri, and Jorik Stoop for fruitful discussions. This work was supported by the NCI of the NIH under Award Number 5R01EB024989. Research reported in this publication was supported by the NIH under Award Number T32GM144291. Computing support for this work came from the Argonne National Laboratory (ANL) Aurora Early Science program. An award of computer time was provided by the INCITE program. This research used resources of the Argonne Leadership Computing Facility, which is a DOE Office of Science User Facility supported under Contract DE-AC02-06CH11357.

References

1. Ames, J., Puleri, D.F., Balogh, P., Gounley, J., Draeger, E.W., Randles, A.: Multi-GPU immersed boundary method hemodynamics simulations. J. Comput. Sci. **44**, 101153 (2020)
2. Cui, J., Liu, Y., Xiao, L., Chen, S., Fu, B.M.: Numerical study on the adhesion of a circulating tumor cell in a curved microvessel. Biomech. Model. Mechanobiol. **20**, 243–254 (2021)
3. Dabagh, M., Gounley, J., Randles, A.: Localization of rolling and firm-adhesive interactions between circulating tumor cells and the microvasculature wall. Cell. Mol. Bioeng. **13**(2), 141–154 (2020)

4. Fedosov, D., Caswell, B., Suresh, S., Karniadakis, G.: Quantifying the biophysical characteristics of plasmodium-falciparum-parasitized red blood cells in microcirculation. Proc. Natl. Acad. Sci. **108**(1), 35–39 (2011)
5. Hammer, D.A., Apte, S.M.: Simulation of cell rolling and adhesion on surfaces in shear flow: general results and analysis of selectin-mediated neutrophil adhesion. Biophys. J . **63**(1), 35–57 (1992)
6. Krüger, T., Kusumaatmaja, H., Kuzmin, A., Shardt, O., Silva, G., Viggen, E.M.: The Lattice Boltzmann Method - Principles and Practice (2016). https://doi.org/10.1007/978-3-319-44649-3
7. Peskin, C.S.: Numerical analysis of blood flow in the heart. J. Comput. Phys. **25**(3), 220–252 (1977)
8. Puleri, D.F., Martin, A.X., Randles, A.: Distributed acceleration of adhesive dynamics simulations. In: Proceedings of the 29th European MPI Users' Group Meeting, pp. 37–45 (2022)
9. Puleri, D.F., Randles, A.: The role of adhesive receptor patterns on cell transport in complex microvessels. Biomech. Model. Mechanobiol. **21**(4), 1079–1098 (2022)
10. Puleri, D.F., et al.: High performance adaptive physics refinement to enable large-scale tracking of cancer cell trajectory. In: 2022 IEEE International Conference on Cluster Computing (CLUSTER),. pp. 230–242. IEEE (2022)
11. Qian, Y.H., d'Humières, D., Lallemand, P.: Lattice BGK models for Navier-stokes equation. Europhys. Lett. **17**(6), 479 (1992)
12. Randles, A., Draeger, E.W., Bailey, P.E.: Massively parallel simulations of hemodynamics in the primary large arteries of the human vasculature. J. Comput. Sci. **9**, 70–75 (2015)
13. Randles, A.P., Kale, V., Hammond, J., Gropp, W., Kaxiras, E.: Performance analysis of the lattice Boltzmann model beyond navier-stokes. In: 2013 IEEE 27th International Symposium on Parallel and Distributed Processing, pp. 1063–1074. IEEE (2013)
14. Roychowdhury, S., et al.: Enhancing adaptive physics refinement simulations through the addition of realistic red blood cell counts. In: Proceedings of the International Conference for High Performance Computing, Networking, Storage and Analysis, pp. 1–13 (2023)
15. Shabbir, F., Mujeeb, A.A., Jawed, S.F., Khan, A.H., Shakeel, C.S.: Simulation of transvascular transport of nanoparticles in tumor microenvironments for drug delivery applications. Sci. Rep. **14**(1), 1764 (2024)
16. Takeishi, N., Imai, Y., Ishida, S., Omori, T., Kamm, R.D., Ishikawa, T.: Cell adhesion during bullet motion in capillaries. Am. J. Physiol. Heart Circulatory Physiol. **311**(2), H395–H403 (2016)
17. Twigg, C.: Catmull-rom splines. Computer **41**(6), 4–6 (2003)
18. Ward, R., Molteno, C.: Table of linear feedback shift registers. Technical report 2012-1, University of Otago. https://www.physics.otago.ac.nz/reports/electronics/ETR2012-1.pdf
19. Ye, H., Shen, Z., Li, Y.: Cell stiffness governs its adhesion dynamics on substrate under shear flow. IEEE Trans. Nanotechnol. **17**(3), 407–411 (2017)

Microfluidic Digital Twin for Enhanced Single-Cell Analysis

Samreen T. Mahmud[1][(✉)], Wentao Ma[1], Taylor Thomsen[2,3],
Chang Chen[2], Rachel Rex[2], Andre Lai[2,3], Lydia L. Sohn[2,3],
and Amanda Randles[1]

[1] Department of Biomedical Engineering, Duke University, Durham, NC 27705, USA
`{samreen.mahmud,wentao.ma,amanda.randles}@duke.edu`
[2] Department of Mechanical Engineering, University of California at Berkeley,
Berkeley, CA 94720-1740, USA
`{taylorthomsen,chang.chen,rachelrex,andre.lai,sohn}@berkeley.edu`
[3] UC San Francisco Graduate Program in Bioengineering, UC Berkeley, Berkeley,
CA 94720, USA

Abstract. Advancing single-cell analysis requires tools that not only enable precise experimental measurements but also offer predictive capabilities to guide device optimization and expand experimental possibilities. This study addresses this need by developing a digital twin framework for mechano-node-pore sensing (mechano-NPS), a high-throughput microfluidic platform for single-cell analysis. By creating a virtual replica that integrates models of fluid dynamics and cellular behavior, the digital twin serves as a critical tool for both device development and hypothesis exploration. The foundation of the digital twin was established by accurately modeling the fluid dynamics within the mechano-NPS device, with simulations at various inlet pressures verified against analytical solutions. To ensure biological relevance, cellular models were rigorously tested to replicate key behaviors within the platform. The digital twin's performance was validated against experimental data, focusing on cell velocity and whole cell deformation index (wCDI). While variances in cell velocity highlighted systematic biases, the strong agreement of simulated wCDI with experimental results underscores the digital twin's reliability. This framework not only demonstrates the potential to enhance the mechano-NPS platform but also exemplifies how digital twins can transform experimental approaches in cellular biology.

Keywords: Digital twin · Microfluidic device · Single-cell analysis

1 Introduction

Microfluidic platforms have revolutionized cellular biology by enabling precise single-cell analysis and integrating multiple processes within a single device, paving the way for micro total analysis systems (μTAS) [1,2]. However, the

M. H. Lees et al. (Eds.): ICCS 2025, LNCS 15903, pp. 283–297, 2025.
https://doi.org/10.1007/978-3-031-97626-1_20

traditional strategy to developing these platforms is often highly involved, time-intensive, and experimentally limiting, restricting their optimization and overall impact [2,3]. A digital twin framework for microfluidic devices addresses this critical gap by creating a virtual replica that not only accelerates design and optimization but also provides detailed insights into physical phenomena that are difficult or impossible to measure in conventional experimental setups. Computational modeling enables the exploration of key parameters such as velocity and pressure fields, shear stresses and forces on the cell surface, and mass and heat transfer [2]. For single-cell analysis, where microfluidic platforms currently offer high throughput but limited measurable parameters, digital twins dramatically expand the scope of accessible data. This integration enhances our understanding of device functionality and the underlying pathophysiology, transforming experimental microfluidic platforms into more versatile and informative tools for cellular biology.

Optimizing device geometry and operating parameters is crucial for improving microfluidic platform performance. Digital twins facilitate rapid assessment of numerous variables while ensuring efficient functionality. Computational studies have been utilized to identify optimal pillar spacings to improve device performance [4], investigate the impact of pillar cross-sections and alignments on fluid distribution in the device [5], and analyze the effect of inlet conditions for the separation of circulating tumor cells (CTC) in a microfluidic chip [6]. In addition to optimizing the operation parameters of a microfluidic device, computational models are invaluable for investigating fluid flow properties that influence the functionality of a microfluidic device, such as velocity distribution and fluid-induced shear stress. These models have been used to replicate different shear flow conditions in microfluidic devices [7] and to study the effects of shear stress in cell cultures [8,9].

Beyond fluid flow, computational models have played a significant role in investigating cellular mechanics within microfluidic environments. For instance, Esposito et al. employed three-dimensional (3D) numerical simulations to examine the influence of fluid inertia on cell softness in cylindrical and rectangular microchannels, aiding in cell sorting based on mechanical characteristics [10]. Tan et al. used immersed boundary methods to examine how cells squeeze through micropores of varying sizes under different pressures, establishing cell deformability as a potential biomarker [11]. Similarly, Hynes et al. modeled CTCs as rigid and deformable spheres within a bioprinted vascular chip to assess mechanical interactions [12]. Despite these advances, experimental limitations persist in measuring critical cellular parameters such as stiffness and forces exerted on the cells. Computational tools provide a crucial bridge, supplementing experiments with in silico models to extract insights that would otherwise be inaccessible. For example, Deng et al. integrated computational modeling with an inertial microfluidic cell stretcher (iMCS) to measure the isoshear modulus of cell membranes, an elusive parameter in purely experimental setups [13]. Similarly, Sadaat et al. combined microfluidic experiments with simulations to determine the shear modulus of red blood cells (RBCs) [14]. However, these approaches often require high-speed imaging systems, making them costly and less accessible. The integration of digital twins with microfluidic devices presents a scalable and cost-effective alternative, enhancing experimental capabilities while reducing expensive hardware.

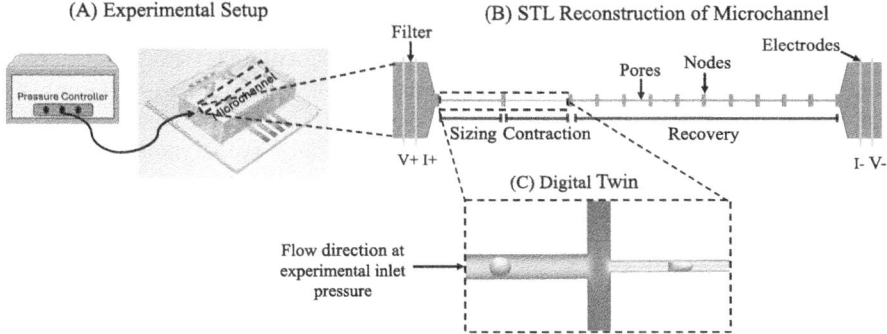

Fig. 1. (A) Fabricated microfluidic device for driving cells through a channel under constant pressure. (B) Reconstructed geometry of the mechano-NPS platform with segmented channels, DC voltage applied via outer electrodes, and current measured by inner electrodes. (C) Digital twin simulating deformable cells and fluid flow using the STL geometry and experimental pressure inputs.

This work establishes a digital twin framework for the microfluidic device mechano-node-pore sensing (mechano-NPS), enhancing its single-cell analysis capabilities while maintaining high fidelity. Mechano-NPS is a mechanophenotyping platform based on the Coulter-counter technique of particle counting (Fig. 1(A)) [15]. Unlike traditional high-speed optical techniques, mechano-NPS achieves a remarkable throughput of 300-500 cells per minute without relying on high-speed optical instruments or costly cameras [16]. Mechano-NPS operates by measuring the modulated electrical currents across a microfluidic channel segmented into nodes and pores (Fig. 1(B)). The device includes a narrow contraction region, forcing cells to squeeze through, allowing measurement of key properties such as cell diameter, resistance to compressive deformation, transverse deformation, and recovery time after deformation. A novel whole cell deformability index (wCDI) quantifies cell stiffness, enabling differentiation of cell lineage, chronological age, and stage of malignant progression in human epithelial cells [16,17]. Given its high throughput, cost-effectiveness, and multiparametric capabilities, mechano-NPS is ideally suited for integration with a digital twin to enhance its performance and expand its analytical capabilities.

To build the digital twin, we first verified the fluid flow within the device by comparing simulation results with analytical solutions, ensuring accuracy in modeling a fundamental aspect that directly influences cell behavior. This robust fluid flow model served as the foundation for further development. Next, we verified the computational representation of cellular behavior by comparing simulated cell velocities against analytical solutions, demonstrating the model's ability to replicate experimental conditions and predict outcomes reliably. Finally, we validated the digital twin against experimental data comparing cell velocity in different device segments and wCDI under various inlet conditions.

Through this work, we present a comprehensive digital twin framework for mechano-NPS that extends its experimental reach and analytical power. By enabling measurement of additional parameters, rapid exploration of design modifications, and optimization of the device performance, this digital twin transforms mechano-NPS into an even more powerful tool for single-cell analysis. More broadly, this approach highlights the potential of digital twins in microfluidic research, paving the way for scalable, data-driven advancements in cellular biomechanics.

2   Methods

2.1   Experimental Setup

The mechano-NPS platform is fabricated using soft lithographic techniques, as previously published [16,17]. Briefly, polydimethylsiloxane (PDMS) mold of the mechano-NPS channel is cast from a silicon negative-relief master mold, the height of which is approximately 20 μm. Once excised, the PDMS mold is bonded to a glass substrate with pre-defined platinum electrodes and gold contact pads that were fabricated using traditional lithography and electron-gun evaporation (Fig. 1(A)). As shown in the schematic of Fig. 1(B), the mechano-NPS channel is comprised of three regions: sizing, contraction, and recovery. The sizing and the contraction segments are 800 μm in length, and the recovery node-pores are each 285 μm in length. The nodes segment the overall channel to provide spatio-temporal resolution and are 85 μm wide, and 50 μm long. Filters are included at the inlet reservoir to exclude cell aggregates and cellular debris. The dimensions of the channel were chosen to provide sufficient transit time while maintaining a high signal-to-noise ratio (SNR).

Human promyelocytic (HL60) cells were introduced into the microfluidic channel under two separate non-pulsatile pressures of 11 and 15 kPa. The modulated current pulse produced by a cell transiting the mechano-NPS was measured using a four-point probe. Current with respect to time was recorded, low-pass filtered, and then processed with a custom-written code [18] to extract information such as magnitude and duration for each pulse event for further analysis.

2.2   Computational Model

To develop a digital twin of mechano-NPS, we utilized the massively parallel computational fluid dynamics solver HARVEY [19,20] to simulate fluid dynamics within a microfluidic device. Cells were explicitly modeled in the microfluidic channel as shown in Fig. 1(C) to represent the HL60 cell line, incorporating five different cell sizes to account for cellular heterogeneity. HARVEY implements the lattice Boltzmann method (LBM) to solve the governing fluid equations, whereas cells are modeled using the finite element method (FEM) and are coupled to surrounding fluid using the immersed boundary method (IBM).

LBM for Fluid Flow. LBM is a deterministic, mesoscopic approach that numerically solves the Navier-Stokes equations by modeling fluid with a particle distribution function. Fluid behavior is discretized using a fixed Cartesian lattice, where the probability function $f_i(\mathbf{x}, t)$ determines the probability of finding a particle at lattice point $\mathbf{x}$ and time t with a discrete velocity $\mathbf{c}_i$ [21]. The evolution of the particles with an external force field is governed by [22]:

$$f_i(\mathbf{x} + \mathbf{c}_i, t + 1) = \left(1 - \frac{1}{\tau}\right) f_i(\mathbf{x}, t) + \frac{1}{\tau} f_i^{eq}(\mathbf{x}, t) + F_i(\mathbf{x}, t) \tag{1}$$

where $f_i^{eq}(\mathbf{x}, t)$ is the Maxwell-Boltzmann equilibrium distribution and F_i is the external force field. HARVEY employs a D3Q19 velocity discretization model with the Bhatnagar–Gross–Krook (BGK) collision operator $\Omega = 1/\tau$, where τ is the relaxation time which determines the relaxation of f_i towards the equilibrium distribution function f_i^{eq}. The kinematic viscosity, ν is linked to τ by $\nu = c_s^2(\tau - 1/2)$ with a lattice speed of sound $c_s = 1/\sqrt{3}$. The density ρ and the velocity $\mathbf{v}$ are calculated respectively as the 0th and the 1st moment of the distribution function, which are then used to calculate the equilibrium distribution function $f_i^{eq}(\mathbf{x}, t)$. The external force $F_i(\mathbf{x}, t)$ that accounts for the body force imparted by the cell on the fluid is calculated by applying Guo's forcing scheme [23]. At the walls, no-slip condition is enforced using the halfway bounce-back boundary conditions whereas a constant density at the inlets and outlets is applied using a Zou-He like algorithm adapted to the D3Q19 velocity discretization [24]. The fluid is simulated with a density of 1000 kgm^{-3} and a viscosity of 0.89 mPa.s at a lattice grid spacing of 0.125 µm.

FEM for Deformable Cells. Deformable cells are modeled as fluid-filled capsules with a triangulated membrane of zero thickness having an initial spherical shape [20,25] using FEM. For simplicity, the cytoplasm is considered an incompressible Newtonian fluid having the same kinematic viscosity as the ambient fluid representing the cytoskeleton while neglecting the nucleus. The cell membrane is modeled to be isotropic and hyperelastic which follows the Skalak constitutive law for resisting shear and area dilation [26]. The strain energy function is given by:

$$W_s = \frac{G_s}{4}[(I_1^2 + 2I_1 - 2I_2) + CI_2^2] \tag{2}$$

where G_s is the shear elastic modulus, I_1 and I_2 are the strain invariants of the Green strain tensor, and C is the ratio of dilation to shear modulus. The membrane's resistance to bending is implemented using the Helfrich formulation [27]. The HL60 cells were modeled with a shear elastic modulus, G_s, of 2.25 ×

10^{-4} Nm^{-1} and a bending modulus, E_b, of 1×10^{-18} J to match the behavior of the cell observed experimentally.

IBM for Coupling. To account for the interaction of the cell with the ambient fluid, the Lagrangian grid of the FEM cell model is coupled to the Eulerian grid of LBM by applying IBM [28]. Here, three components of IBM are implemented with the following sequence: interpolation, updating, and spreading. At first, to determine the cell membrane deformation, Lagrangian membrane velocity $\mathbf{V}$ is interpolated from the Eulerian velocity $\mathbf{v}$ with a three-dimensional Dirac delta function δ having four-point support as follows:

$$\mathbf{V}(\mathbf{X}, t) = \sum_{\mathbf{x}} \mathbf{v} \delta(\mathbf{x} - \mathbf{X}(t)) \tag{3}$$

where $\mathbf{X}$ is the vertex location of the Lagrangian grid and $\mathbf{x}$ is the fluid lattice location in the Eulerian grid. Next, we update the position of the cell vertex with a no-slip condition, assuming unit timesteps. Lastly, the forces calculated at each Lagrangian vertex $\mathbf{G}$ are spread onto the surrounding Eulerian grid using the same delta function:

$$\mathbf{g}(\mathbf{x}, t) = \sum_{\mathbf{X}} \mathbf{G}(\mathbf{X}, t) \delta(\mathbf{x} - \mathbf{X}(t)) \tag{4}$$

3   Results and Discussions

3.1   Verification of Fluid Flow in Microfluidic Digital Twin Against Analytical Solution

Accurate fluid flow modeling within mechano-NPS is essential for ensuring the validity of the digital twin, as fluid dynamics directly influence cell velocity and deformation. As an initial verification step, we simulated fluid-only flow within the microfluidic device using our LBM-based model, excluding device filters from the simulation. As these filters do not induce a pressure drop and only serve to prevent cell aggregates or debris from clogging the entrance of the sizing pore, their omission does not affect the core flow dynamics. The fluid flow was analyzed at two different inlet pressures: 11 and 15 kPa. Upon convergence of the flow, we assessed the magnitudes of pressure and velocity along the centerline of the device, as shown in Figs. 2(A) and 2(B). The pressure profiles demonstrated a steady decline from the inlet pressure to zero at the outlet, with a consistent slope throughout each section. As expected, the most significant pressure drop occurred within the contraction pore, where the channel narrows, while the wider node sections exhibited minimal pressure variations. This behavior is characteristic of Poiseuille flow, confirming the expected pressure distribution.

The velocity profile demonstrated a consistent magnitude within each section of the device, aligning qualitatively with Poiseuille flow characteristics. At both 11 kPa and 15 kPa inlet pressures, the sizing and recovery pores exhibited similar

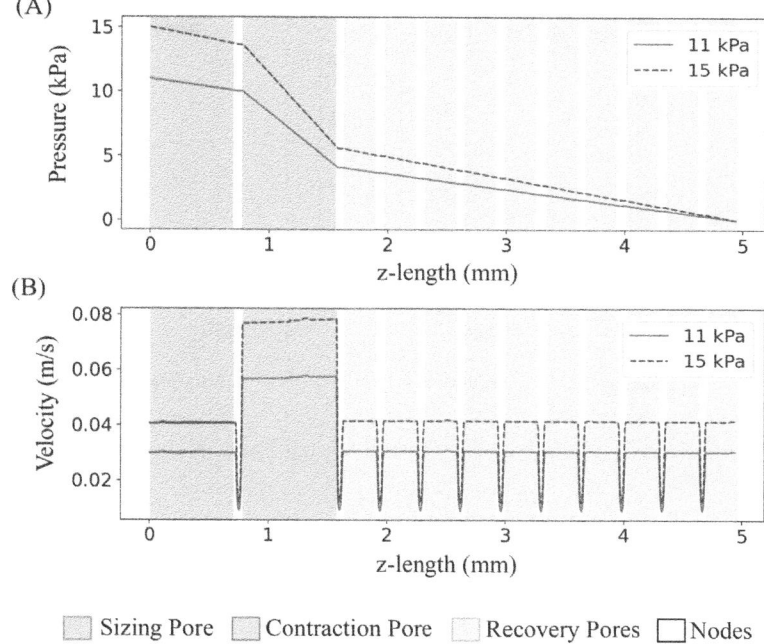

(A)

(B)

Sizing Pore Contraction Pore Recovery Pores Nodes

Fig. 2. (A) Pressure and (B) Velocity profiles across the mechano-NPS platform modeled with the digital twin at inlet pressures of 11 kPa and 15 kPa.

velocities due to their identical dimensions. In contrast, the contraction pore, being the narrowest section, exhibited the highest velocities, while the nodes showed minimal velocity, as illustrated in Fig. 2(B). To quantitatively verify the simulated fluid velocity, we compared our results to the analytical velocity profile of Poiseuille flow. Our analysis concentrated on two critical segments: the sizing pore, where cells move freely, and the contraction pore, where significant cell deformations occur. Given that the recovery pores mirror the sizing pore in design, they exhibit identical velocity characteristics, making the sizing pore analysis representative of both regions. The longitudinal velocity profile for a rectangular channel is given by [29]:

$$u_x(y,z) = \frac{16a^2}{\mu\pi^3}\left(-\frac{dp}{dx}\right)\sum_{i=1,3,5,...}^{\infty}(-1)^{(i-1)/2}\left[1 - \frac{\cosh(i\pi z/2a)}{\cosh(i\pi b/2a)}\right]\frac{\cos(i\pi y/2a)}{i^3}.$$

(5)

In Eq. (5), a and b are the two widths of the rectangular cross-section, with $-a \leq y \leq a$ and $-b \leq z \leq b$, so the centerline velocity can be found by setting $y = z = 0$. x is the flow direction, so $-dp/dx$ is the pressure gradient driving the flow. The above equation gives a relation between the pressure gradient and the longitudinal velocity. Therefore, after obtaining the pressure gradient from Fig. 2(A), we could calculate the analytical centerline velocity using Eq. (5) for

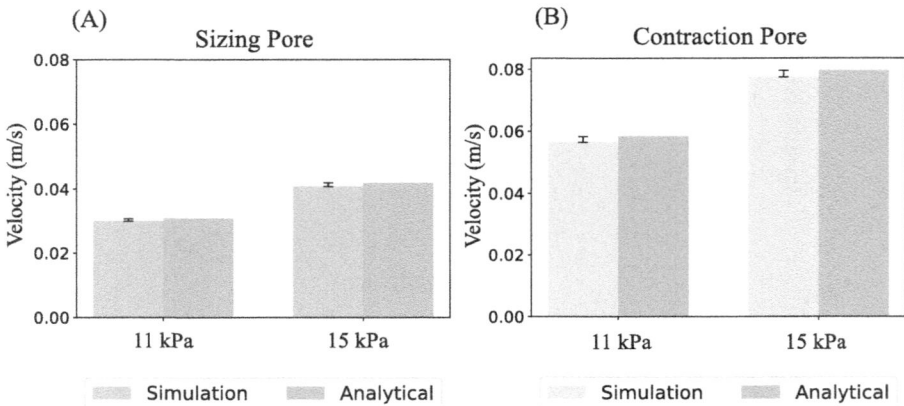

Fig. 3. Comparison of simulated velocities in (A) sizing and (B) contraction pores with analytical results.

both the sizing and the contraction pores. The simulated velocity profiles closely matched the analytical solutions, with a percentage error of less than 4% at both 11 kPa and 15 kPa (Fig. 3). This strong agreement confirms the accuracy of our fluid flow model, establishing a robust foundation for subsequent simulations involving cellular transport and deformation.

3.2   Verification of Simulated Cell Velocity in a Square Channel and the Sizing Pore

To verify the accuracy of the cell velocity modeling, we performed simulations of cells moving through a fluid-filled square channel and compared the results with analytical solutions from [30], which were previously validated against experimental data. For this study, we constructed a square channel with a side length of 20 μm and a longitudinal length of 350 μm, replicating the dimensions used in the experimental validation. Three cell diameters: 10, 14, and 17 μm were studied in the square channel. Given the fluid and cell parameters, the Reynolds number and the size ratio between the cells and the channel closely resemble those within the sizing pore of the microfluidic device. Therefore, simulations conducted in the square channel serve as appropriate validation tests for assessing the digital twin's ability to accurately capture the physics within the sizing pore.

Figure 4 compares the simulated and analytical results for cell mobility. The horizontal axis represents the cell-to-channel size ratio, while the vertical axis shows the ratio of cell velocity to undisturbed centerline fluid velocity for each cell size. The results demonstrate a strong agreement between the simulation and the analytical solutions. Although the analytical solution is based on rigid beads, it effectively captures the physics of cell mobility in both the simulated square channel and the sizing pore. This agreement is due to the small cell-to-channel

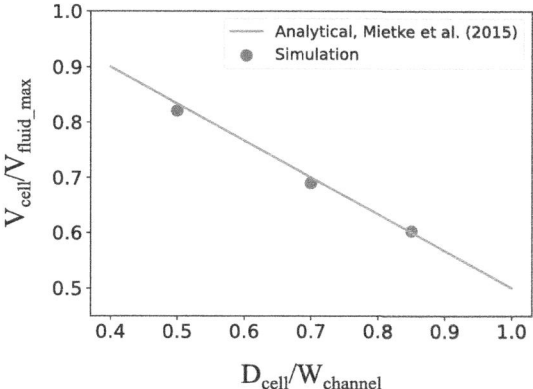

Fig. 4. Verification of cell model through comparison with analytical solutions in a square channel.

size ratio (<0.9), where cell deformation (quantified by the non-dimensional capillary number) is negligible. In particular, Ahmmed et al. [31] measured the velocity of various cancer cells in a square microchannel and found that their results also aligned with analytical solutions for rigid beads. Similarly, Kuriakose et al. [32] reported that the capillary number–and thus cell deformation–had minimal influence on cell mobility in their experiments.

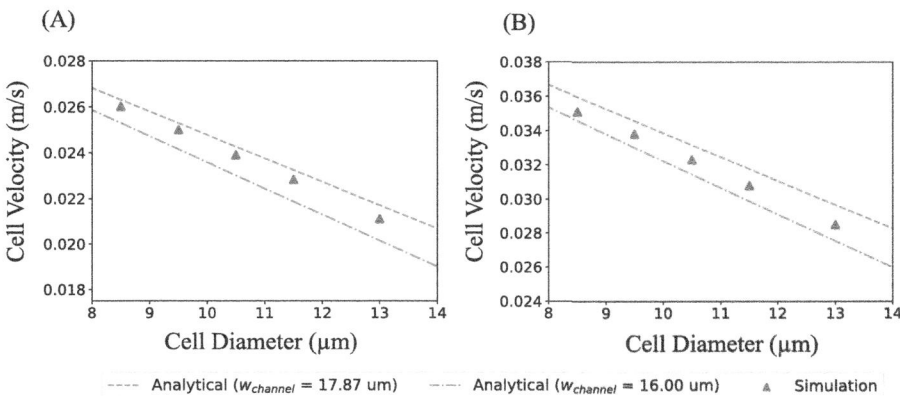

Fig. 5. Verification of cell model through comparison with analytical solutions in the sizing pore at (A) 11 kPa and (B) 15 kPa inlet pressures.

Building on these analytical solutions for square channels, we infer the expected analytical cell velocity within the sizing pore, which has a rectangular cross-section. In Fig. 5, the two lines in each subplot represent the analytical cell velocity for square channels with widths of 17.87 and 16.0 μm, corresponding to the two edges of the sizing pore cross-section. Since the dimensional difference between these edges is small, the actual analytical cell velocity is expected to fall between these two bounds, aligning well with the simulated data points. Thus, our simulations successfully validate cell velocity in the sizing pore, further reinforcing the accuracy of the digital twin model.

3.3   Validation of the Simulated Cell Velocities and wCDI Against Experimental Data

After successfully verifying our fluid and cell models, we proceeded to validate the digital twin using experimental data from HL60 cell screening with mechano-NPS. Given the substantial computational demands of simulating the cell transiting throughout the entire device, we focused our analysis on a key region of interest, depicted in Fig. 1(C). This section consists of the sizing and contraction pores, where the cell first moves undeformed through the sizing pore before undergoing significant deformation in the contraction pore. To ensure consistency, input conditions were carefully tuned to match the flow characteristics throughout the device. To account for cellular heterogeneity, we selected five distinct cell sizes from the experimental data set, reflecting the natural size variation of HL60 cells. Validating the digital twin against experimental results is essential for ensuring its reliability and accuracy in predicting real-life scenarios. Our validation process focused on key performance metrics, that include cell velocities within the sizing and contraction pores which are critical components of the mechano-NPS platform. Additionally, we assessed the wCDI by comparing simulation-derived values with experimental measurements across a range of cell diameters, demonstrating the digital twin's capability to replicate experimental observations. The wCDI serves as an indicative measure of cell stiffness, which can be assessed using the mechano-NPS device and is defined as:

$$wCDI = (\frac{V_c}{V_0})(\frac{d_0}{h}) \tag{6}$$

where V_c is the cell velocity in the contraction pore, V_0 is the average cell velocity for all the different cell sizes in the sizing pore, d_0 is the cell's initial diameter, and h is the height of the channel. Cell velocities at an inlet pressure of 11 kPa were analyzed in both the sizing and contraction pores and compared with experimental data, as illustrated in Figs. 6(A) and 6(B), respectively. The results revealed a consistent trend across both datasets: cell velocity decreased as cell diameter increased in both the sizing and contraction pores, as expected. However, a constant bias was observed between the simulation and experimental data, with identical slopes but differing intercepts due to this bias. To investigate the source of the discrepancy, we conducted an additional analysis at an input pressure of 15 kPa, as shown in Fig. 7. The results reproduced the same bias, again showing identical slopes across datasets. Given the prior validation of the cell model and the consistency of this bias across inlet pressures, the discrepancy likely stems from a systematic difference between the experimental and simulation setups rather than an issue with the digital twin itself.

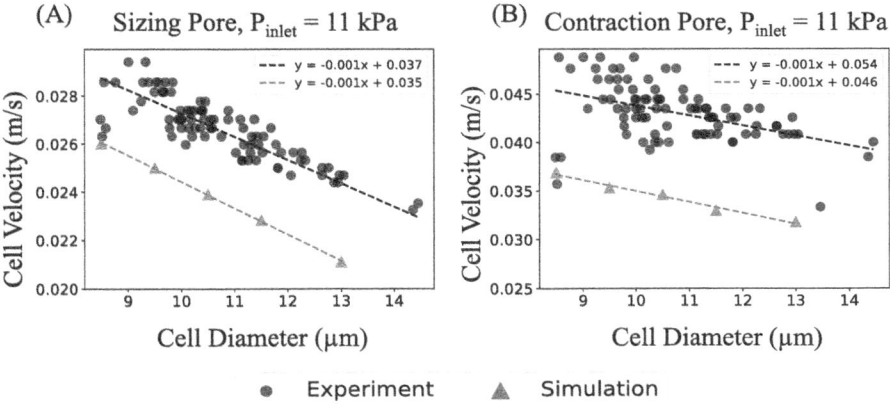

Fig. 6. Comparison of experimental and simulated cell velocities at 11 kPa inlet pressure in (A) sizing and (B) contraction pores.

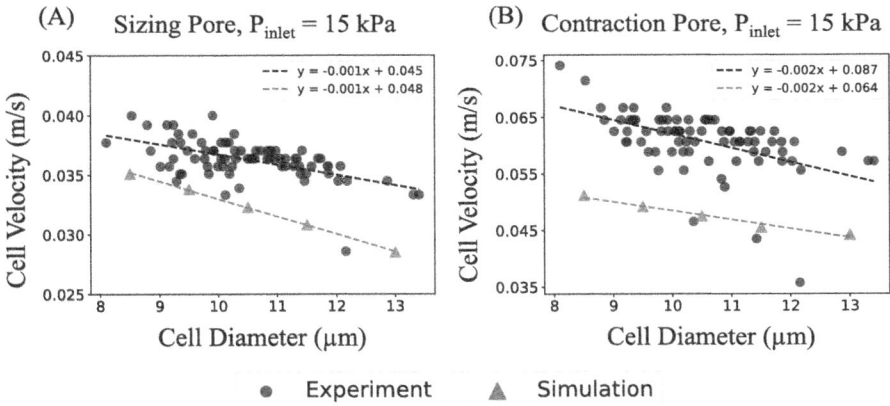

Fig. 7. Comparison of experimental and simulated cell velocities at 15 kPa inlet pressure in (A) sizing and (B) contraction pores.

In validating the digital twin, a key objective was to accurately replicate the experimental wCDI, a novel parameter used by mechano-NPS to characterize cell stiffness. As wCDI is derived from the ratio of velocities in the sizing and contraction pores (Eq. 6), its validation required first confirming the velocities in both regions. Despite the presence of a systematic bias in velocity measurements, this bias was effectively neutralized in the wCDI calculations due to the use of velocity ratios, leading to strong alignment between the experimental and simulated wCDI values. Figures 8(A) and 8(B) compare experimental and simulated wCDI at inlet pressures of 11 kPa and 15 kPa, respectively. At 11 kPa, the experimental and simulated wCDI values agreed closely, with a percent difference of less than 4% for all cell sizes except for the 8.5 μm cell, which showed a slightly higher 7% difference. This deviation can be attributed to the presence of outliers in cell velocity measurements at that size, as observed in Fig. 6. For an inlet pressure of 15 kPa, the percent difference between the experimental and simulated wCDI values remained below 5%, further demonstrating strong agreement between the two data sets. To further assess these differences and evaluate the digital twin's capability in capturing cell behavior within the mechano-NPS device, a Bland-Altman analysis was conducted. The results revealed a bias close to zero, indicating no systematic difference between the two measurement approaches. The 95% limits of the agreement confirmed that the simulated wCDI values closely matched the experimental results. These findings underscore the reliability of the digital twin in accurately replicating cell behavior across varying test conditions within the mechano-NPS device, reinforcing its potential as a robust tool for in silico cell analysis and device optimization.

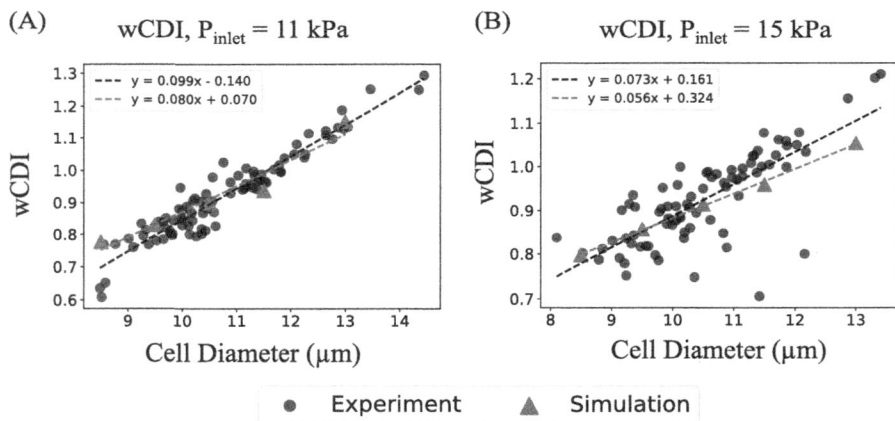

Fig. 8. Comparison of experimental and simulated wCDI at (A) 11 kPa and (B) 15 kPa inlet pressures.

4   Conclusion

Microfluidic devices have become indispensable in single-cell analysis, enabling precise investigations into live-cell states, mechanical properties, and molecular components. Integrating these platforms with digital twins presents a transformative opportunity to accelerate cost-effective device optimization, providing access to otherwise unmeasurable parameters, and enhancing throughput without compromising accuracy. Beyond improving efficiency, digital twins also deepen our understanding of the fundamental physical principles governing microfluidic devices and the complex biological phenomena they interrogate.

In this work, we introduce a digital twin framework for mechano-NPS, a high-throughput microfluidic platform designed for single-cell mechanophenotyping. The framework replicates the experimental setup by integrating fluid dynamics and cellular behavior models to provide an accurate in silico representation of the system. Development began with rigorous fluid flow modeling, ensuring that simulated pressure and velocity distributions closely matched analytical solutions, thereby providing a robust foundation. Building on this, we developed and verified a cellular model that accurately represents the mechanical behavior of the cell within the device, ensuring the reliability of the digital twin for future predictions. Finally, the digital twin was validated against experimental data, focusing on key performance metrics such as cell velocity and wCDI. Although minor discrepancies in cell velocity were observed, the simulated wCDI closely matched the experimental results, underscoring the accuracy and robustness of the digital twin. While this study focuses on a specific platform, the proposed algorithm is generalizable and can be adapted to simulate a wide range of microfluidic devices in silico. A key challenge in implementing such digital twin models lies in the significant computational cost. Finite element method (FEM)-based modeling of cellular behavior demands high spatial resolution to capture accurate physical interactions, particularly when simulations are customized for specific experimental configurations. This requirement can lead to a trade-off between model fidelity and computational efficiency. Nevertheless, this limitation can be partially alleviated by leveraging scalable cloud computing resources, enabling broader accessibility and faster turnaround for high-fidelity simulations.

This work highlights the transformative potential of digital twins in experimental microfluidics, demonstrating their ability to extend capabilities, optimize performance, and unlock new avenues of discovery. By bridging computational modeling with experimental biology, digital twins redefine how microfluidic devices are designed, tested, and utilized, paving the way for the next generation of high-throughput, data-driven cellular analysis platforms.

Acknowledgments. The authors thank Jorik Stoop for fruitful discussions. This work was supported by NIH 5R01EB024989. The content does not necessarily represent the official views of the NIH. An award of computer time was provided by the INCITE program. This research used resources of the Argonne Leadership Computing Facility, which is a DOE Office of Science User Facility supported under Contract DE-AC02-06CH11357.

References

1. Yin, H., Marshall, D.: Microfluidics for single cell analysis. Curr. Opin. Biotechnol. **23**(1), 110–119 (2012)
2. Carvalho, V., Rodrigues, R.O., Lima, R.A., Teixeira, S.: Computational simulations in advanced microfluidic devices: a review. Micromachines **12**(10), 1149 (2021)
3. Sheidaei, Z., Akbarzadeh, P., Kashaninejad, N.: Advances in numerical approaches for microfluidic cell analysis platforms. J. Sci. Adv. Mater. Dev. **5**(3), 295–307 (2020)
4. Jun-Shan, L., Zhang, Y.-Y., Zhong, W., Jia-Yi, D., Xuan, Y., Ri-Ye, X., et al.: Design and validation of a microfluidic chip with micropillar arrays for three dimensional cell culture. Chin. J. Anal. Chem. **45**(8), 1109–1114 (2017)
5. Chen, W.-X., Li, J.-G., Wan, X.-H., Zou, X.-S., Qi, S.-Y., Zhang, Y.-Q., et al.: Design of a microfluidic chip consisting of micropillars and its use for the enrichment of nasopharyngeal cancer cells. Oncol. Lett. **17**(2), 1581–1588 (2019)
6. Zhang, X., Xu, X., Ren, Y., Yan, Y., Wu, A.: Numerical simulation of circulating tumor cell separation in a dielectrophoresis based YY shaped microfluidic device. Sep. Purif. Technol. **255**, 117343 (2021)
7. Kou, S., Pan, L., van Noort, D., Meng, G., Wu, X., Sun, H., et al.: A multishear microfluidic device for quantitative analysis of calcium dynamics in osteoblasts. Biochem. Biophys. Res. Commun. **408**(2), 350–355 (2011)
8. Calibasi Kocal, G., Güven, S., Foygel, K., et al.: Dynamic microenvironment induces phenotypic plasticity of esophageal cancer cells under flow. Sci. Rep. (2016)
9. Wong, J.F., Young, E.W., Simmons, C.A.: Computational analysis of integrated biosensing and shear flow in a microfluidic vascular model. AIP Adv. **7**(11) (2017)
10. Esposito, G., Romano, S., Hulsen, M.A., D'Avino, G., Villone, M.M.: Numerical simulations of cell sorting through inertial microfluidics. Phys. Fluids **34**(7) (2022)
11. Tan, J., Sohrabi, S., He, R., Liu, Y.: Numerical simulation of cell squeezing through a micropore by the immersed boundary method. Proc. Inst. Mech. Eng. C J. Mech. Eng. Sci. **232**(3), 502–514 (2018)
12. Hynes, W., et al.: Examining metastatic behavior within 3D bioprinted vasculature for the validation of a 3D computational flow model. Sci. Adv. **6**(35), eabb3308 (2020)
13. Deng, Y., Davis, S.P., Yang, F., Paulsen, K.S., Kumar, M., Sinnott DeVaux, R., et al.: Inertial microfluidic cell stretcher (iMCS): fully automated, high-throughput, and near real-time cell mechanotyping. Small **13**(28), 1700705 (2017)
14. Saadat, A., Huyke, D.A., Oyarzun, D.I., Escobar, P.V., Øvreeide, I.H., Shaqfeh, E.S., et al.: A system for the high-throughput measurement of the shear modulus distribution of human red blood cells. Lab Chip **20**(16), 2927–2936 (2020)
15. Don, M.: The Coulter principle: foundation of an industry. JALA: J. Assoc. Lab. Autom. **8**(6), 72–81 (2003)
16. Kim, J., Han, S., Lei, A., Miyano, M., Bloom, J., Srivastava, V., et al.: Characterizing cellular mechanical phenotypes with mechano-node-pore sensing. Microsyst. Nanoeng. **4**(1), 1–12 (2018)
17. Lai, A., Rex, R., Cotner, K.L., Dong, A., Lustig, M., Sohn, L.L.: Mechano-nodepore sensing: a rapid, label-free platform for multi-parameter single-cell viscoelastic measurements. JoVE (J. Visualized Exp.) (2022)
18. Saleh, O., Sohn, L.: Correcting off-axis effects in an on-chip resistive-pulse analyzer. Rev. Sci. Instrum. **73**(12), 4396–4398 (2002)

19. Randles, A.P., Kale, V., Hammond, J., Gropp, W., Kaxiras, E.: Performance analysis of the lattice Boltzmann model beyond Navier-Stokes. In: 2013 IEEE 27th International Symposium on Parallel and Distributed Processing, pp. 1063–1074 (2013)
20. Gounley, J., Draeger, E.W., Randles, A.: Numerical simulation of a compound capsule in a constricted microchannel. Procedia Comput. Sci. **108**, 175–184 (2017)
21. Chen, S., Doolen, G.D.: Lattice Boltzmann method for fluid flows. Annu. Rev. Fluid Mech. **30**(1), 329–364 (1998)
22. Krüger, T., Kusumaatmaja, H., Kuzmin, A., Shardt, O., Silva, G., Viggen, E.M.: The Lattice Boltzmann Method. Springer, Cham, vol. 10. no. 978–3, pp. 4–15 (2017)
23. Guo, Z., Zheng, C., Shi, B.: Discrete lattice effects on the forcing term in the lattice Boltzmann method. Phys. Rev. E **65**(4), 046308 (2002)
24. Hecht, M., Harting, J.: Implementation of on-site velocity boundary conditions for D3Q19 lattice Boltzmann simulations. J. Stat. Mech: Theory Exp. **2010**(01), P01018 (2010)
25. Krüger, T., Varnik, F., Raabe, D.: Efficient and accurate simulations of deformable particles immersed in a fluid using a combined immersed boundary lattice Boltzmann finite element method. Comput. Math. Appl. **61**(12), 3485–3505 (2011)
26. Skalak, R., Tozeren, A., Zarda, R., Chien, S.: Strain energy function of red blood cell membranes. Biophys. J. **13**(3), 245–264 (1973)
27. Zhong-Can, O.-Y., Helfrich, W.: Bending energy of vesicle membranes: general expressions for the first, second, and third variation of the shape energy and applications to spheres and cylinders. Phys. Rev. A **39**(10), 5280 (1989)
28. Peskin, C.S.: The immersed boundary method. Acta Numer **11**, 479–517 (2002)
29. Zou, Q., He, X.: On pressure and velocity boundary conditions for the lattice Boltzmann BGK model. Phys. Fluids **9**(6), 1591–1598 (1997)
30. Mietke, A., Otto, O., Girardo, S., Rosendahl, P., Taubenberger, A., Golfier, S., et al.: Extracting cell stiffness from real-time deformability cytometry: theory and experiment. Biophys. J . **109**(10), 2023–2036 (2015)
31. Ahmmed, S.M., Suteria, N.S., Garbin, V., Vanapalli, S.A.: Hydrodynamic mobility of confined polymeric particles, vesicles, and cancer cells in a square microchannel. Biomicrofluidics **12**(1) (2018)
32. Kuriakose, S., Dimitrakopoulos, P.: Motion of an elastic capsule in a square microfluidic channel. Phys. Rev. E-Stat. Nonlinear, Soft Matter Phys. **84**(1), 011906 (2011)

Energy-Efficient Neural Network Training for Scientific Datasets with Advanced Similarity Analytics and Orchestration

Kin Wai NG[1], Orcun Yildiz[2], Tom Peterka[2], Florence Tama[3,4], Osamu Miyashita[3], Catherine Schuman[1], and Michela Taufer[1(✉)]

[1] University of Tennessee, Knoxville, TN 37996, USA
mtaufer@utk.edu
[2] Argonne National Laboratory (ANL), Lemont, IL 60439, USA
[3] Center for Computational Science, RIKEN, Kobe, Hyōgo, Japan
[4] Nagoya University, Nagoya, Aichi, Japan

Abstract. Scientific computing increasingly depends on neural architecture search (NAS) to identify accurate neural networks (NNs) that facilitate breakthroughs in various fields, from protein classification to material discovery. However, conventional NAS workflows face challenges due to excessive training times and inefficient energy consumption resulting from redundant computations and inflexible orchestration. In this paper, we present A4NN.2, the next generation of the Analytics for Neural Network (A4NN) workflow, which overcomes these challenges by introducing a structural similarity engine and advanced orchestration using the Wilkins framework. These enhancements eliminate redundant training and enable modular high-performance workflow executions. A4NN.2 accelerates NN training, reduces energy consumption, and demonstrates broad applicability across benchmark datasets and scientific domains. When used to train NNs to classify protein configurations from X-ray images, A4NN.2 achieves significant efficiency gains by reducing computational costs while maintaining high accuracy, thus accelerating scientific discovery in structural biology.

Keywords: Workflows · Flow control · Energy efficiency · CIFAR10 · CIFAR100 · Protein XFEL diffraction dataset

1 Introduction

Scientific computing increasingly depends on neural architecture search (NAS) to identify accurate neural network (NN) configurations, driving advancements in fields such as protein classification and material discovery. However, long training times and inefficient energy use often hinder traditional NAS workflows due to redundant computations and inflexible orchestration frameworks. This paper presents A4NN.2, the latest version of the Analytics for Neural Network (A4NN) workflow, designed to overcome these challenges. A4NN.2 integrates a structural similarity engine and advanced orchestration capabilities,

M. H. Lees et al. (Eds.): ICCS 2025, LNCS 15903, pp. 298–312, 2025.
https://doi.org/10.1007/978-3-031-97626-1_21

using the Wilkins framework to improve efficiency. Building on its predecessor, A4NN.1 [4], which featured a prediction engine to estimate NN accuracy and terminate underperforming networks early, A4NN.2 introduces two key enhancements. First, it automatically identifies and eliminates structural similarities and redundancies among NNs, reducing unnecessary computations. Second, it optimizes workflow flexibility through advanced orchestration using the Wilkins framework [19], enhancing the modularity of tasks and overall efficiency. By reducing redundant computations, A4NN.2 significantly lowers the total FLOPs during the NAS search process—a key proxy for energy consumption—thereby enhancing the sustainability of NAS. We present our findings on the efficiency of A4NN.2, particularly in terms of its reduced energy consumption, while maintaining the accuracy level of NNs. This is demonstrated across two widely recognized benchmarks, CIFAR10 and CIFAR100, in addition to a dataset involving protein diffraction generated by X-ray Free Electron Laser (XFEL).

Identifying Structural Similarity and Redundancy. A4NN.1 relied on a prediction engine that used parametric modeling methods to determine when to stop training early, but it did not address potential structural redundancies among candidate NNs. This oversight resulted in unnecessary training of similar networks, thereby increasing both computational demands and energy consumption. A4NN.2 introduces a new similarity engine that uses graph edit distance techniques to assess and identify structural similarities among NNs. By omitting networks that are structurally similar, we reduce redundant computations, optimize resource usage and improve energy efficiency.

Optimizing Workflow Flexibility with Advanced Orchestration. A4NN.1 used a tightly coupled workflow, which limited the flexibility to integrate new tasks or modify existing components without substantial rework. This rigidity hindered the ability to adapt the workflow to different NAS algorithms or evolving research needs. In A4NN.2, we integrate the Wilkins framework [19], a modular workflow orchestrator that decouples NAS algorithms from the task engines used for prediction and similarity. This modularization enhances flexibility, allowing for easy swapping of components, such as replacing the prediction engine with a more advanced model or adding new analytics tasks. The result is a flexible workflow specification, now configured through a high-level specification file, making it easier to customize and adapt to various research needs.

Demonstrate Efficiency in Energy Consumption Although A4NN.1 reduced training time by early stopping, it did not fully optimize energy consumption, as redundant models were still being trained. Furthermore, it lacked a comprehensive evaluation of energy proxies beyond FLOPs. We transform A4NN.2 to achieve significant efficiency gains through the early termination of converging models combined with the exclusion of redundant models, minimizing both training time and energy usage, and comprehensive evaluation of energy proxies, including training time per epoch, total training time per model, and FLOPs per model. We validate the gains of A4NN.2 by applying the workflow to two benchmark datasets (CIFAR10 and CIFAR100) to evaluate its effectiveness across varying complexity levels. The experimental results show up to 64.3% reduction in FLOPs and 59% reduction in training time in CIFAR10 and 53.4% reduction in FLOPs and 45% reduction in training time in CIFAR100. The well-known benchmarks are used to generate trust in the outcome of A4NN.2. More importantly, we demonstrate the impact of A4NN.2 in searching for accurate NNs for spectroscopy databases and in identifying protein types from X-ray diffraction images, reducing FLOPs by as much as 47.4% to 53.5% across varying beam intensities while preserving accurate solutions.

A4NN.2 addresses the reproducibility concerns of the AI community by open-source workflow configuration files, task codes, and orchestration scripts for full transparency and reproducibility, and integration with Data Commons to store and share training metadata, fostering community collaboration and validation. The code is available at: https://github.com/TauferLab/A4NN_workflow.

2   The A4NN Workflow

We design and implement an ecosystem for accelerating NAS by integrating modular analytic engines that enhance search efficiency and reduce computational overhead. This modularization enables A4NN.2 to support a suite of engines, allowing for extensibility and incorporating additional analytical tasks beyond the core workflow. Figure 1 shows the A4NN.2 workflow orchestrated by the Wilkins workflow system. This workflow consists of three main tasks, the NAS algorithm (red), the parametric prediction engine (blue), and the similarity engine (green).

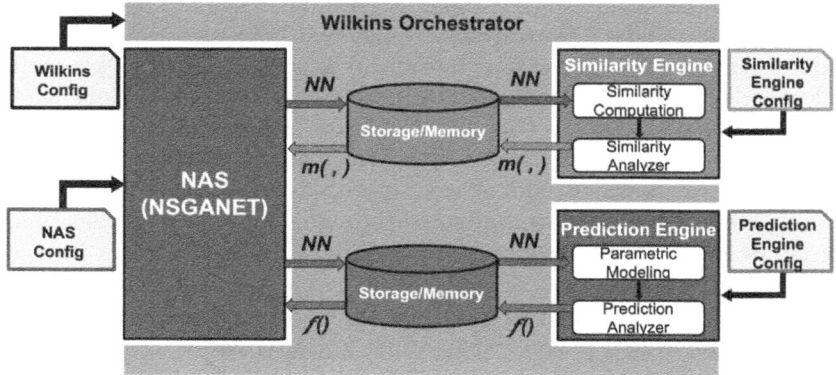

Fig. 1. Main components of the A4NN.2 workflow. Wilkins orchestrates the workflow, managing communication between three main tasks: NSGANET NAS (red), the similarity engine (green), and the prediction engine (blue). The similarity engine identifies and removes redundant NNs based on structural similarity, while the prediction engine estimates NN fitness to enable early termination. The entire workflow is configurable through high-level specification files. (Color figure online)

A4NN.2 is compatible with various NAS frameworks, allowing flexibility in selecting search algorithms. In this study, we use NSGANET [12], a multi-objective approach that optimizes NNs for accuracy and computational efficiency. NSGANET evolves NN architectures iteratively, generating an initial population and refining it across generations through mutation and crossover operations. It prioritizes FLOPs minimization to encourage energy-efficient models. FLOPs are estimated by the number of floating-point operations needed to perform a single forward pass through the NN architecture. A4NN.2 accelerates NSGANET by introducing two key areas. The prediction engine estimates the fitness trajectory of NNs, enabling early termination when performance stabilizes. The similarity engine detects and eliminates structurally similar NNs, preventing redundant training and improving search efficiency. The Wilkins orchestrator facilitates seamless communication between these components, ensuring efficient data exchange and workflow execution.

To support diverse configurations with different NAS strategies, A4NN.2 provides configurable settings for each component. Users can specify parameters that control the NAS settings, the A4NN engines (i.e., prediction and similarity), and the Wilkins orchestrator settings. NAS and A4NN engines configurations are specified in JSON format, and Wilkins in YAML.

NAS Conf. File. Users set the NSGANET parameters via a JSON file, which includes paths for input datasets, the initial population size, nodes per phase, offspring per generation, total generations, and training epochs.

A4NN Engine Files. Users set up the A4NN engines via JSON files, defining parameters for the prediction and similarity engines. For the prediction engine, the configuration outlines the parametric function selection, the number of required data points, the epoch for predicting fitness, the count of predictions for early stopping, and the permitted prediction variance. The similarity engine settings specify the similarity metric and threshold for structural comparisons.

Wilkins Conf. File. Users define data and resource needs in a YAML configuration file. Input and output file needs for each task are outlined, and data transfers occur via files. Resource allocation includes one process per task.

2.1   Identification of Structural Similarity and Redundancy

A4NN.2 augments the workflow with a similarity engine that identifies and removes structurally similar architectures before training begins. This prevents redundant computation and improves the overall efficiency of the search. The similarity engine is designed for comparison across NNs. It operates alongside NAS, analyzing architectures and signaling whether a given NN should proceed to training or be ignored. The similarity engine consists of two steps: the structure similarity computation and the similarity analyzer.

Following the encoding scheme of NSGANET, NNs are represented as directed acyclic graphs. Each NN consists of a sequence of phases, where computational operations (e.g., convolution, pooling, or batch normalizations) within a phase are encoded as a binary string. We decode the binary string into a graph, where vertices are computational operations, and edges are connections between them. To quantify structural similarity, we use the approximated graph edit distance (GED) metric [1]. GED measures the minimum number of edit operations (e.g., vertex or edge insertions or deletions) required to transform one graph in to the other. A lower GED indicates a high degree of similarity between two NN graphs, while a higher GED suggests less similarity. GED has been widely used in many applications requiring graph comparisons [3,6]. The similarity engine is designed for extensibility, and in future work, we can explore more advanced metrics, including kernel-based [9], or embedding-based approaches [13].

The similarity analyzer step evaluates whether a NN scheduled for training is structurally similar to an already trained network. If the similarity measure indicates redundancy, the analyzer signals the NAS to discard the architecture, thus avoiding redundant computation. If no similar architecture is found, training proceeds as usual, with the prediction engine determining whether early termination is possible. By filtering similar NNs, the similarity engine accelerates the NAS process, reducing computational cost of model search without compromising diversity in candidate NNs.

2.2 Decoupling and Modularization of Workflows for Flexibility

A4NN.1 used a plug-in that was not optimized for scalable communication, which led to rigid communication patterns. This limitation constrained the efficient execution of decoupled tasks in distributed environments. To address this, we empower A4NN.2 with the Wilkins orchestrator, introducing three key features: (i) High-performance HDF5-based data transport for scalable communication between NAS and decoupled tasks; (ii) Adaptive flow control mechanisms to manage data dependencies dynamically, minimizing idle time and optimizing resource utilization; and (iii) Automatic communication channel creation by matching data requirements between NAS and workflow tasks, streamlining execution. These features significantly improve scalability and reduce execution overhead, enabling the workflow to leverage larger HPC systems efficiently.

As shown in Fig. 1, Wilkins orchestrates A4NN.2 by launching its tasks concurrently, and managing their communication and dependencies transparently to the user. In this workflow, the NSGANET task sends NN architecture information to the similarity engine, and receives similarity data, $m(,)$, from previously trained NNs to determine whether the NN should be eliminated from training. Next, the NSGANET task sends NN architecture information to the prediction engine, and receives fitness predictions, $f()$, to assess the potential for early termination. Data exchanges occur over HDF5 files, ensuring interoperability and structured storage of A4NN's generated data. Wilkins also allows tasks to communicate in situ using MPI message passing; however, we chose file-based communication to ensure reproducibility and facilitate the integration of the results with Data Commons. Wilkins leverages the LowFive data model [16], an HDF5-VOL plugin, which enables seamless integration with existing A4NN task codes with minimal modification. By decoupling workflow tasks and managing their execution through a modular framework, Wilkins provides the flexibility needed to adapt A4NN to diverse NAS implementations and research objectives.

3   Evaluating A4NN on Benchmark Datasets

We evaluate A4NN on benchmark datasets to assess its efficiency, accuracy, and overall performance. Our evaluation is structured around four key questions: (i) Runtime performance: How much does A4NN reduce the time required for NAS compared to NSGANET; (ii) Energy efficiency proxies: How effectively does A4NN reduce training epochs, training time, and FLOPs compared to NSGANET?; (iii) Impact of model complexity: How does model complexity influence energy efficiency metrics such as FLOPs per epoch?; and (iv) Balancing accuracy and efficiency: How does A4NN balance model accuracy and computational efficiency, and what trade-offs exist between these metrics? We present our results on CIFAR10 and CIFAR100 datasets.

3.1   Benchmark Datasets and Evaluation Settings

We evaluate A4NN using the CIFAR10 and CIFAR100 datasets, two widely used computer vision benchmarks that differ in complexity, allowing for a thorough assessment across varying task difficulties. Both datasets were introduced in 2009 [10] as a subset of the 80 Million Tiny Images dataset [18].

CIFAR10: It contains 60,000 color images (32×32 pixels) divided into 10 categories, each with 6,000 images. It is divided into 50,000 training images and 10,000 test images, providing a balanced and straightforward classification task.

CIFAR100: This dataset is a more intricate version, consisting of 60,000 images divided into 100 detailed classes, each with 600 images. It is partitioned into 50,000 for training and 10,000 for testing. The increased class count and variability within classes present added classification difficulties.

We configure our experiments based on the NSGANET setup from our previous work [4], initializing a population of 10 NNs, generating 10 offspring per generation, and evolving over 10 generations, resulting in 100 trained models. Each model is trained for 25 epochs.

For the prediction engine, we use a concave function of the form $\mathcal{F}(x) = a - b^{c-x}$ to extrapolate a candidate fitness prediction at future epochs. We require three predictions within a variance threshold of 0.5 to determine convergence. For the similarity engine, we use GED with a threshold of 2 to determine structural similarity. We maintain consistent experimental parameters across all runs and repeat each workflow five times to assess variability. Experiments were run on the DARWIN HPC cluster at the University of Delaware, using one NVIDIA Tesla V100 GPU and four CPU cores from a 32-core AMD EPYC 7002 Series processor. The cluster features a high-performance Lustre filesystem.

3.2   Runtime Performance

To answer the question *"How much does A4NN accelerate NAS compared to NSGANET?"*, we evaluate the total runtime required to explore and identify best-performing NNs. Our analysis focuses on two key aspects: overall wall time and detailed runtime breakdowns to pinpoint efficiency gains.

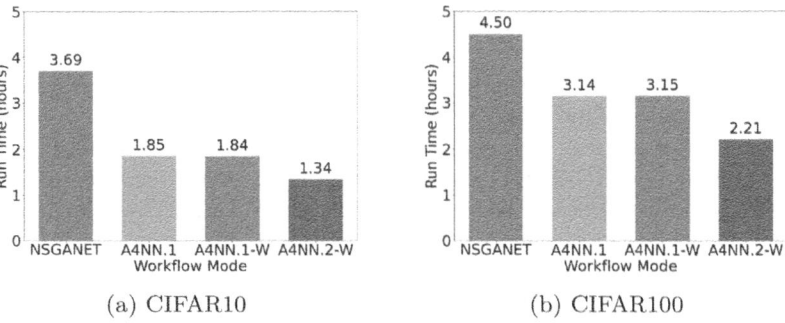

(a) CIFAR10   (b) CIFAR100

Fig. 2. Wall times for NNs trained with standalone NAS and A4NN (with and without Wilkins orchestrator) on CIFAR10 and CIFAR100 datasets.

Figure 2 shows the total wall times for the CIFAR10 and CIFAR100 datasets, comparing NSGANET (independently) with the three A4NN configurations (i.e., A4NN.1 using only the prediction engine, A4NN.1-W using Wilkins and the prediction engine, and A4NN.2-W using Wilkins, the prediction engine, and the similarity engine). Our results demonstrate a substantially reduced runtime for all A4NN configurations compared to NSGANET. On CIFAR10 (Fig. 2a), A4NN.1-W achieves a 2x speedup, while A4NN.2-W, which incorporates both the prediction and similarity engines, improves further with a 2.7x speedup. Similarly, for CIFAR100 (Fig. 2b), A4NN.1-W provides a 1.4x speedup, and A4NN.2-W achieves a 2x improvement over NSGANET.

Table 1 provides a detailed decomposition of the runtime, highlighting the impact of the prediction and similarity engines of A4NN and the integration of Wilkins on overall efficiency. The results highlight two key points. First, Wilkins does not add overhead to the A4NN workflow, as shown by the minimal difference in execution time between A4NN.1 and A4NN.1-W. Second, as expected, NN training accounts for most of the total runtime, while the time spent on A4NN's task engines is relatively minimal. The results also confirm that A4NN can effectively accelerate NAS by reducing redundant computations, with the similarity engine providing additional runtime savings. The integration of Wilkins also ensures efficient and scalable orchestration without penalizing performance.

3.3   Energy-Efficiency Proxies

To answer the question *How effectively does A4NN reduce training epochs, training time, and FLOPs compared to NSGANET?*, we evaluate energy efficiency using three key proxies: the number of training epochs required per model, the total training time per model, and the computational cost in FLOPs per model. Our analysis focuses on comparing the distributions of these metrics to quantify reductions in training effort and computational cost.

Figures 3a and 3d show the distribution of epoch counts per model for CIFAR10 and CIFAR100, where NSGANET trains models to the maximum

Table 1. Runtime breakdown per workflow component (in hours) for each method, comparing standalone NSGANET and A4NN variants across the CIFAR10 and CIFAR100 benchmarks.

Dataset	Method	NAS Time	A4NN Task Time	Other Time	Total Time
CIFAR10	NSGANET	3.6 ± 0.01	N/A	0.09 ± 0.0	3.69 ± 0.01
	A4NN.1	1.78 ± 0.01	0.004 ± 0.0	0.06 ± 0.0	1.85 ± 0.01
	A4NN.1-W	1.77 ± 0.02	0.004 ± 0.0	0.06 ± 0.0	1.84 ± 0.02
	A4NN.2-W	1.25 ± 0.01	0.03 ± 0.0	0.06 ± 0.0	1.34 ± 0.01
CIFAR100	NSGANET	4.4 ± 0.02	N/A	0.1 ± 0.0	4.5 ± 0.01
	A4NN.1	3.05 ± 0.01	0.007 ± 0.0	0.09 ± 0.0	3.14 ± 0.01
	A4NN.1-W	3.05 ± 0.04	0.006 ± 0.0	0.09 ± 0.0	3.15 ± 0.04
	A4NN.2-W	2.1 ± 0.03	0.03 ± 0.0	0.09 ± 0.0	2.21 ± 0.04

epoch limit (i.e., 25) while A4NN methods adaptively terminate training earlier. We observe that A4NN methods significantly reduce the number of training epochs compared to NSGANET. A4NN.2-W achieves the lowest median epoch count, reducing training epochs by 58% on CIFAR10 and 44% on CIFAR100. Similarly, A4NN achieves substantial reductions in training time (Figs. 3b and 3e), with A4NN.2-W reducing the median training time per model by 59% on CIFAR10 and 45% on CIFAR100. We observe a similar trend for FLOPs (Figs. 3c and 3f), where A4NN.2-W achieves the highest efficiency, reducing FLOPs by 55% on CIFAR10 and 47% on CIFAR100 compared to NSGANET.

Moreover, the shape of the distributions in Fig. 3 provides further insights into the variability between workflows. The NSGANET distributions are narrow across all metrics, unsurprisingly, since all models train for the maximum number of epochs, resulting in consistently high computational costs. In contrast, A4NN methods exhibit wider distributions, particularly A4NN.2-W, which shows a concentration of models near zero due to the elimination of redundant training for similar models. Differences in distribution shapes across datasets further highlight the impact of task complexity. On CIFAR10, many models fall in the lower range of the distributions, indicating a tendency to train less complex architectures. Conversely, CIFAR100 distributions show a greater proportion of models in the upper range, suggesting a preference for more complex architectures. Given the similar reduction trends and the consistent distribution shapes across energy proxies, we conclude that FLOPs serve as a reliable proxy for energy consumption and efficiency. Our results demonstrate that A4NN methods, especially A4NN.2-W, achieve substantial efficiency gains by reducing training epochs, training time, and FLOPs compared to NSGANET. These observed trends hold on both CIFAR10 and CIFAR100, despite differences in dataset complexity.

3.4   Impact of Model Complexity on Energy Proxies

To answer the question *How does model complexity influence energy efficiency metrics such as FLOPs per epoch?*, we evaluate the relationship between model

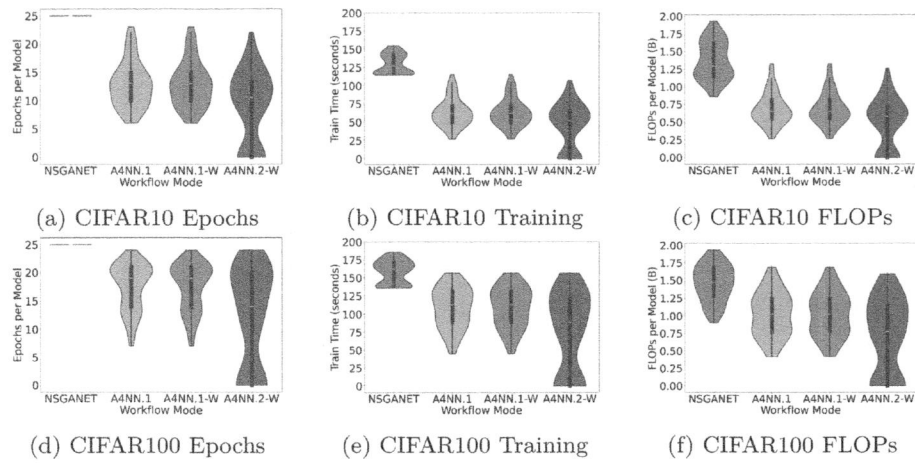

Fig. 3. Distribution of training epochs per model, training time (in seconds) per model, and FLOPs per model for four workflow methods.

complexity, measured by the number of trainable parameters, and FLOPs per epoch, our proxy for energy consumption. Our analysis focuses on identifying correlation between these variables to provide insights into training dynamics and their impact on reducing computational costs in NAS.

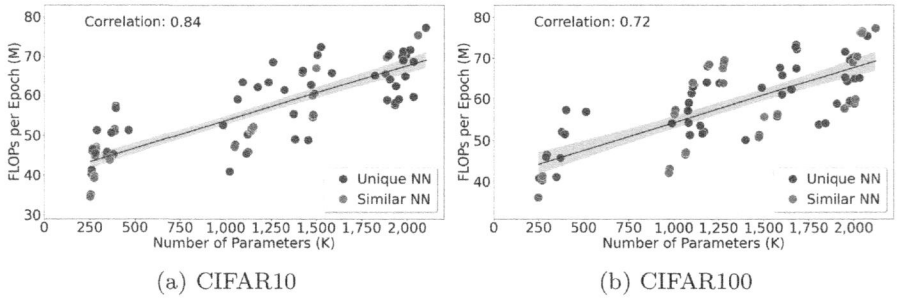

Fig. 4. Model complexity, measured by parameter numbers, correlates with epoch-wise FLOPs. Models are colored to denote structural uniqueness: blue dots for unique models, and red dots for models similar to past trained ones. (Color figure online)

Figure 4 shows the relationship between model complexity and FLOPs per epoch for both CIFAR10 and CIFAR100 datasets. Each point represents a trained NN, color coded to distinguish between unique models (blue) and similar models (red). For CIFAR10, there are 28 similar models out of 100. For CIFAR100, there are 31 similar models out of 100. We observe a strong positive correlation between model complexity and FLOPs per epoch. For CIFAR10, the correlation coefficient is 0.84, while for CIFAR100 is 0.72. These results indicate that computational cost increases with model complexity, which is expected. However, despite these strong correlations, models with similar parameter counts exhibit considerable variability in FLOPs. This variability suggests that models with the same parameter count can have significantly different architectures (e.g., differing in layer arrangement, types, and connectivity patterns) directly impacting computational cost.

Furthermore, we observe that similar models are not concentrated in a specific region but are instead dispersed across the parameter space. This distribution indicates that structurally similar networks can have different computational costs despite having a similar number of parameters. These findings highlight the importance of our similarity engine in identifying redundant models, and reducing unnecessary computational costs associated with training them.

3.5   Balancing Accuracy and Efficiency

To answer the question "How does A4NN balance model accuracy and computational efficiency, and what trade-offs exist between these metrics?", we analyze the relationship between validation accuracy and FLOPs across trained NNs. Our analysis focuses on two key aspects: constructing Pareto frontiers to capture optimal trade-offs and quantifying FLOPs reductions achieved by A4NN to pinpoint efficiency gains.

Figure 5 presents the Pareto optimal solutions identified by A4NN variants and NSGANET, where each point represents a model, with symbol shape indicating the workflow method and symbol size corresponding to the number of parameters. By analyzing these frontiers, we evaluate A4NN's ability to balance accuracy and efficiency compared to NSGANET. For CIFAR10, A4NN models (blue and red) achieve validation accuracy comparable to NSGANET (gray) while requiring fewer FLOPs. It is important to note that A4NN.2 preserves most of the solutions identified by A4NN.1, even when similar models are dropped from training. Furthermore, A4NN models often reach high accuracy models with fewer parameters than NSGANET. We observe a similar pattern for CIFAR100 (Fig. 5b), where A4NN models maintain efficiency gains over NSGANET, achieving similar accuracy at a lower computational cost. This is evident from the leftward shift of Pareto optimal solutions generated by A4NN compared to those of NSGANET. We did not tune hyper-parameters or apply data augmentation, keeping settings aligned with A4NN.1 for fair comparison. Despite this, A4NN.2 reliably selects competitive architectures.

We further quantify these improvements in Fig. 6 by aggregating total FLOPs across all 100 evaluated architectures. A4NN.2-W achieves a 64.3% reduction in FLOPs in CIFAR10 and a 53.4% reduction on CIFAR100 compared to NSGANET. Overall, these findings demonstrate that A4NN.2 effectively balances accuracy and efficiency, reducing computational costs significantly while retaining high-performing solutions.

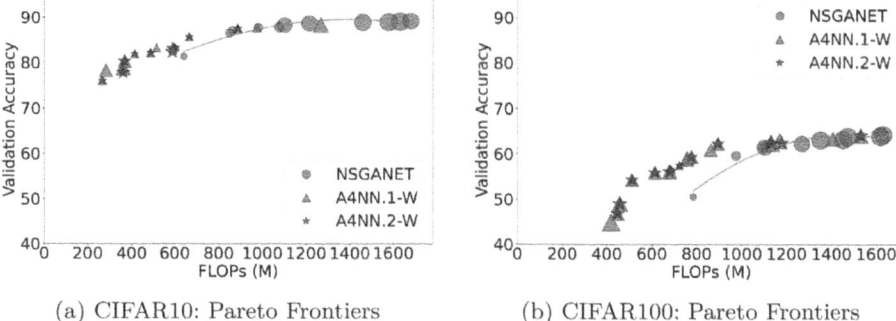

(a) CIFAR10: Pareto Frontiers   (b) CIFAR100: Pareto Frontiers

Fig. 5. Pareto-optimal frontiers for NNs illustrating the trade-off between validation accuracy and total FLOPs. Models produced by NSGANET (gray circles), A4NN.1-W (red triangles), and A4NN.2-W (blue stars) are shown as markers. Each marker is a Pareto-optimal model where enhancing one metric (accuracy or FLOPs) reduces the other. Marker size denotes the model's parameter count. (Color figure online)

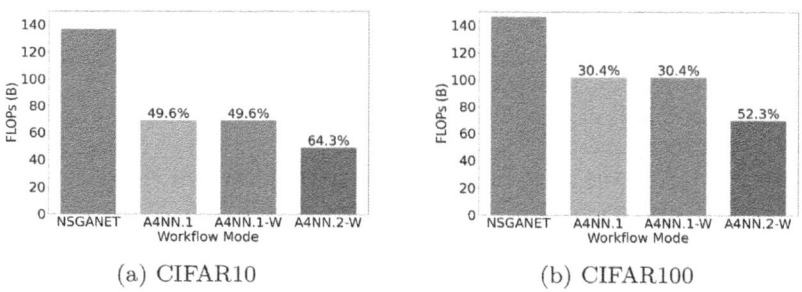

(a) CIFAR10   (b) CIFAR100

Fig. 6. Total FLOPs for training 100 NN architectures and FLOPs percentages saved by A4NN vs. standalone NSGANET for CIFAR10 and CIFAR100.

4   Applying A4NN to Scientific Datasets

We compare A4NN's performance to NSGANET to evaluate the trade-off between accuracy and computational cost in a real-world scenario. Specifically, we use A4NN with the Protein XFEL Diffraction dataset [14] to assess its ability to reduce power consumption while training NNs for classifying protein conformations within the dataset. The dataset consists of diffraction patterns generated by XFEL experiments, where proteins are exposed to intense laser beams, producing photo scattering patterns that capture structural information. The *spsim* simulator was used to generate different diffraction patterns for two conformations of EF2 with PDB ID 1n0u and 1n0v from the Protein Data Bank. In this study, we generate, train, and evaluate NNs to classify these protein conformations (i.e., differentiate between 1n0u and 1n0v). The dataset includes three subsets generated by varying intensities of the XFEL beam on the same proteins: Low (1×10^{14} photons/μm^2/pulse), Medium (1×10^{15} photons/μm^2/pulse), and

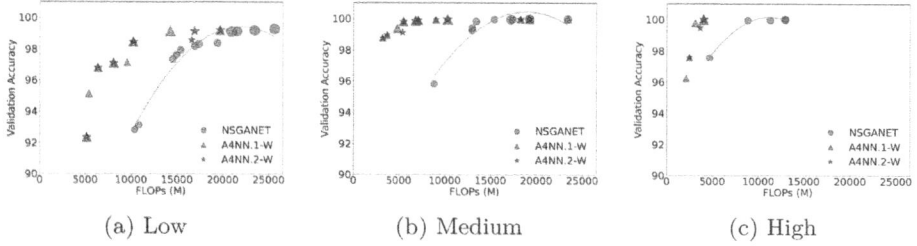

(a) Low   (b) Medium   (c) High

Fig. 7. Pareto-optimal frontiers for NNs using (a) Low, (b) Medium, and (c) High beam intensity protein diffraction datasets, illustrating the balance between validation accuracy and total FLOPs. Models by NSGANET (gray circles), A4NN.1-W (red triangles), and A4NN.2-W (blue stars) are depicted as markers. Each marker is a Pareto-optimal model, where enhancing one metric affects the other. Marker size indicates the number of model parameters. (Color figure online)

High (1×10^{16} photons/μm^2/pulse). The XFEL beam's intensity directly affects the resultant images' signal-to-noise ratio and low beam intensities are a proxy for noise. The lower the intensity, the higher the noise.

Figure 7 shows the FLOPs of the Pareto optimal models selected by each workflow method across all intensity levels, where lower values indicate better performance. At Low beam intensity, A4NN.2-W achieves comparable validation accuracy to NSGANET while reducing FLOPs by approximately 47.4%, demonstrating a more efficient search for optimal architectures even in noisy data scenarios. Specifically, A4NN.2-W models consistently require fewer than 10,000 MFLOPs, compared to NSGANET models that exceed 15,000 MFLOPs. This reduction not only accelerates training, but also conserves energy, highlighting the capability of A4NN to handle low signal-to-noise ratios efficiently. At Medium beam intensity, A4NN.2-W continues outperforming NSGANET by achieving the same high validation accuracy with up to 48.8% fewer FLOPs. The Pareto front shows that A4NN models cluster around 10,000 MFLOPs, whereas NSGANET models require up to 20,000 MFLOPs. This efficiency gain accelerates the convergence to high accuracy, highlighting A4NN's adaptability to medium noise levels. Even at High beam intensity, where data quality is improved and training naturally converges faster, A4NN.2-W maintains a FLOPs reduction of 53.3% compared to NSGANET. The Pareto frontiers reveal that A4NN models consistently achieve high accuracy with FLOPs below 5,000 MFLOPs, while NSGANET models demand higher computational costs.

Figure 8 shows the total FLOPs required for training 100 NN architectures and percentages of FLOPs saved by A4NN compared to standalone NSGANET for the protein diffraction datasets. Figure 8 shows the total FLOPs required for training 100 NNs and the corresponding percentages of FLOPs saved by A4NN compared to standalone NSGANET for protein diffraction datasets. Lower FLOPs indicate better energy efficiency and reduced computational cost. At low beam intensity, A4NN.2-W achieves the highest savings, reducing total

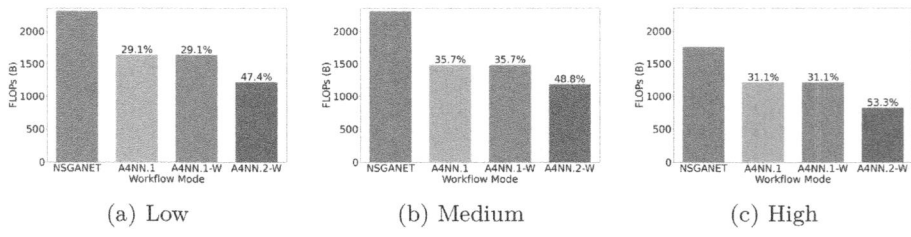

(a) Low   (b) Medium   (c) High

Fig. 8. Total FLOPs for training 100 NN architectures and FLOPs percentages saved by A4NN vs. standalone NSGANET for the protein diffraction datasets.

FLOPs by 47.4% compared to NSGANET. This significant reduction is due to the effective elimination of redundant models and the early termination of converging architectures. Both A4NN.1 and A4NN.1-W achieve a 29.1% reduction, highlighting the impact of early termination even without the similarity engine. The considerable savings in FLOPs at low beam intensity demonstrate the ability of A4NN to efficiently handle noisy data sets where signal-to-noise ratios are low and training is typically more resource intensive. For medium intensity, A4NN.2-W demonstrates superior performance by reducing total FLOPs by 48. 8%, maintaining high precision while requiring fewer computations. Both A4NN.1 and A4NN.1-W achieve a 35.7% reduction, showcasing the benefits of modular orchestration and early stopping. The medium-intensity efficiency gains highlight A4NN's adaptability to datasets with moderate noise levels, optimizing training cycles without sacrificing model performance. At high beam intensity, where data quality improves and training naturally converges faster, A4NN.2-W achieves a remarkable 53.3% reduction in total FLOPs compared to NSGANET, the highest savings across all intensity levels. In contrast, both A4NN.1 and A4NN.1-W yield a 31.1% reduction. The superior performance of A4NN.2-W at high beam intensity illustrates its capability to maximize resource utilization even under optimal data conditions. This efficiency gain is attributed to the synergy between the structural similarity engine and advanced orchestration, which collectively eliminate redundant computations.

Figures 7 and 8 demonstrate that A4NN.2-W consistently achieves significant energy savings across all intensity levels, reducing total training costs while maintaining high accuracy.

5   Related Work

This study builds upon the previous works of Olaya et al. [14], Patel et al. [15], Rorabaugh et al. [8], and Channing et al. [4]. We also take inspiration from other studies leveraging NAS for scientific datasets, such as Kandasamy et al. [7] and Balaprakash et al. [2]. Olaya et al. [14] introduced the XPSI framework to predict protein types and orientations from 2D diffraction patterns, but required significant human intervention and did not address computational efficiency. Patel et al. [15] expanded XPSI by employing NSGANET for NAS, reducing manual tuning. However, their approach still faced long runtimes and lacked distribution. Rorabaugh et al. [8] introduced the PENGUIN fitness prediction engine to decouple search and prediction strategies, improving efficiency in NAS workflows. Channing et al. [4] proposed the A4NN workflow which laid the foundation for a more efficient workflow in NN training. These studies, however, do not address potential architectural redundancies among candidate NNs in NAS. Other NAS applications for scientific datasets, like DENSE [7] and cancer modeling on HPC machines [2], face similar challenges of time and resource consumption, limiting accessibility for domain scientists. Efforts to improve NAS efficiency have led to methods like early stopping [11], learning curve extrapolation [5], and training speed estimation [17], which reduce computation time and resource usage. This work addresses these challenges by augmenting the A4NN workflow to further reduce wall times and energy consumption.

6 Conclusion

This paper demonstrates the effectiveness of A4NN.2 in accelerating NAS while significantly reducing energy consumption across diverse datasets. Integrating a structural similarity engine and advanced orchestration, A4NN.2 eliminates redundant training and optimizes resource usage. Our experiments on a Protein XFEL Diffraction dataset show that A4NN.2 achieves up to 47.4% FLOPs savings at Low beam intensity, 48.8% at Medium beam intensity, and 53.3% at High beam intensity compared to NSGANET, while maintaining high accuracy. These gains are especially prominent in noisy datasets, where A4NN.2 accelerates convergence with fewer training epochs. Overall, A4NN.2 balances accuracy and efficiency, paving the way for sustainable scientific computing.

Acknowledgments. This work was supported by the National Science Foundation (NSF) under grant numbers 2331152 and 2223704. This work was supported by Advanced Scientific Computing Research, Office of Science, U.S. Department of Energy, under Contract DE-AC02-06CH11357.

References

1. Abu-Aisheh, Z., et al.: An exact graph edit distance algorithm for solving pattern recognition problems. In: 4th International Conference on Pattern Recognition Applications and Methods (2015)
2. Balaprakash, P., et al.: Scalable reinforcement-learning-based neural architecture search for cancer deep learning research. In: Proceedings of the International Conference for High Performance Computing, Networking, Storage and Analysis (2019)
3. Bunke, H., Allermann, G.: Inexact graph matching for structural pattern recognition. Pattern Recogn. Lett. (1983)
4. Channing, G., et al.: Composable workflow for accelerating neural architecture search using in situ analytics for protein classification. In: Proceedings of the 52nd International Conference on Parallel Processing (2023)

5. Domhan, T., et al.: Speeding up automatic hyperparameter optimization of deep neural networks by extrapolation of learning curves. In: Proceedings of the 24th International Conference on Artificial Intelligence (2015)

6. Ferrer, M., Bunke, H.: Graph edit distance–theory, algorithms, and applications. Image Process. Anal. Graphs: Theory Pract. (2012)

7. Kasim, M.F., et al.: Building high accuracy emulators for scientific simulations with deep neural architecture search. Sci. Technol. Mach. Learn. (2021)

8. Keller Rorabaugh, A., et al.: Building high-throughput neural architecture search workflows via a decoupled fitness prediction engine. IEEE Trans. Parallel Distrib. Syst. (2022)

9. Kriege, N.M., et al.: A survey on graph kernels. Appl. Netw. Sci. (2020)

10. Krizhevsky, A., Hinton, G.: Learning Multiple Layers of Features from Tiny Images. University of Toronto, Toronto, Ontario, Technical report (2009)

11. Li, L., et al.: Hyperband: a novel bandit-based approach to hyperparameter optimization. J. Mach. Learn. Res. **18**(185), 1–52 (2018)

12. Lu, Z., et al.: NSGA-Net: neural architecture search using multi-objective genetic algorithm. In: Proceedings of the Genetic and Evolutionary Computation Conference, pp. 419–427 (2019)

13. Makarov, I., et al.: Survey on graph embeddings and their applications to machine learning problems on graphs. PeerJ Comput. Sci. **7**, e357 (2021)

14. Olaya, P., et al.: Identifying structural properties of proteins from x-ray free electron laser diffraction patterns. IEEE 18th International Conference on e-Science (e-Science) (2022)

15. Patel, R., et al.: A methodology to generate efficient neural networks for classification of scientific datasets. IEEE 18th International Conference on e-Science (e-Science) (2022)

16. Peterka, T., et al.: Lowfive: in situ data transport for high-performance workflows. In: IEEE International Parallel and Distributed Processing Symposium (IPDPS) (2023)

17. Ru, R., et al.: Speedy Performance estimation for neural architecture search. In: Advances in Neural Information Processing Systems (2021)

18. Torralba, A., et al.: 80 million tiny images: a large dataset for nonparametric object and scene recognition. IEEE Trans. Pattern Anal. Mach. Intell. (2008)

19. Yildiz, O., et al.: Wilkins: HPC in situ workflows made easy. Front. High Perform. Comput. (2024)

Author Index

The manufacturer's authorised representative in the EU is Springer
Nature Customer Service Centre GmbH, Europaplatz 3, 69115 Heidelberg,
Germany. If you have any concerns regarding our products, please
contact ProductSafety@springernature.com

Printed and bound by CPI Group (UK) Ltd, Croydon, CR0 4YY

28/04/2026

02098522-0002